T0326258

Determination of Metals in Natural Waters, Sediments, and Soils

Determination of Metals in Natural Waters, Sediments, and Soils

T. R. Crompton
Chemical Analysis Department
Water Authority, Leeds
Yorkshire, UK

ELSEVIER

AMSTERDAM • BOSTON • HEIDELBERG • LONDON • NEW YORK • OXFORD
PARIS • SAN DIEGO • SAN FRANCISCO • SINGAPORE • SYDNEY • TOKYO

Elsevier
Radarweg 29, PO Box 211, 1000 AE Amsterdam, Netherlands
The Boulevard, Langford Lane, Kidlington, Oxford OX5 1GB, UK
225 Wyman Street, Waltham, MA 02451, USA

ISBN: 978-0-12-802654-0

British Library Cataloguing in Publication Data
A catalogue record for this book is available from the British Library

Library of Congress Cataloging-in-Publication Data
A catalog record for this book is available from the Library of Congress

For Information on all Elsevier publications
visit our website at http://store.elsevier.com/

Contents

Preface

This book is concerned with a discussion of methods currently available in the world literature up to early 2014 for the determination of metals, and metalloids in natural waters, sediments, and soils. The occurrence of metals, many of which are toxic, can have profound effects on the ecosystem.

In the case of soils, the presence of deliberately added or adventitious metallic compounds can cause contamination of the tissues of crops grown on the land or animals feeding on the land and, consequently, can cause adverse toxic effects on man, animals, birds, and insects. Also drainage of these substances from the soil can cause pollution of adjacent streams rivers and eventually the oceans. Some of the substances included in this category are fertilizers, crop sprays, sheep dips, etc. In the case of river sediments, a major input of metals is a consequence of industrial activity. Many industries discharge metal-containing effluents to rivers or directly to sea or discharge of such wastes by ship dumping.

Surface water drainage is another input. The presence of metallic compounds in river sediments is due, in part, to man-made pollution and monitoring the levels of these substances in the sediment and sediment cores provides an indication of the time dependence of their concentration over large time spans. Contamination of sediments is found not only in rivers but also in estuarine and oceanic sediments and thus sediment analysis provides a means of tracking metallic compounds from their source through the ecosystem.

Sediments have the property of absorbing such contaminants from waters within their bulk (accumulation) and, indeed, it has been shown that the concentration, for example, for some types of metallic contaminants in river sediments can be up to one million times greater than it occurs in the surrounding water.

Most of the metallic elements in the periodic table have been found to be present in natural waters, and soil sediments and sludges, some naturally occurring and others as a result of industrial activity of one kind or another. To date, insufficient attention has been given to the analysis of sediments and one of the objects of this book is to draw attention of analysts and others concerned to the methods available and their sensitivity and limitations.

The use of correct sampling and sample preservation procedures is mandatory in the analysis of solid materials and this is discussed fully in Chapters 1 and 2 which discuss aspects of sampling including sample homogeneity, comminution and grinding of samples, sample digestion procedures. The application of nondestructive methods of analysis to solid samples is also discussed.

Chapters 3 to 5, respectively, deal with the determination of trace metal concentrations in river waters, groundwaters, and aqueous precipitation.

Chapter 6 deals with the sampling of sediments and Chapters 7 and 8, respectively, deal with the determination of "traces of metals in sediments and soils."

Examination of solid samples for metals combines all the exciting features of analytical chemistry. First, the analysis must be successful and in many cases must be completed quickly. Often the nature of the substances to be analyzed for is unknown, might occur at exceeding low concentrations and might, indeed, be a complex mixture. To be successful in such an area requires analytical skills of a high order and the availability of sophisticated instrumentation.

The work has been written with the interests of the following groups of people in mind: management and scientists in all aspects of the water industry, river management, fishery industries, sewage effluent treatment and disposal, land drainage, and water supply; also management and scientists in all branches of industry. It will also be of interest to agricultural chemists, agriculturalists concerned with the ways in which metals used in crop or soil treatment permeate through the ecosystem, the biologists and scientists involved in fish, plant, insect, and plant life, and also to the medical profession, toxicologists and public health workers and public analysts. Other groups or workers to whom the work will be of interest include oceanographers, environmentalists and, not least, members of the public who are concerned with the protection of our environment.

Finally, it is hoped that the work will act as a spur to students of all subjects mentioned and assist them in the challenge that awaits them in ensuring that the pollution of the environment is controlled so as to ensure that into the new millennium we are left with a worthwhile environment to protect.

Chapter 1

Metals in Natural Water Samples

Sampling Techniques

Chapter Outline

1.1 INTRODUCTION

Heavy metals are among the most toxic and persistent pollutants in freshwater systems. Many research and monitoring efforts have been conducted to determine sources, transport, and fate of these metals in the aquatic environment. However, studies have shown that contamination artifacts have seriously compromised the reliability of many past and current analyses,[1] and in some cases, metals have been measured at 100 times true concentration.[2] These induced errors are of great concern, since artifact-free data are necessary to detect trends and to identify factors that control the transport and fate of toxic heavy metals. In addition, without accurate and reliable data, it is impossible to accurately monitor the effect of costly regulations aimed at reducing metal emissions. To avoid these problems, and to enhance the quality of trace metal data, laboratories are putting substantial effort into improving protocols for sample collection, handling, and analysis.[3] This greater level of effort devoted to clean methods is costly in both money and time.

The problems caused by contamination when measuring trace metals were first brought to the attention of the scientific community by Patterson in his investigation of stable isotopes in the 1960 and 1970s.[4–6] Largely through his influence, clean methods became part of the standard operating procedures used by chemical oceanographers starting in the mid-1970s.[5,6] Freshwater chemists were slow to adopt these same techniques, with the notable exception of the long series of investigations on lead cycling by Patterson and co-workers.[7–11]

With few exceptions,[12–15] limnologist have begun to use clean techniques only in the last 10 years. This was spurred, in part, by oceanographers who began to study freshwater systems such as the Mississippi River,[16–18] Great Lakes,[19] and Amazon River.[20] The result of this activity has been to cast serious doubt on earlier routine measurements. Thus, Flegal and Coale[21] have questioned surveys of lead in surface waters,[22,33] and Windom et al.[2] disputed the reliability of the United States Geological Survey (USGS) National Stream Quality Accounting Network. Ahlers et al.[4] and Benoit[24] implied that most previously reported results may be in error because of failure to follow appropriate clean protocols. A parallel can be drawn with chemical oceanography, where virtually all uncensored (i.e., other than nondetects) trace metal data from before about 1975 are considered invalid.

Common to all analytical methods is the need for correct sampling. It is still the most critical stage with respect to risks to accuracy in aquatic trace metal chemistry, owing to the potential introduction of contamination. Systematic errors introduced here will make the whole analysis unreliable.

Surface-water samples are usually collected manually in pre-cleaned polyethylene bottles (from a rubber or plastic boat) from the sea, lakes, and rivers. Sample collection is performed in front of the bow of the boat, against the wind. In the sea or in larger inland lakes, a sufficient distance (about 500 m) in an appropriate wind direction has to be kept between the boat and the research vessel to avoid contamination. The collection of surface water samples from the vessel itself is impossible, considering the heavy metal contamination plume surrounding each ship. Surface-water samples are usually taken at 0.3 to 1 m depth, in order to be representative and to avoid interference by the air–water interfacial layer in which organics and consequently bound heavy metals accumulate. Usually, sample volumes between 0.5 and 2 L are collected. Substantially larger volumes could not be handled in a sufficiently contamination-free manner in the subsequent sample pre-treatment steps.

Reliable deep-water sampling is a special and demanding art. It usually has to be done from the research vessel. Special devices and techniques have been developed to provide reliable samples.

Samples for mercury analysis should preferably be taken in pre-cleaned glass flasks. If, as required for the other ecotoxic heavy metals, polyethylene flasks are commonly used for sampling, then an aliquot of the collected water sample for the mercury determination has to be transferred as soon as possible into a glass bottle, because mercury losses with time are to be expected in polyethylene bottles.

Luettich et al.[25] have described an instrument system for remote measurement of physical and chemical parameters in shallow water.

Martin et al.[26] compared surface grab and cross-sectionally integrated sampling and found that very similar concentrations of dissolved metals were obtained by these procedures.

Cox and McLeod[27] have discussed difficulties associated with preserving the integrity of chromium in water samples.

1.2 SAMPLING DEVICES

The job of the analyst begins with taking the sample. The choice of sampling gear can often determine the validity of the sample taken; if contamination is introduced in the sampling process itself, no amount of care in the analysis can save the results.

Sampling the subsurface waters present some not completely obvious problems. For example, the material from which the sampler is constructed must not add any metals or organic matter to the sample. To be completely safe, then, the sampler should be constructed either of glass or of metal. All-glass samplers have been used successfully at shallow depths; these samplers are generally not commercially available.[28,29] To avoid contamination from material in the surface film, these samplers are often designed to be closed while they are lowered through the surface, and then opened at the depth of sampling. The pressure differential limits the depth of sampling to the upper 100 m; below this depth, implosion of the sampler becomes a problem.

Implosion at greater depths can be prevented either by strengthening the container or by supplying pressure compensation. The first solution has been applied in the Blumer sample.[30] The glass container is actually a liner inside an aluminum pressure housing; the evacuated sampler is lowered to the required depth, where a rupture disc brakes, allowing the sampler to fill. Even with the aluminum pressure casing, however, the sampler cannot be used below a few 1000 m without damage to the glass liner.

Another approach to the construction of glass sampling containers involves equalization of pressure during the lowering of the sampler. Such a sampler has been described by Bertoni and Melchiorri-Santolini.[31] Gas pressure is supplied by a standard diver's gas cylinder, through an automatic delivery valve of the type used by scuba divers. When the sampler is opened to the water, the pressurizing gas is allowed to flow out as the water flows in. The sampler in its original form was designed for used in Lake Maggioŕe, where the maximum depth is about 200 m; but in principle it can be built to operate at any depth.

Stainless steel samplers have been devised, largely to prevent organic contamination. Some have been produced commercially. The Bodega-Bodman sampler and the stainless steel Niskin bottle, formerly manufactured by General Oceanics, Inc, are examples. These bottles are both heavy and expensive. The Bodega-Bodman bottle, designed to take very large samples, can only be attached to the bottom of the sampling wire; therefore, the number of samples taken at a single station is limited by the wire time available, and depth profiles require a great deal of station time.

The limitations of the glass and stainless steel samplers have led many workers to use the more readily available plastic samplers, sometimes with a full knowledge of the risks and sometimes with the pious hope that the effects resulting from the choice of sampler will be small compared with the amounts of organic matter present.

Smith[32] has described a device for sampling immediately above the sediment water interface of the ocean. The device consists of a nozzle supported by a benthic sled, a hose and a centrifugal deck pump, and is operated from a floating platform. Water immediately above the sediment surface is drawn through the nozzle and pumped through the hose to the floating platform, where samples are taken. The benthic sled is manipulated by means of a hand winch and a hydrowire.

Basu[33] has described a device for collecting water samples at depth comprising a plastic cylinder (60×60cm) attached to a light-alloy bracket and shaft. The lower end of the cylinder terminates in an inverted cone and nozzle (0.5cm diameter). The upper end, fitted with a sealing ring, is closed by a plastic ball (6.3cm diameter) attached to the inside wall of the cylinder by an elastic cable. The ball is held in the open position by one arm of a pivoted lever; a wire attached to the bracket passes through a hole in the other arm of the lever. Weights at the lower end of the cylinder keep it steady at the sampling depth. The apparatus is lowered into position with both ends of the cylinder open; a lead weight then slides down the wire, strikes the lever arm, and releases the ball, which closes the top of the cylinder; and the sampler is carefully withdrawn. A small orifice, closed by a thumb-screw, controls the discharge of water from the sampler.

Bhagat et al.[34] Evans and Edgar,[35] and Noth[36] reviewed sampling in the aquatic environment, automated sampling, and sampling of frozen water, respectively.

Gibs et al.[37] used a multiport sampler with seven screened intervals to study vertical variations in water chemistry.

Powell and Puls[38] studied differences in ground water chemistry between the casing and screened interval values of four wells. Tracer experiments were used to study the differences in natural flushing between the casing and screened interval volumes.

Benoliel[39] has reviewed the storage and preservation of natural water samples.

Salbu and Oughton[40] have reviewed strategies for the sampling, fractionation, and analysis of natural waters.

Droppo and Jaskot[41] have studied the effect of river transport characteristics on contaminant sampling.

Hall et al.[42] have studied the effects of four different filter membranes on concentrations of 28 dissolved elements in five different natural water matrices.

Benoit et al.[43] have demonstrated that clean techniques are necessary for reliable measurement of trace metals in freshwaters at ambient, though not necessarily regulatory, concentrations. These workers compared conventional sample-handling methods to clean techniques for 35 individual steps used in protocols for analysis of filtrate and filter-retained forms of silver, cadmium, copper, and lead. Approximately two-thirds of all steps contributed statistically significant amounts of contamination in the measurement of dissolved and particulate cadmium, copper, and lead. Average contamination for a single contributing step was 300%, 141%, and 200% for the three metals, respectively (where 100% represents no added contamination). Relative copper contamination tended to be lower, partly because real levels in water are higher for this metal. Contamination generally was not a problem for silver when it was present in water at higher than background levels. With that exception, it does not seem possible to develop clean technique protocols, even when measuring trace metals in polluted freshwaters where levels are moderately high. The expectation of Benoit et al.[43] is that most other metals (e.g., zinc, chromium or nickel) will have contamination behavior that is similar to cadmium, copper, and lead, rather than the rare metal silver.

Brite et al.[23] have described a downhole groundwater sampler to reduce bias and error due to sample handling and exposure while introducing minimal distribution to natural flow conditions in the formation and well. This in situ sealed (ISS) or "snap" sampling device includes removable/flab-ready sample bottles, a sampler device to hold double end-opening sample bottles in an open position, and a line for lowering the sampler system and triggering closure of the bottles downhole.

Before deploying, each bottle is set open at both ends to allow flow-through during installation and equilibration downhole. Bottles are triggered to close downhole without well purging; the method is therefore passive or nonpurge. The sample is retrieved in a sealed condition and remains unexposed until analysis. Data from six field studies comparing ISS sampling with traditional methods indicate ISS samples typically yield higher volatile organic compound (VCO) concentrations; in one case, significant chemical-specific differentials between sampling methods were discernible. For arsenic, filtered and unfiltered purge results were negatively and positively biased, respectively, compared to ISS results. Inorganic constituents showed parity with traditional methods. Overall, the ISS is versatile, avoids low VOC recovery bias, and enhances reproducibility while avoiding sampling complexity and purge water.

1.3 FILTRATION OF WATER SAMPLES FOR TRACE METAL DETERMINATION

Filtration and centrifugation are the principal methods available for the separation of dissolved and undissolved trace metal fractions in waters; in situ dialysis has also been proposed as a means of effecting such a separation,[44] but interpretation of the results is not straightforward and the technique requires further evaluation.

Since it is essential that the separation should be carried out immediately after sample collection (to avoid trace metal redistribution on storage), centrifugation is generally inconvenient and filtration is usually the only practicable procedure.

Cheeseman and Wilson[45] have presented a general discussion of filtration and filter media in relation to trace metals. The use of membrane filters with a pore size of about 0.5 μm is generally considered to give a separation of practical utility, but glass fiber filters and membrane filters of other pore sizes have also been used for trace-metal studies.

The U.S. Environmental Protection Agency has chosen 0.45-μm membrane filters as the basis of its standard separation technique,[46] and similar filters have been recommended in a number of authoritative texts.[47–49]

Paper filters are not recommended for trace metal studies,[45] although their use for determination of dissolved iron appears to be endorsed by the American Society for Testing and Materials.[50] However, the latter publication recommends the use of 0.45-μm membranes for the determination of dissolved copper and manganese.

Hunt[51] have discussed the effect of filtration of water samples prior to trace metal determination. From this work it is evident that separation of the dissolved and undissolved fractions of trace metals in water samples is not simple. Filtration, usually the only practicable procedure, may be associated with problems of contamination and adsorption, in addition to being subject to the basic difficulty of incomplete separation.

The adsorption of trace metals during sample filtration has received very little study. It does not appear to be mentioned as a possible cause of error in a number of analytical manuals. The limited information available suggests, however, that it can cause serious difficulties.

Thus, it is not possible at present to recommend a filtration procedure known to be suitable for all metals. This position is clearly unsatisfactory.

Hunt[51] gives the following guidelines for conducting and reporting tests of filtration systems for trace metal analysis.

1. Tests should be designed to assess both contamination *and* adsorption. For this reason, tests involving only filtration of blanks to assess contamination are inadequate.
2. As a minimum, a blank and a solution containing the maximum dissolved determinand concentration expected in routine samples should be used in the tests. It is desirable that a low-concentration solution should also be used, since adsorption may be more important at lower concentrations.
3. Ideally, actual samples should be used for the tests, in which case prior filtration is necessary to ensure that the determinand is present only in a filterable form. This filtration should be carried out using filters of smaller effective pore-size than those to be tested.

However, if wide variations in a sample matrix are expected, preliminary studies using solutions of defined composition and appropriate determinand concentrations may be more useful, particularly for screening a number of filters and filter holders. Such solutions should not contain matrix components likely to produce wholly atypical dissolved metal specification patterns, because trace metal adsorption may be influenced by the adsorbate speciation, and should be undersaturated with respect to all solid phases (to avoid retention by the filters of determinand contained in, or adsorbed on, precipitated material). A test with actual samples should, however, precede routine application of the procedure.

4. It is essential, in view of the likely effects of pH on trace metal adsorption, that test samples or solutions should cover the anticipated pH range of samples. Tests at the extremes of the natural water pH range are inadequate, since adsorption may be greatest at intermediate pH values.

5. If natural samples are used, it is desirable that the major ion composition (or salinity of seawater samples) be known.
6. The experimental procedure should be defined unambiguously. This definition should include details of any pretreatment (e.g., acid washings) applied to the filters and filtration apparatus.
7. The filtrate should be collected into acid[45] to minimize adsorption losses subsequent to filtration.
8. The analytical technique adopted for the tests, if different from that to be used for analysis of the filterable fraction in routine operation, should be capable of giving results of adequate accuracy.
9. Appropriate statistical methods[52] should be used in assessment of test results.

REFERENCES

1. Taylor HE, Shiller AM. *Environ Sci Technol* 1995;**29**:1313.
2. Windom HL, Byrd JT, Smith RG, Huan F. *Environ Sci Technol* 1991;**25**:1137.
3. Horowitz AJ, Demas CR, Fitzgerald KK, Miller TL, Ricker DA. *Open-file rep.* US Geological Survey; 1994. No. 94–539.
4. Ahlers WW, Reid MR, Kim JP, Hunter KA. *Aust J Mar Freshwater Res* 1990;**41**:713.
5. Participants of the lead in seawater workshop. *Mar Chem* 1974;**2**:69.
6. Participants of the lead in seawater workshop. *Mar Chem* 1976;**4**:388.
7. Hirao Y, Patterson CC. *Science* 1974;**184**:989.
8. Shirahata H, Elias RW, Patterson CC, Koide M. *Geochim Cosmochim Acta* 1980;**44**:149.
9. Ng A, Patterson CC. *Geochim Cosmochim Acta* 1981;**45**:2109.
10. Erel T, Patterson CC, Scott MJ, Morgan J. *J Chem Geol* 1990;**85**:383.
11. Erel Y, Morgan JJ, Patterson CC. *Geochim Cosmochim Acta* 1991;**55**:707.
12. Sigg L. In: Stumm W, editor. *Chemical processes in lakes*. New York: John Wiley; 1985. p. 283.
13. Sigg L, Strum M, Kistler D. *Limnol Oceanogr* 1987;**32**:112.
14. Nriagu JO, Gaillard JF. In: Nriagu JO, editor. *Environmental impact of smelters*. New York: John Wiley & Sons; 1984. p. 349.
15. Morfett K, Davison W, Hamilton-Taylor J. *J Environ Water Sci* 1988;**11**:107.
16. Shiller AM, Boyle EA. *Nature* 1985;**317**:49.
17. Shiller AM, Boyle EA. *Geochim Cosmichim Acta* 1987;**51**:214.
18. Trefry JH, Metz S, Trocine RP, Nelson TA. *Science* 1985;**230**:439.
19. Coale KH, Flegal AR. *Sci Total Environ* 1989;**87/88**:297.
20. Boyle EA, Huested SS, Grant B. *Deep-Sea Res* 1982;**29**:1355.
21. Flegal AR, Coale KS. *Water Res Bull* 1989;**25**:1275.
22. Smith RA, Alexander RB, Wolman MG. *Science* 1987;**235**:1607.
23. Brite SL, Parker BL, Cherry JA. *Environ Sci Technol* 2010;**44**:4917.
24. Benoit G. *Environ Sci Technol* 1994;**28**:1987.
25. Leuttich RA, Kirby Smith WW, Hunnings W. *Estuaries* 1993;**16**:190.
26. Martin GR, Smoot JL, White KD. *Water Environ Res* 1992;**64**:866.
27. Cox AG, McLeod CW. *Michrochim Acta* 1992;**109**:161.
28. Gump BH, Hertz HA, May WE, Chesler SN, Dyszel SM, Enagonio DP. *Anal Chem* 1975;**47**:1223.
29. Keizer PD, Gordon Jr DC, Dale J. *J Fish Res Board, Can* 1977;**34**:347.
30. Clark Jr RC, Blurner M, Raymond SO. *Deep-Sea Res* 1967;**14**:125.

31. Bertoni R, Melchiorri-Santolini U. *Mem Inst Ital Hydrobiol* 1972;**29**:97.
32. Smith KL. *Limnol Oceanogr* 1971;**16**:675.
33. Basu AK. *Trib Cebedeau* 1969;**22**:272.
34. Bhagat SK, Proctor DE, Funk WH. *Water Sewage Works* 1971;**118**:180.
35. Evans MR, Edgar R. *Water Pollut Control Lond* 1971;**70**:111.
36. Noth TC. *Water Sewage Works* 1971;**118**:179.
37. Gibs J, Brown GA, Turner KS, McLeod CL, Jelinski JC, Koehnlein SA. *Groundwater* 1993;**31**:201.
38. Powell RM, Puls RW. *J Contam Hydrol* 1993;**12**:51.
39. Benoliel JL. *Int J Environ Anal Chem* 1994;**57**:197.
40. Salbu B, Oughton DH. *Trace Elem Nat Waters* 1995;**41**:69.
41. Droppo EG, Jaskot C. *Environ Sci Technol* 1995;**29**:161.
42. Hall GEM, Bonham Carter GF, Horowicz AJ, et al. *Appl Geochem* 1996;**11**:243.
43. Benoit G, Hunter KS, Rozan TF. *Anal Chem* 1997;**69**:1006.
44. Benes P, Steinnes E. State of trace elements in natural waters. *Water Res* 1974;**8**:947.
45. Cheeseman RV, Wilson AL. *Technical memorandum TM 78*. Water Research Association; 1973.
46. United States Environmental Protection Agency. *Methods for chemical analysis of water and wastes*. EPA–625–/6–74–003. Washington: Office of Technology Transfer; 1974. 298 p. (p. 81).
47. Strickland JDH, Parsons TR. *Practical handbook of sea water analysis*. Ottawa: Fisheries Research Board of Canada; 1968. Bull. No. 167, 311 p.
48. American Public Health Association, et al. *Standard methods for the examination of water and waste water*. 14th ed. Washington: The Association, xxxix; 1976. 1193 p. (p. 147).
49. Association of Official Analytical Chemists. In: Horwtiz W, editor. *Official methods of analysis of the association of official analytical chemists*. 12th ed. Washington: The Association, xxi; 1975. 1094 p. (p. 619).
50. American Society for Testing and Materials. *Annual book of ASTM standards, 1977. Part 31, water*. Philadelphia: The Society, xxi; 1977. 1110 p. (pp. 323, 364, 386).
51. Hunt DTE. *Filtration of water samples for trace metal determinations*. Technical Report TR 104. Medmenham, U.K: Water Research Centre; February 1979.
52. Cheesman RV, Wilson AL. *Manual on analytical quality-control for the water industry*. Technical Report TR66. Water Research Centre. Medmenham, U.K: Water Research Centre; 1978. 167 p. 1978.

Chapter 2

Water Sample Preservation

Chapter Outline

2.1 INTRODUCTION

The problems that can be encountered in sample preservation and analysis have been discussed by King and Ciacco,[1] Grice et al.[2] Hume,[3] Riley,[4] and Grasshoff.[5,6]

Mart[7] have described a typical sample-bottle cleaning routine for use when taking samples for very-low-level metal determination.

Sampling bottles and plastic bags, both made of high-pressure polyethylene, were rinsed by the following procedure. First clean with detergent in a washing machine, rinse with deionized water, soak in a hot (about 60 °C) acid bath, beginning with 20% hydrochloric acid, reagent grade, followed by two further acid baths of lower concentration, the last being of Merck, Suprapur quality or equivalent. The bottles are then filled with dilute hydrochloric acid, Merck, Suprapur, this operation being carried out on a clean bench; rinse and fill them up with very pure water (pH 2). Bottles are wrapped into two polyethylene bags. For transport purposes, lots of 10 bottles are enclosed hermetically into a larger bag.

The determination of traces of heavy metals in natural waters can be greatly affected by contamination (positive or negative) during filtration and storage of samples.[13–23] Until the work of Scarponi et al.[14] this problem, especially as regards filtration acid had not been studied adequately and systematically with respect to the determination of cadmium lead and copper.

The same filter can release or adsorb trace metals depending on the metal concentration level and the main constituents of the sample; therefore, claimed results must be considered with caution in working with natural samples. To avoid contamination, the following procedures have often been used. Filters have been cleaned by soaking them in acids[11,25–29] or complexing agents[7–10] Sometimes, however, the washing procedure has not been found to be fully satisfactory. For example, it has been reported that strong adsorption of cadmium and lead occurs on purified unconditioned membrane filters when triple-distilled water is passed through the filter, while there is no change in the concentration with a river water sample after filtration of 500 mL.[24] Some investigators prefer to avoid filtration when the particulate matter does not interfere with the determination (in which case the analyses must be complete soon after sampling)[31] or when open seawater is analyzed; in the latter case, filtered and unfiltered samples do not seem to differ significantly in measurable metal content.[7,32,33]

Heavy metals are generally present at very low concentrations in water, and hence their determination is extremely susceptible to problems of contamination. The first sources of potential error are associated with the sampling and subsequent storage of the sample prior to analysis. The sample container must be carefully selected to avoid contamination due to leaching of metals into the sample and also to minimize losses of metal from the solution by adsorption onto the walls of the container.

In their review of the literature, Batley and Gardner[10,25] concluded that polythene and Teflon containers are suitable for sample collection and storage. Indeed, polythene sample containers have now gained widespread acceptance for routine use. Selection of a cleaning method for the sample container is, nevertheless, still somewhat arbitrary. Many methods are reported in Table 2.1, but no attempt appears to have been made to compare the different methods. The analyst therefore either has to hope that the method chosen is adequate or to perform a series of tests to establish the adequacy of the method Furthermore, some of the methods currently recommended require cleaning periods of several weeks and may be consuming an undue amount of the analyst's time, as well as tying up large numbers of sample containers "under preparation."

The method chosen to clean or prepare the sample container must fulfil two requirements: It must reduce contamination to an acceptable level and it must minimize or prevent adsorption losses to the container wall. It is widely accepted that adsorption losses may be reduced by acidifying the sample, usually to 0.5% nitric acid. This acidification step is, nevertheless, more likely to leach metals from an improperly cleaned container. In addition, there is a growing requirement for samples to be maintained at their natural pH in order to perform analyses that help elucidate the chemical form of the metal.[44] These analytical schemes for speciation studies can be time-consuming and hence necessitate the storage of samples at their natural pH for several days prior to analysis.[47] Such storage conditions favor adsorption losses from the sample.

TABLE 2.1 Methods Reported for Cleaning Sample Containers*

Basis of Method	References
Hot conc. HNO	34
50% HNO_3	54
50% HCl, then 50% HNO_3	35
Ca 40% HCl	25
25% HNO_3	36
20% HNO_3	37,38
Ca 15% HCl	10
10% HNO_3	39–42,51
2.5% $HClO_4$	43
2% HNO_3	48
1% HNO	44
Detergent only	45
Sample rinse only	10

**After initial cleaning with 15% HCl.*

Indeed, Subramian et al.[48] report significant losses of metals from river water samples (pH 6–8) during the first 10 days of storage in polythene containers.

In marked contrast, Florence[47] report no detectable change in the metal concentration of a freshwater sample (pH 6.1) stored for 23–26 days at either 26 or 4 °C, also in polythene containers.

A possible explanation for these contradictory results may be that different cleaning methods were used to prepare the sample container. For instance, acids used for cleaning may activate adsorption sites.[10,49] Pretreatment of the sample containers by soaking with a concentrated salt solution (1 g of calcium sulfate plus 1 g of magnesium sulfate L^{-1})[54] or a portion of the water to be sampled has been recommended to overcome adsorption losses. It is still not certain, however, whether such time-consuming pretreatment is necessary.

Some examples of sample preservation procedures are given in the next sections.

2.2 CLEANING METHODS FOR POLYETHYLENE CONTAINERS PRIOR TO THE DETERMINATION OF LEAD, COPPER, ZINC, AND CADMIUM IN FRESHWATER

The methods used by Laxen and Harrison[50] to clean and prepare the sample containers are summarized in Table 2.2. Certain of the methods were chosen

TABLE 2.2 Cleaning Procedures for Sample Containers

Designation	Procedure	References
A	No treatment.	
B	Rinse with water.[a]	
C	Detergent wash; water rinse; 10% HNO, wash; water, rinse: drain dry.	51
D	10% HNO, 48 h soak; water rinse; drain dry.	39
E	10% HNO, rinse water, rinse; drain dry.	39
F	50% HNO, rinse water rinse; 0.2% Aristar HNO soak, minimum 2 weeks, water rinse before use.	52
G	2.5% Perchloric acid (36 mL of 70%, $HClO_4$ to ca 1000 mL of water in glass container), 24-h soak water rinse, 2.5% $HClO_4$, 24-h soak; water rinse; 2.5% $HClO_4$ 24-h soak, water rinse; drain dry.	43
H	5% Decon 90,[b] 24-h soak, water rinse; 10% HNO_3, soak 2 weeks; water rinse; water soak minimum 1 week with change of water.	41
J	5% Decon 90,[b] 24-h soak; water rinse; 10% HNO_3, soak 2 weeks; water rinse; 0.2% Aristar HNO_3, soak 1 week; water rinse; water, soak minimum 1 week with change of water.	50
K	5% Decon 90,[b] 24-h soak; water rinse; drain dry.	50
L	Immediately before use: detergent (household liquid) wash; water rinse; 2% HNO_2, 24-h soak; tap water rinse 6 times; water rinse 6 times.	48
M	15% HCl, soak minimum 3 days; water rinse before use	10
N	10% HNO_3, 24-h soak; water rinse; 1 g of $CaSO_4$ + 1 g of $MgSO_4$ L^{-1} of water, soak minimum 24 h; water rinse before use.	53

[a] *Water means deionized and distilled water.*
[b] *Decon 90 is a mixture of anionic and nonionic surface active agents, from Decon Laboratories, Conway Street, Hove, Sussex, England.*

because they are specifically recommended for the preparation of new containers; others were selected for routine cleaning. Two sources of pure water were used; one was distilled and then deionized, and the other deionized and then double distilled from a Fistream FSL/DMC/4DBD system. The blank metal values were respectively [Zn] = 0.3 and 0.18 ng mL^{-1}, [Cd] = <0.01 and <0.01 ng mL^{-1}, [Pb] = 0.09 and 0.09 ng mL^{-1}, and [Cu] = 0.25 and 0.08 ng mL^{-1} in the two waters. The higher- quality water only becomes available towards the end of the study. In the description of the cleaning methods, both pure waters are merely termed "water." All reagents were AnalalR grade unless otherwise indicated.

Duplicate bottles were cleaned by methods A, C, D, F, G, H, and J (Table 2.2) and were then filled with water, and 5 mL of 'Aristar' nitric acid was added. The concentration of zinc, cadmium, lead, and copper were then determined by graphite furnace atomic absorption spectrometry analysis immediately and after 2 weeks.

The outcome of this work[50] was the recommendation that a 48 h soak with 10% nitric acid (i.e., method D, Table 2.2) be used for both the preliminary cleaning of new bottles and for routine cleaning. The results further confirm the suitability of polyethylene sample containers for the collection and storage of freshwater samples prior to analysis for zinc, cadmium, lead, and copper.

2.3 PROLONGED STORAGE OF NATURAL-WATER SAMPLES CONTAINING IRON, CHROMIUM, NICKEL, THALLIUM, COBALT, MANGANESE, SILVER, COPPER, CADMIUM, LEAD, AND ZINC IN POLYETHYLENE CONTAINERS IN PRESENCE OF AQUEOUS NITRIC AND PRESERVATION REAGENT

In work described by Marchant and Klopper,[36] unfrozen samples of pure water and nitric acid, stored in high-density polyethylene containers, were used as contamination blank controls for hydro-geochemical samples preserved by deep-freezing in similar vessels. The maximal levels of 11 metal contaminants in the blanks after four years of storage are reported. These levels were below the anomalous threshold concentrations established in most hydro-geochemical prospecting surveys, and it is concluded that thoroughly cleaned polyethylene containers can be used to store samples for several years, provided that the natural trace element content of the waters can be effectively stabilized.

Typical results obtained in the work are noted in Table 2.2.

After 4 years, the maximum levels of metal contaminants in these blanks were (in μg L^{-1}) Fe, 0.4; Cr, 0.6; Ni, 1.0; T1, 0.6; Co, 1.5; Mn, 0.14; Ag, 0.17; Cu, 1.5; Cd, 0.17; Pb, 2.4; Zn, 2.4. Values slightly lower than these are found for Co, Mn, Ag, Cu, Zn, and Pb in the pure water samples. Approximately 0.5–1.0 μg L^{-1} of both Zn and Pb are derived from the nitric acid.

2.4 PRESERVATION AND STORAGE OF SURFACE-WATER SAMPLES CONTAINING ZINC, LEAD, COPPER, CADMIUM, MANGANESE, AND IRON IN GLASS VIALS IN THE PRESENCE OF NITRIC ACID

Henricksen and Balmer[37] have developed a method for preserving lake-water samples in which the samples are collected in 24-mL glass vials with polyethylene snap-caps that have been washed in succession with solutions of 12% Na-EDTA and 5% Deconex detergent, soaked in nitric acid, and thoroughly rinsed with distilled water. After arrival at the laboratory, the samples are preserved with 0.25 mL 3.5 N nitric acid, and the concentrations of heavy metals determined by flameless atomic-absorption spectrophotometry. Statistical evaluation of results from samples collected from 18 lakes in Norway indicate that neither the time of addition of acid nor the length of time between sample preservation and analysis greatly affects the measured concentrations of zinc, lead, copper, manganese, and iron.

Evaluation of lead and cadmium was precluded because the concentrations of these metals were at or below the analytical limit in most of the lake samples.

2.5 PRESERVATION OF MERCURY-CONTAINING SAMPLES

Much work has been carried out on the preservation and storage of samples intended for mercury analysis.[55–64] Several workers have discussed the storage of samples in polyethylene containers.[58–62] Pyrex glass has also been considered as a storage material. Acetic acid-formaldehyde,[58] nitric acid,[57,63] and potassium dichromate[64] have all been considered as preservatives. Nitric acid (to pH 1.0) was found to be moderately effective in preventing loss of mercury provided that the acid is placed in the container before the sample.[63] However, with this preservation, 80% of the mercury originally associated with particulate matter will within one week's storage pass into the solution phase. There is no significant loss to the container. If the sample is not acidified, most of the mercury is retained by the particulate matter, 15% is absorbed by the containers, and only 10% remains in solution.

This has serious implications when determination of total and soluble mercury is required. A relationship has been found between compositional differences in different sources of commercial polyethylene bottles and the stability of mercury solutions stored therein.[59] It has also been found that pretreatment of polyethylene bottles with chloroform suppresses losses from stored 1 $\mu g\,L^{-1}$ mercury solutions in natural and distilled water.[65] Addition of humic acid at the 50 $\mu g\,L^{-1}$ level reduces losses from 1 $\mu g\,L^{-1}$ mercury solutions stored in commercial polyolefin bottles for more than 15 days to less than 10%. Humic acid is simple to use and was found to be more effective than preservatives based on nitric acid and oxidants.

The uptake of mercury (II) by two humic acids varied with pH, with more than 98% being absorbed at pH 4.5. Addition of increasing amounts of chloride reduced uptake by 10–20% and shifted the region of maximal sorption to higher pH values. Calcium, magnesium, and ammonium ion promoted near total adsorption and reduced the influence of pH at 10 mM concentrations and had a slight effect at 01 mM concentrations.

Craig[62] showed that the passage of mercury vapor from ambient air through container walls, made of conventional polyethylene, linear polyethylene, or Teflon, can seriously contaminate the samples of distilled and natural water stored inside. When the samples contain oxidizing preservatives such as nitric acid or potassium permanganate, the rate of mercury contamination is greatly increased. Freezing the samples in plastic containers prevented such mercury contamination, and storage in glass minimized it.[63]

Ambe and Suwabe[63] found that the addition of sodium chloride to dilute mercury solutions in Pyrex glass ampoules greatly improves the stability of the solutions. As a result of their experimental work, these researchers chose 3% sodium chloride adjusted to pH 0.5–1.0 with sulfuric acid as a preservative, and the solutions were sealed in Pyrex glass ampoules. In the concentration range 1–1000 $\mu g L^{-1}$, mercury concentrations did not change in 18 months.

Mahon and Mahon[64] studied mercury losses in untreated water samples to develop a means of minimizing mercury loss without treatment with strong acids or oxidants. They recommended that if polyethylene containers are used, they should be rinsed with water characteristic of the sample. After the samples are collected and filtered, they must be agitated frequently and vigorously, to maintain oxygen saturation, and the sample must be analyzed as soon as possible, preferably within a few hours of collection and certainly on the same working day.

Samples that received no agitation after being introduced into the polyethylene containers lost mercury much more rapidly than samples handled in the same way except for vigorous and frequent agitation. Soaking and rinsing sample containers in river water were done in order to passivate the wall surfaces.

For the preservation of mercury in aqueous samples by acidification, a pH of less than 2 is unsatisfactory because of the relatively rapid loss of mercury vapor to the headspace. Strong oxidizing agents have been used to preserve the mercury in the mercury (II) form. Zhang et al.[65] describe a method where potassium dichromate is added to water samples and results in preservation for up to 2 months. Five consecutive determinations of mercury in water in the 2.0–26 $\mu g L^{-1}$ range over a period of 6–20 days had relative standard deviations of 2.2–11.0%.

Ramelhusa et al.[66] studied the effect of organic matter in contaminated water samples on the determination of mercury.

2.6 PRESERVATION OF SELENIUM-CONTAINING SAMPLES

Cheam and Agemian[30] studied the stability of inorganic selenium(IV) and selenium(VI) species at levels of 1 and 10 $\mu g L^{-1}$ under various conditions of pH, type of water, and type of container. Use of polyethylene containers and adjustment to pH 1.5 will provide optimum conditions of preservation for both distilled and natural water samples up to 125 days. Algal growth is detrimental to solution stability at natural pH values of pH 5.4–7.2, but adjustments to sulfuric acid with a pH of 1.5 successfully prevents this effect. Storage of samples at 4 °C gave satisfactory stability but is less practicable. However, sulfuric acid interferes with some analytical methods, and if natural waters are to be tested it may be preferable to use unacidified natural water stored in polyethylene bottles at 4 °C. It was noted that selenium(VI) is generally more stable than selenium(IV) in aqueous solutions. An inter-laboratory quality-control study[67] involving adjustment to pH 1.5 with sulfuric acid used the preferred conditions or acidifications of 0.02% v/v sulphuric acid and storage at room temperature in polyethylene bottles gave excellent recoveries over the selenium concentration range 0–1000 $\mu g\,L^{-1}$ confirmed the effectiveness of this preservation method. Cutter[46] stated that acidification can be used for the preservation of water samples, but that it should not be excessive because the speciation can be changed. A sample spiked with selenium(VI) and stored in 4 M hydrochloric acid for 7 days showed 60% conversion into selenium(IV). Sample storage in 1 M hydrochloric acid preserves the selenium(VI) as well as the selenium(IV). Freezing the samples avoids the introduction of contaminants and in addition, the frozen samples can be analyzed for volatile compounds—the methyl mercury species, even in airtight containers, are completely lost from liquid samples within a day.

Massee et al.[68] studied sorption losses for selenium from distilled water, during storage in containers made of borosilicate glass, high-pressure polyethylene, or Teflon. The effects of pH and storage times were studied and special attention was paid to the effect of the ratio of inner container surface area to sample volume. For selenium(IV) at the 8 $\mu g\,L^{-1}$ level, losses were insignificant in all three container materials, irrespective of the water matrix composition. This is probably because the selenium is present as oxy-acids, which are partly dissociated, and the anions do not adhere to the container walls.

Thompson et al.[69] spiked acidified river water with selenium at the 50 $\mu g\,L^{-1}$ level. They observed losses starting after 16 days of storage at 20 °C. Robberecht and Van Grieken,[70] using tracer experiments, showed that losses of both selenium(IV) and selenium(VI) from simulated natural waters onto Pyrex and polyethylene containers were negligible even after prolonged storage. Elemental selenium, however, appeared to be rapidly lost on both materials.

Reamer et al.[12] studied selenium adsorption on several hydride generation systems and construction materials by use of 75 Se as a radiotracer (100 $\mu g\,L^{-1}$). Polypropylene, two types of Teflon, and both silaned and unsilaned glass were evaluated. Glass and polypropylene cause the highest adsorption losses (23–32%) and silanols glass the lowest (2–8%). The adsorption decreased as a function of the number of reductions done and adsorbed selenide was leached from the material only by 2 N nitric acid.

The various data in the literature indicate that adsorption losses of selenium will depend on factors such as element concentration, chemical form, container material, contact time, pH, salinity, and suspended matter with microorganisms. Reduction of contact time and acidification with a strong acid will generally prevent or minimize the losses. Acidification, however, can change the initial composition of the aqueous sample, but storage of the unacidified water at 4 °C in pre-cleaned polyethylene bottles can prevent this problem. For the study of volatile organo-selenium compounds, freezing the samples is the best method of storage.

REFERENCES

1. King DI, Ciacco LL. *Sampling of natural waters and waste effluent*. New York: Marcel Dekker; 1997. pp. 451–481.
2. Grice GD, Harvey GR, Brown VT, Bachus RH. *Bull Environ Contam Toxicol* 1972;**7**:125.
3. Hume DN. Fundamental problems in oceanographic analysis in analytical methods. In: Gidd RP, editor. *Oceanography*. Washington: American Chemical Society; 1975. p. 1–8.
4. Riley JP. In: Riley JP, Skirrow G, editors. *Chemical oceanography*, vol. 2. London: Academic Press; 1975. p. 193–215.
5. Grasshoff K. *Fresenius Z Für Anal Chem* 1966;**220**:89.
6. Grasshoff K. *Verl Chem* 1976;**1**:50.
7. Mart L. *Fresenius Z für Anal Chem* 1979;**296**:350.
8. Riley JP, Robertson DE, Dutton SWR. In: 2nd ed. Riley JP, Skirrow G, editors. *Chemical oceanography*, vol. 3. London: Academic Press; 1975. p. 193.
9. Zief M, Michell JW. Contamination control. In: *Trace element analysis*. New York: Wiley; 1976.
10. Batley GE, Gardner D. *Water Res* 1975;**9**:517.
11. Robertson DE. In: Zief M, Speight R, editors. *Ultra purity, methods and techniques*. New York; Dekker; 1972. p. 207.
12. Reamer DC, Veilton C, Tokonsbalides T. *Anal Chem* 1981;**53**:245.
13. Truit RE, Weber JH. *Anal Chem* 1979;**51**:2057.
14. Scarponi G, Capodaglio G, Oescon P. *Anal Chim Acta* 1982;**135**:268.
15. Weber JH, Truitt RE. *Research report 21, water resource research centre*. Durham, N.H: University of New Hampshire; 1979.
16. Marvin KT, Proctor Jr RR, Neal RA. *Limnol Oceanogr* 1970;**15**:320.
17. Gardiner J. *Water Res* 1974;**8**:157.
18. Spencer DW, Manheim FT. *US Geol Surv Prof Pap* 1969;**650-D**:288.
19. Spencer DW, Brewer PG, Sachs PL. *Geochim Cosmochim Acta* 1972;**36**:71.
20. Dams R, Rahn KA, Winchester JW. *Environ Sci Technol* 1972;**6**:441.
21. Wallace Jr GT, Fletcher IS, Duce RA. *J Environ Sci Health* 1977;**A12**:493.
22. Duychserts G, Gillian G. In: Wanninen E, editor. *Essays on analytical chemistry*. Oxford: Pergamon; 1977. p. 417.
23. Smith RG. *Talanta* 1978;**25**:173.
24. Nurnberg HN, Valenta P, Maat L, Raspor B, Sipos L. *Fresenius Z für Anal Chem* 1976;**382**:357.
25. Bately GE, Gardner D. *Estuar Coast Mar Sci* 1978;**7**:59.
26. Bruland GW, Franks RP, Knauer GA. *Anal Chim Acta* 1979;**105**:233.
27. Burrell DC. *Mar Sci Commun* 1979;**5**:283.
28. Fukai R, Huynh-Ngoc H. *Mar Pollut Bull* 1976;**7**:9.
29. Figura P, McDuffie B. *Anal Chem* 1980;**52**:1433.
30. Cheam V, Agemian H. *Anal Chim Acta* 1980;**113**:237.
31. DeForest A, Pettis RW, Fabris G. *Aust J Mar Freshw Res* 1978;**29**:193.

32. Fuaki R, Huynh-Ngoc L. *Anal Chim Acta* 1976;**83**:375.
33. Zirino A, Leiherman SH, Clavelle C. *Environ Sci Technol* 1978;**12**:73.
34. Patterson CC, Settle DM, National Bureau of Standards Special Publication US; 1976, **422**: 321.
35. Moody JR, Whitfield M. *Anal Chem* 1977;**49**:2264.
36. Marchant JW, Klopper BC. *J Geochem Explor* 1978;**9**:103.
37. Henricksen A, Balmer K. *Vatten* 1977;**77**:33.
38. Calibrese EJ, Tuthill RW, Sieger FL, Klar JM. *Bull Environ Contam Toxicol* 1979;**33**:107.
39. Department of the Environment. *Lead in potable waters by atomic adsorption spectrometry*. London: HMSO; 1976. p. 6.
40. Sugai SF, Healey ML. *Mar Chem* 1978;**6**:291.
41. Mill AJB. PhD Thesis, Imperial College of Science and Technology, London; 1976.
42. King WG, Rodriguez JM, Wai CM. *Anal Chem* 1974;**46**:771.
43. *Trace metal decontamination procedures for sample containers sheet M-27*. Environmental Science Associates Inc., Burlington, MA.
44. Meranger JC, Subramanian KS, Chalifoux C. *Environ Sci Technol* 1979;**13**:707.
45. Shendrikar AD, Dharmarajan V, Walker-Merrick H, West PW. *Anal Chim Acta* 1976;**84**:409.
46. Cutter GA. *Anal Chim Acta* 1978;**98**:59.
47. Florence TM. *Water Res*; 197(11):681.
48. Subramian KS, Chakrabarti CL, Sueras JE, Mainse LS. *Anal Chem* 1978;**50**:444.
49. Whiteside P, editor. *Atomic absorption and electrothermal atomization*. Cambridge, UK: Pye Unicam Ltd; 1977. p. 14.
50. Laxon DPH, Harrison RM. *Anal Chem* 1981;**53**:345.
51. King WG, Rodrigues JM, Wai CM. *Anal Chem* 1974;**46**:771.
52. Poldoski JE, Glass GE. *Anal Chim Acta* 1978;**101**:79.
53. Nurnberg HW, Valenta P, Mart L, Raspor B, Sipos L. *Fresenius Z für Anal Chem* 1976;**282**:357.
54. Karin RW, Buone JA, Fashini JL. *Anal Chem* 1975;**47**:2296.
55. Jenne EA. *J Environ Qual* 1975;**4**:427.
56. Carron J, Agen H. *Anal Chim Acta* 1977;**92**:61.
57. Carr RA, Wilkniss PE. *Environ Sci Technol* 1973;**7**:62.
58. Coyne RV, Collins JA. *Anal Chem* 1972;**44**:1093.
59. Heiden RW, Aikens DA. *Anal Chem* 1977;**49**:68.
60. Heiden RW, Aikens DA. *Anal Chem* 1979;**51**:151.
61. Heiden RW, Aikens DA. *Anal Chem* 1983;**55**:2327.
62. Craig JH. *Anal Chim Acta* 1979;**110**:313.
63. Amber M. *Anal Chim Acta* 1977;**92**:55.
64. Mahon KI, Mahon SE. *Anal Chem* 1977;**49**:662.
65. Zhang GZ, Xu M, Fan Huanjing M. *Kexue* 1987;**8**:10.
66. Ramelhosa E, Segarder SR, Pereira E, Vale C, Duart A. *Int J Environ Anal Chem* 2003;**83**:81.
67. Cheam V, Aspila KI. *Interlaboratory quality control study, no. 26, arsenic and selenium in water report, no. 68*. Burlington, Ontario: Inland Water Directorate Water Quality Branch; 1980.
68. Massee R, Maessen FJMJ, Geotj JJM. *Anal Chim Acta* 1981;**37**:181.
69. Thompson M, Pahlavanpour B, Thorne LT. *Water Res* 1981;**15**:407.
70. Robberecht HJ, Van Grieken RE. *Anal Chem* 1980;**52**:449.

Chapter 3

Metals in River Water

Chapter Outline

3.1 ALUMINUM

3.1.1 Spectrophotometry

Various workers have described methods based on the use of pyrocatechol violet for the determination of down to 5 $\mu g\,L^{-1}$ aluminum in natural water.[1–4]

Seip et al.[5] have investigated two spectrophotometric procedures for determining various forms of aluminum in river and stream waters in connection with the investigations of the effects of acid rain in Norway. These methods employ ferron-orthophenanthroline[6] and pyrocatechol violet as chromogenic reagents.[7]

To isolate different aluminum fractions they used the fractionation scheme described by Driscoll.[8] Three aluminum fractions are determined; nonlabile monomeric aluminum, labile monomeric aluminum, and acid-soluble aluminum.

Interferences appear to cause considerable problems in the ferron orthophenanthroline method, and much less in the pyrocatechol violet method. The sensitivity of the latter method is also better. The pyrocatechol violet method is therefore recommended.

Wiganowski et al.[9] used bromopyrogallol red as chromogenic reagent in a flow-injection method for the determination of aluminum in river waters. The reagent solution contains bromopyrogallol red, *n*-tetradecyltrimethyl-ammonium bromide, and hexamine in 60% ethanolic solution, and the carrier solution contains acetate buffer, 1,10-phenanthroline and hydroxyl-ammonium chloride. Sample solutions acidified by sulfuric acid are injected and the peak absorbance at 623 nm is recorded. The detection limit is about 0.1 $\mu g\,L^{-1}$ and calibration plots are linear for the ranges up to 300 $\mu m\,L^{-1}$ aluminum.

Bromopyrogallol red reacts with metal ions such as iron, zinc, lead, manganese, cobalt, nickel, copper, cadmium, etc. In the above this method, 10^{-5} M iron(II) and iron(III) interfere if no masking agent is added.

Wiganowski et al.[9] pointed out that river water may contain several forms of aluminum that do not react with bromopyrogallol red, e.g., complexed or undissolved species.

To convert all forms of aluminum in the sample to a form that is included in this method, the sample (fixed with 5 mL L^{-1} concentrated sulfuric acid when sampled) is concentrated to 10 mL by boiling, cooled, diluted with water to 50 mL, filtered, and then injected into the analyzer. This treatment converted 85–110% of the various forms of aluminum present in river water samples to an ionic form which is included in the bromopyrogallol red procedure.

Other reagents that have been used for the spectrophotometric determination of aluminum include chrome azurol L,[9,10] eriochrome cyanine R,[11–13] ferron, orthophenanthroline quinolin-8-ol, and 1,10-phenanthroline.[14] Sampson and Fleck[15] described a method having a detection limit for aluminum of 5 μg L^{-1}. This method was used routinely for the 25–250 μg L^{-1} range. Interferences from other cations were corrected or avoided altogether. The basis of the method is to form a complex of aluminum, chrome azurol S, and cetylpyridinium chloride. Absorbance was measured at 640 nm.

3.1.2 Fluorometry

Hydes and Liss[16] describe a fluorometric method utilizing lumogallion for determining down to 0.05 μg L^{-1} aluminum in river waters. In waters abnormally rich in dissolved organic matter, competition for the dissolved aluminum between natural organic ligands and the Lumogallion reagent is overcome by ultraviolet irradiation prior to analysis. All forms of aluminum in filtered water may be detected except when the aluminum occurs in stable mineral structures, such as clay particles, small enough to pass through the filter.

The precision of the method is 0.05 μg L^{-1} aluminum at the 1 μg L^{-1} aluminum level and 0.6 μg L^{-1} aluminum at the 22 μg L^{-1} aluminum level.

De Pablos et al.[17] have given details of the development of a procedure for spectrofluorimetric determination of aluminum in a water/dimethylformamide medium of pH 3.7, using *N*-oxalylamine (salicylaldehyde hydrazone) as the fluorometric reagent. Under these conditions, the complex had excitation and emission maxima at 387 and 474 nm, respectively, and the detection limit was 5 μg L^{-1}. The effects of various possible interfering substances are discussed.

Sanchez Rojas et al.[18] described a fluorometric procedure for determining aluminum in natural waters based on the reaction of aluminum with *N*-(3-hdroxy-2-pyride)salicylaldimine. The reaction is carried out in the pH range of 4–6 in an aqueous *N,N′*-dimethylformamide medium and has been applied to aluminum in the range of 3.5–400 μg of aluminum L^{-1}. The limit of detection is 1.4 μg L^{-1}, and the relative standard deviation is 1.9%.

Yuan[19] used cetyltrimethylammonium bromide and aluminon in an acetic acid-sodium acetate buffer to form an aluminum complex for the spectrophotometric determination of aluminum in natural waters. Beer's law was obeyed in the range of 0–700 μg L^{-1} of aluminum. Iron interferes with the determination of aluminum. The recovery ranged from 96% to 102% and the relative standard deviation of the determination ranged from 1.4% to 28.3%.

Vilchez et al.[20] used solid-phase spectrofluorimetry to determine down to 20 ptt of aluminum in natural waters.

Sutheimer and Cabaniss[21] determined down to 3.7 nm aluminum in natural water by flow-injection analysis followed by fluorescence detection.

Other reagents that have been used in the fluorimetric determination of aluminum include ferron.[22] Morin,[23] 8-hydroxyquinoline-5-sulfonic acid,[24] eriochrome cyanine R,[25] stilbazo and zephiramine,[26] and lumigallon.[27] Zhu et al. reviewed absorption and fluorometric methods for the determination of aluminum.[28]

3.1.3 Ultraviolet Emission Spectrometry

Uchino et al.[29] used the 167.1-nm vacuum ultraviolet emission line to determine aluminum in lakewater by inductively coupled plasma atomic emission spectrometry. Improved sensitivity over more commonly used lines was demonstrated and 60 elements were used to assess the selectivity, iron having the nearest emission line to aluminum. High-quality mirrors, gratings, and lenses were required to ensure high tranmissivity at such a short wavelength. Water samples from the unpolluted Mashu Lake were analyzed without preconcentrations and aluminum concentrations of 2–3 $\mu g L^{-1}$ were detected.

3.1.4 Flow-Injection Analysis

Various chromogenic reagents have been used for the determination of aluminum by flow-injection analysis. These include eriochrome cyanine,[30,31] pyrocatechol violet,[31,32] aluminon,[31] and chrome azurol S.[33] Using eriochronic cyanine R with cetyltrimethylammonium bromide as a cationic surfactant, Roeyset[30,31] was able to determine down to 0.1 $\mu L L^{-1}$ aluminum in natural waters. He reported a sampling rate of 120 samples per hour. Roeyset[30] also compared pyrocatechol violet, aluminum, eriochrome cyanine R, and eriochrome cyanine R with cetylmethylammonium bromide as chromogenic reagents. The effects of acidity of the samples on sensitivity were examined. Interferences from iron, fluoride, and phosphate in the determination of aluminum were studied. The detection limits were 5 μg aluminum for the pyrocatechol violet and eriochrome cyanine R method, 1 $\mu g L^{-1}$ aluminum for the eriochrome cyanine-cetyltrimethylammonium bromide method, and 50 $\mu g L^{-1}$ aluminum for the aluminon method. The pyrocatechol violet method was recommended for the determination of aluminum in natural waters.

Zolzer and Schwedt[33] have described continuous-flow and flow-injection analysis methods for the spectrophotometric determination of traces of labile aluminum in water and soil. These are based on complex formation with chrome azurol S.[33] Interference from ferric ions was suppressed by reduction with ascorbic acid. The use of glass equipment must be avoided to eliminate adsorption of aluminum ions. For equilibrium solutions with a pH value of 4.0–8.0, only the labile fraction is determined by the method—in contrast to other analytical methods such as atomic adsorption spectrometry. Detection limits are 10 $\mu g L^{-1}$ aluminum. Labile aluminum is between 0.9% and 32% of total aluminum content, depending on the pH of the sample.

Chung and Ingle[34] have discussed kinetic data obtained from individual peak profiles corrected for dispersion in the application of the flow-injection analysis to the determination of down to 0.13 $\mu g\,L^{-1}$ of aluminum in natural waters.

3.1.5 Atomic Absorption Spectroscopy

Brueggemeyer and Fricke[35] compared furnace atomization behavior of aluminum from standard and thorium-treated L'vov platforms. Pretreatment with thorium causes a considerable sharpening of the peaks as well as improved thermal stability during charring.

Takahashi et al.[36] employed electrothermal atomic-absorption spectrophotometry to the determination of aluminum in geothermal waters. Matrix interferences were reduced by the addition of ammonium nitrate for the sample. The detection limit was 1 ppb, and the precision was 4.1% at the 20 ppb aluminum level.

Electrothermal atomic absorption spectrophotometry with matrix modification by ammonium in EDTA-magnesium chloride was described by Huang et al.[37] for the determination of aluminum in natural waters. The modifier was effective in removing interferences and enhancing the absorption signal. The detection limit was 0.9 μg of aluminum L^{-1}, recoveries ranged from 93.7% to 104.3%, and the relative standard deviation was 2.6%.

3.1.6 Ion Chromatography

The ion chromatography of mixtures of aluminum and iron has been discussed.[38] A sulfosalicylic acid-ethylene diamine medium (pH 5) is used, and the separation is achieved on a low-capacity cationic-exchange resin (Dionex CS-1). Postcolumn derivatization with chrome azurol S-acetyl methyl ammonium bromide (Triton X-100) was used.

Satiroglu and Tokgaz[39] have discussed the cloud point determination of aluminum in water samples and its determination by electrothermal atomic-absorption spectrometry, flame atomic absorption spectrometry, and UV-visible spectrophotometry.

3.1.7 Ion-Selective Electrodes

Using a fluoride-selective electrode, Redic[40] determined nanomole quantities of aluminum in natural water. He studied the rate and mechanism of the reaction between aluminum and fluoride ions in buffered aqueous solution. Potential time curves were recorded during the reaction, using a lanthanum fluoride electrode in conjunction with a reference electrode. The initial rates of decrease of the concentration of free fluoride ion were calculated and shown to be proportional to the amount of aluminum in solution. The procedure was used to determine aluminum in the range of 8–3000 nmol.

3.1.8 Differential Pulse Polarography

Ritchie et al.[41] have used differential pulse polarography for the direct determination of aluminum in natural waters. If the pH is carefully controlled to 4.00 ± 0.01, there is a linear relationship between the peak height of the polarographic wave and the aluminum concentration up to 600 $\mu g L^{-1}$ aluminum. The coefficient of variation is about 4% at the 250 $\mu g L^{-1}$ level. With increasing aluminum concentrations, the relationship ceases to be linear, and above 300 $\mu g L^{-1}$, the peak splits, probably because of hydrolysis and polymerization. Sodium, ammonium, magnesium, and calcium interfere at levels 100 times greater than that of aluminum, whereas iron(III), iron(II), copper(II), zinc(II), nickel(II), nitrate, perchlorate, chloride, and sulfate do not interfere.

3.1.9 Miscellaneous Techniques

Other methods that have been used to determine aluminum in water include inductively coupled plasma atomic-emission spectrometry, which has a detection limit of 0.4 ppb[42]; emission spectrometry[43]; photodiode ionization[44–46]; and cathodic stripping voltammetry.[47]

Various workers have discussed the speciation of aluminum in water.[46,48–51]

Lu et al.[52] fractionated dissolved aluminum in the field and the laboratory using a cation-exchange method. Although absolute differences between results obtained from field and laboratory fractionations were generally small, relative differences, expressed as the ratio between labile aluminum determined after laboratory fractionation and that obtained after field fractionation, could be large. The differences found were not statistically significant, although this may simply reflect the spread in the results. The ratio of laboratory fractionation to field fractionation had no apparent relationship with the temperature difference between the field and the laboratory. Although some significant correlations were found between the ratio at laboratory fractionation to field fractionation and hydrogen ion concentration, no significant correlations were found with the equivalent relative difference in hydrogen ion concentration between the laboratory and the field; nor was any significant correlation found with dissolved organic carbon.

Goenaga et al.[53] have shown that a progressive removal of particles from freshwater samples by filtration using various pore-diameter polycarbonate capillary membranes (0.4, 0.1, 0.05, and 0.015 μm) caused a reduction in the levels of labile aluminum (9–23%), as detected with pyrocatechol violet, in the filtrates. Removal of aluminum adsorbed onto suspended solids and aluminum losses through adsorption onto the membranes are thought to be responsible for these observations. Losses of aluminum during filtration of freshwater samples were evaluated by filtration of particle-free synthetic solutions and found to be <10%.

Driscoll[54] described a procedure for the fractionation of dilute acidic waters. Acid-soluble aluminum, nonlabile monomeric aluminum, and labile monomeric aluminum could be determined. The inorganic speciation of aluminum may be calculated by using labile monomeric aluminum, pH, fluoride and sulfate data with a chemical equilibrium model. Nonlabile monomeric aluminum is thought to contain colloidal aluminum. In natural water, levels of the labile monomeric aluminum increased exponentially with decreases in solution pH, while the nonlabile species were strongly correlated with organic carbon concentration.

3.2 AMMONIUM

3.2.1 Spectrophotometry

The Berthelot reaction[55–57] or modifications of it are commonly used for the spectrophotometric determination of ammonia. In such methods, the sample containing dissolved ammonia is made to react with a phenolic compound and a chlorine-donating reagent at a high pH in the presence of a catalyst. A complexing agent is usually also added to prevent the precipitation of metal hydroxides and other compounds. This reaction results in the formation of an intensely colored blue indophenol dye. The procedure is reported to be highly specific for ammonia as against other nitrogen-containing compounds (e.g., urea, amino acids), and is more sensitive than other commonly used methods, such as that based on Nessler's reagent.

Stewart[58] has shown that an increase in reaction temperature increased the rate of formation of indophenol blue, but also significantly reduced the sensitivity of the method. He recommended that the final solutions should be brought to the same temperature before mixing.

Verdouw et al.[59] have also used salicylic instead of phenol in the Berthelot reaction.

As a result of detailed studies on a modified Berthelot reaction using salicylate and dichloroisocyanurate instead of phenol and sodium hypochlorite, Kron[60] concluded that it is necessary to optimize the pH value for each combination of reagents used.

The implications of the work of Kron[60] on the choice of reagents and conditions for the determination of ammonia are as follows. There are no practical or theoretical disadvantages attached to the use of dichloroisocyanurate and sodium salicylate that detracted from the known advantages—namely, the increased stability of dichloroisocyanurate compared with sodium hypochlorite and the lesser toxicity of sodium salicylate compared with phenol.

Sodium nitrosylpentacyanoferrate(II) catalyst has been used with considerable success as the catalyst by a number of workers.

Crowther and Evans[61,62] have described an automated[63] distillation spectrophotometric method for determining ammonium in water.

Nickels[64] described a device for minimizing sample contamination during automated ammonium analyses.

Carson and Gross[65] have described a method for determining low $mg L^{-1}$ quantities of ammonia in natural water. The method involves reaction with a large excess of 2,5-dimethoxyoxolane in 1,2-dichloroethane. The resultant pyrrole is then determined spectrophotometrically following reaction with

(E)-*p*-dimethylaminocinnamaldehyde to form an intensely blue complex. Only primary amines interfere with the determination.

3.2.2 Continuous-Flow Fluorometry

Aoki et al.[66,67] have described a continuous-flow fluorometric method for the determination of ammonium in river water. The sample, in an outer tube, is treated with sodium hydroxide to release molecular ammonia which passes through a microporous PTFE membrane into an inner tube containing a buffered *o*-phthaldialdehyde reagent stream; the reaction product then enters a fluorimeter, and fluorescence is measured at 486 nm. The percentage recovery ranged from 95% to 105% over the concentration range 1–10 µm.

Evidence was found for slight interference in this method by inorganic salts (sodium chloride, sodium nitrate, and sodium sulfate), but only when their concentration exceeded 100 times that of ammonium. Some amines interfere and some do not.

3.2.3 Flow-Injection Analyses

Krug and Kizicka[68] and others[69] have described a rapid turbidimetric Nessler method for the automated determination of ammonia in natural water using flow injection analysis.

Krug et al.[70] used zone trapping in flow-injection analysis for the modified Berthelot spectrophotometric determination of low levels of ammonia in natural waters. Interference of metals is prevented by the addition of EDTA.

Van Son et al.[71] determined down to 1 $\mu mol\,L^{-1}$ total ammonia and nitrogen (ammonia plus ammonium ions) in water by flow-injection analysis and a gas diffusion membrane. In this method, the sample is injected into an alkaline stream in which the ammonium ions present are converted to ammonia molecules. This solution is fed to a cell with a gas-permeable membrane in which a reproducible exchange of ammonia with a flowing solution of the acidic form of bromothymol blue occurs. The resulting absorbance change is measured spectrophotometrically in a flow-through cell.

In a series of application notes, workers at Tecator Ltd[72–76] have described a series of flow inspection methods for the determination of between 50 $\mu g\,L^{-1}$ and 1000 $mg\,L^{-1}$ of ammonium nitrogen in natural waters.

Gailani et al.[77] described a miniaturized flow injection analysis system with online chemiluminescence detection based on the luminol–hypochlorite reaction for the determination of ammonium in river water.

3.2.4 Ammonia Electrodes

Hara et al.[78] have described a simple gas dialysis concentrator, consisting of a microporous poly(tetrafluoroethylene) membrane and polymer nets, combined with an ammonia selective gas electrode. This limited the detection limit of the electrode to about 3 $\mu g\,L^{-1}$ ammonium.

Hara et al.[79] have also described a computer-controlled automatic switching of a four-way valve that enabled the alternate introduction of the sample solution and distilled water to wash the electrode membrane. Rapid and precise determination of 0.1–10 mg L^{-1} ammonium ions was possible.

3.2.5 Molecular-Emission Cavity Analysis

Molar-emission cavity analysis has been applied[80] to the determination of ammonium (and nitrate) nitrogen in river water samples. The water sample is injected onto solid sodium hydroxide in a small vial. The ammonia generated is swept by nitrogen into a molecular emission cavity analyzer oxy-cavity and the intensity of the NO–O continuum is measured at 500 nm. The method is rapid, precise, and free from interference.

3.2.6 Ion Chromatography

Mizobuchi et al.[81] have described a method for the determination of ammonia in water based on reaction with fluorescamine to form a fluorophore, which is then chromatographed on trichrosorb RP-18 using a mobile phase consisting of 0.05 m phosphate buffer solution (pH 2.0) and acetonitrite (65:25 v/v). Average recoveries were from 95.8% to 100.7% and their standard deviations were from 1.9 to 5.4.

Fluorimetric determination of ammonia in rainwater and river water has been done by the Mizobuchi American Chemical Society.[81]

Conboy et al.[82] employed ion chromatography coupled with a mass spectrometric detector for the determination of ammonium ions (and sulfate) in natural waters.

3.2.7 Miscellaneous Techniques

Workers at the Department of the Environment in the United Kingdom[83] have compared three methods for the determination of ammonium in river water, namely direct titration, the use of an ammonia-selective electrode, and spectrophotometric methods depending on the formation of a colored complex related to indophenol blue. The upper limit of ammonia concentration for the various methods ranges from 1 to 100 mg L^{-1} with only the selective electrode method applicable to concentrations in excess of 50 mg L^{-1}.

Khadro et al.[84] have described a novel conductiometric sensor based on a PVC membrane containing nonactin for the determination of ammonia. Marquez[85] described a multipumping flow system[39] for chemiluminescent determination of ammonium in natural waters.

3.3 ARSENIC

3.3.1 Spectrophotometry Methods

Arsenic has been determined in river water by the silver diethyldithiocarbamate method.[86] Levels of up to 1.4 μg L^{-1} have been found in river and lake waters.

Stauffer[87] determined arsenic and phosphorus in groundwater by a reduced molybdenum blue spectrophotometric method.

In further work, Stauffer[88] evaluated the reduced molybdenum-blue method for estimating phosphorus, arsenic(V), arsenic(III), and total arsenic in the chemically diverse geothermal waters of Yellowstone National Park, Wyoming, and Steamboat Springs, Nevada. A systematic investigation of the arsenic(V) and phosphorus molybdate complexes in the presence of high levels of silica and fluoride, is reported. The method is very suitable for the determination of arsenic in waters containing these high levels of potentially interfering compounds. Nyamah and Torgbor[89] have described a spectrophotometric method for the determination of arsenic(V) in natural waters. Arsenic(V) reacts quantitatively with potassium iodite in the presence of sulfuric acid, releasing an equivalent amount of free iodine, which is extracted into chloroform, and measured at 515 nm. The detection limit was $0.5\,\mu g\,L^{-1}$.

Li and Wang[90] also used spectrophotometry for the determination of arsenic(III) and total inorganic arsenic in natural waters. Potassium borohydride was used as a reducing agent. Arsenic(III) was determined in an ammonium citrate buffer solution, and the total inorganic arsenic determined in tartaric acid, with arsenic(V) determined by difference. The detection limit was $0.16\,\mu g\,L^{-1}$ for arsenic(III) and $0.36\,\mu g\,L^{-1}$ for inorganic arsenic. The relative standard deviation was $\leq 3.5\%$.

3.3.2 Graphite Furnace Atomic-Absorption Spectrometry

Chakraborthi and Irgolic[91] have considered the interference of phosphate, sodium, sulfate, chloride, aluminum, nitrate, potassium, and selenous acid (singly and in combination) in the determination of arsenic by graphite furnace atomic-absorption spectrometry. The arsenic signals were not only dependent on the phosphate concentration but also on the phosphate/arsenic ratio. Ashing curves (arsenic signals as a function of ashing temperature) showed that arsenite and arsenate in salt-free solutions with $400\,mg\,L^{-1}$ (as nitrate, sulfate, or chloride) can be determined at ashing temperatures of 1100 °C. In the absence of nickel the signals are less intense and the ashing temperature should not be higher than 900 °C.

Solutions of arsenic compounds ($100\,\mu g\,L^{-1}$ arsenic) $400\,mg\,L^{-1}$ nickel, (0.1 M nitric acid) containing various salts were analyzed at an ashing temperature of 900 °C. The results indicate that sodium sulfate and phosphoric acid can certainly interfere with arsenic analysis, whereas there is no such influence of arsenic signals when these substances are present singly. Severe signal reductions were caused by mixtures of these salts.

Pacey and Ford[92] achieved arsenic speciation by ion-exchange chromatographic separation followed by analysis of the separated arsenic species by graphite furnace atomic-absorption spectrometry. The separation of monomethylarsenic acid, dimethylarsenic acid, and the tri- and pentavalent arsenic ions was achieved by anion-exchange chromatography, with the pH maintained between 4 and 10. The absolute detection limit was 0.5 ng. The minimum concentration of arsenic(V) in a sample that could be detected was 0.4 μg L^{-1} and the absolute amount of arsenic(V) detectable in the furnace was 0.2 ng.

Duttemans and Massart[93] described a method of optimization of the matrix modification technique for water samples for the determination of arsenic by graphite furnace atomic-absorption spectroscopy. In the presence of nickel, ashing temperatures of up to 1300 °C could be used without significant losses of arsenic. The addition of 1% nitric acid enhanced the peak height.

Subramanian et al.[94] have described a procedure for the determination of tri- and pentavalent and total arsenic in river waters by a method based on ammonium pyrrolidinedithiocarbamate-methyl isobutylketone extraction–graphite furnace atomic-absorption spectroscopy, and cross-checked the results by analyzing the samples using stripping voltammetry.[95]

Ficklin[96] has described a simple ion-exchange technique that separates arsenic(III) and arsenic(V) in groundwater samples for detection by graphite furnace atomic-absorption spectrophotometry.

Faust et al.[97,98] carried out an assessment of the chemical and biological significance of arsenic compounds in contaminated lake water. They determined arsenic by the graphite furnace atomic-absorption technique using a cation-exchange column to speciate the arsenic compounds (arsenite, mono- and dimethylarsonic acid). Concentrations in the sediments were found to be up to 4327 times higher than in water. Monomethylarsonic acid and arsenate were the predominant forms in water and sediments, respectively.

Tefalidct and Irgum[99] have described a method for volatilization of arsenic for facile and interference-free introduction of the element into a flame atomic spectrometer. The generation is based on conversion of trivalent arsenic in solution into low-boiling arsenic trichloride by strong hydrochloric acid. The arsenic trichloride gas is produced in a flow system and separated from the liquid sample stream by a permeable tube, whose outer side is purged with hydrogen gas. Subsequent detection of arsenic(III) is achieved by using an oxygen-hydrogen flame-in-tube atomizer and atomic absorption spectrometry. In this method the tolerance towards the interferents cobalt, copper, nickel, and iron is improved considerably, when compared with hydride generation methods using sodium tetrahydroborate(III) as reagent.

3.3.3 Hydride-Generation Atomic-Absorption Spectrometry

Various workers have discussed the application of this technique to the determination of arsenic species in natural waters.[100–125]

Shaikh and Tallman[109] determined total arsenic and its speciation in natural waters by flameless atomic-absorption spectrometry with nanogram sensitivity. The arsenic species are reduced to the hydrides and collected in a liquid nitrogen trap. They are selectively vaporized from the trap and directly injected into a graphite furnace. Total arsenic calculated from speciation analysis agreed to within 10% with that determined by a sample digestion procedure. Welz and Melcher[113] studied the effect of the valency state of arsenic on the degrees of signal depression caused by copper, nickel, and iron in the determination of arsenic by hydride-generation atomic-absorption spectrometry.

Sun et al.[121] used the reducing agent potassium iodide-sulfourea in dilute sulfuric acid to reduce arsenic to arsenic(III), followed by sodium borohydride reduction to arsine. The arsine is atomized in a quartz tube at 920 °C and determined by atomic absorption. The sensitivity of the method was $0.2\,\mu g\,L^{-1}$ and the relative standard deviation was ≤10%.

In the determination of arsenic in natural waters, Narasaki[122] collected the generated hydride in a gas-collection device prior to introduction into the quartz tube atomizer of the atomic absorption spectrophotometer.

Pierce et al.[123–125] have described a hydride-generation atomic-absorption spectrometric procedure for determining arsenic in natural waters that, it is claimed, is relatively free from interference. In this automated method, the sample is first carried through a persulfate digestion to convert organo-arsenic species to the inorganic form and all inorganic arsenic to the pentavalent form. The solution is then acidified with hydrochloric acid and reduced to arsine by addition of sodium borohydride; an electrically heated quartz tube is employed as atom cell.

The effects of a wide range of potential interfering substances have also been tested, and none has been found to produce an effect greater than ±10% at the maximum level likely to occur in potable waters.

Chen et al.[104] has described a procedure for the co-precipitation of AsIII and AsV with zirconium hydroxide. AsIII can be volatilized on a zirconium-coated graphite furnace at greater than 1400 °C, leaving AsV behind for subsequent determination at higher temperatures.

Brovko[105] determined both arsenic and selenium in natural water by hydride-generation atomic-absorption spectrophotometry. The hydrides were pre-concentrated on the antinomy-coated inner surface of the electrothermal atomizer. The limits of detection were 1 and 0.5 ng for arsenic and selenium, respectively. Narasaki et al.[106] collected the generated hydride in a gas-collection device prior to introduction into the quartz tube atomizer of the atomic absorption spectrophotometry.

Anderson et al.[115] and Kumar and Ryuzuddin[102] carried out the selective reduction of arsenic species AsIII, AsV, monomethyl arsenic acid, and dimethyl arsinic acid in river water by continuous hydride generation. These workers investigated conditions for using this approach to determine different arsenic species, using sodium tetrahydroborate as reducing agent and atomic absorption spectrometry for detection. The pH value was not the only factor affecting reduction; other factors such as kinetic control and complexation were also involved. Using molecular emission spectrometry with hydride generation, Matsumoto and Fuwa[116] determined subnanogram amounts of arsenic in river water and obtained a detection limit of 0.2 ng. Narasaki and Fuwa[117] determined nanogram levels of arsenic and selenium using a flow-injection batch system where hydride gas stored in a gas liquid separator up to an appropriate pressure, is swept automatically into an atomic absorption furnace. This process resulted in an effective reduction of the elements to their hydrides and minimized reagent consumption. Sensitivities were between 0.01 and 0.004 absorbance ng^{-1}.

Hinners[114] claims that the presence of methylated organoarsenic (such as trimethylarsinic acid) compounds can cause errors in methods based on the direct conversion of arsenic to its hydride followed by atomic absorption spectrometry. Speciation of inorganic arsenic(III) by hydride evolution directly into an atomic absorption system was found to be subject to error. Organic forms of arsenic can produce an underestimation of total arsenic when the hydride response from concentrated acid is quantitated against inorganic arsenic. While the method of Hinners[114] is reliable, when applied to solutions containing only arsenate and arsenite, in the presence of organoarsenic compounds, which commonly co-occur with inorganics in water samples, unreliable analyses will result. It is necessary to separate all four arsenic species from each other before applying atomic absorption spectrometry.

Arbab Zavar and Howard[110] describe the development of optimized conditions for an automated method of determining arsenic down to $0.9\,ng\,L^{-1}$ that uses hydride-generation atomic-absorption spectroscopy.

Aggett and Hayashi[118] have discussed interferences by copper(II), cobalt(II), and nickel(II) on the determination of arsenic by arsine-generation atomic-absorption spectrometry.

Hagen and Lovett[119] have described the use of an iodine-trapping solution in the determination of arsenic by hydride-generation graphite furnace atomic-absorption spectrometry with hydride generation. Due to the equivalence of the absorption for the trapped arsenic in the iodine solution and aqueous arsenic solutions, hydride generation efficiencies can be determined directly. The high trapping efficiencies of the iodine solution ($96.2 \pm 0.9\%$ at the $4\,ng\,As\,mL^{-1}$ level), combined with nickel matrix modification to prevent charring losses of arsenic, provide a simple, rapid, and inexpensive technique for arsenic isolation and determination. A detection limit of 30 pg of injected arsenic ($0.6\,\mu g\,As\,L^{-1}$ of aqueous sample) was obtained.

Pallier et al.[103] adapted voltammetry to qualify As(III) in model water samples with a low mineral content and compared results obtained by this procedure to those obtained by hydride generation atomic absorption spectrometry. The specificity lies in the study of the influence of interfering species on the quantification of samples by coagulation using iron salts. First, a square-wave cathodic stripping voltammetry method at a hanging mercury drop electrode was optimized and compared with the hydride-generation atomic-absorption spectrometric method. Second, the influence of the presence of Fe(II), Fe(III), and As(V) was evaluated. No interference from As(V) was observed, however, for both methods; the As(III) signal decreased by 10% in the presence of iron. A standard addition method was thus required to efficiently quantify As(III) in the complex matrix. The square-wave cathodic stripping voltammetric method (limit of quantification = $0.5\,\mu g\,L^{-1}$)

was preferred to the hydride-generation atomic-absorption spectrometry technique to determine low As(III) levels down to ≤2 μg L^{-1}.

Kumar and Ryuzuddin[102] reviewed nonchromatographic hydride-generation atomic-absorption spectrometric techniques for the determination of arsenic and also antimony, selenium, and tellurium in natural waters.

Hydride generation coupled with AAS, ICP–AES, and AFSA have been used for the speciation of arsenic, antimony, selenium, and tellurium in the environmental water samples. Careful control of experimental conditions such as offline/online sample pretreatment methods employing batch, continuous, and flow-injection techniques, and cryogenic trapping of hydrides, enable the determination of various species of hydride-forming elements without the use of chromatographic separation. Other nonchromatographic approaches include solvent extraction, ion exchange, and selective retention by microorganisms. Sample pretreatment, pH dependency of hydride generation, and control of sodium borohydride/hydrochloric acid/hydride generation concentration facilitate the determination of As(III), As(V), monomethylarsonate, and dimethyl arsinate species. Inorganic species of arsenic are dominant in terrestrial waters, whereas inorganic methylated species are reported in seawater. Selenium and tellurium speciation analysis is based on the hydrides generation only from the tetravalent state. Se(IV) and Se(VI) are the inorganic species mostly reported in environmental samples, whereas speciation of tellurium is rarely reported. Antimony speciation analysis is based on the slow kinetics of hydride formation from the pentavalent state and is mainly reported in seawater samples.

3.3.4 Inductively Coupled Plasma Atomic-Emission Spectrometry

Davies and Kempster[126] studied the potential interferences to the determination of arsenic in natural waters by inductively coupled plasma emission spectrometry using a scanning monochromator. Three of the four arsenic emission lines (189.042, 193.759, and 228.812 nm) have adequate sensitivity for natural waters, but the 197.262 nm line does not have adequate sensitivity. They report significant interferences from aluminum, cadmium, chromium, cobalt, and vanadium on one or several of the arsenic emission lines.

Huang et al.[127] used an in situ nebulizer–hydride generator with an inductively coupled plasma mass spectrometer to determine arsenic down to 3 ptt in natural waters.

Feng and Cao[128] determined AsIII and As(V) by a method based on the reaction kinetics of As(V) with L-cysteine. The arsenic species were converted to hydrides and analyzed by inductively coupled plasma atomic emission spectrometry. The detection limits were 3.4 and 0.7 μg L^{-1} for As(V) and As(III), respectively. The relative standard deviations were 1.45% and 1.21% for As(V) and As(III), respectively, at a concentration of 100 μg L^{-1}.

Davies and Kempster et al.[129] studied the potential of inductively coupled plasma atomic emission spectrometry for determining arsenic. Three of the four arsenic emission lines (189.042, 193.759 and 228.812 nm) have adequate sensitivity for natural waters, but the 197.262-nm lines does not have adequate sensitivity. They reported significant interferences from A1, Cd, Cr, Co and V. An anion exchange polymer-based column has been used to separate arsenic species prior to the generation of hydride,[130] which were subsequently analyzed by inductively coupled plasma atomic emission.

3.3.5 Inductively Coupled Plasma Mass Spectrometry

Various workers have applied inductively coupled plasma mass spectrometry to the determination of arsenic in natural waters.[131–135]

Krystek and Ritsema[134] carried out a validation assessment on the determination of arsenic by inductively coupled plasma mass spectrometry. Vasquez et al.[135] have described a routine ICP-MS method for the determination of arsenic and selenium in fresh and sewage water. After a first phase of optimization, where the torch alignment, flow of gases, and ion optic adjustments were calibrated, the method was validated successfully. The limits of detection, linearity, working range, sensitivity, interferences, precision, and accuracy were studied. Groundwaters were spiked at four different matrices at three levels of concentration. The limits of detection obtained were 0.2 and 0.8 $\mu g L^{-1}$. The percentages of linearity obtained were 99.2% for arsenic and 99.8% for selenium. All recovery values were according to the AOAC intervals, from 95% for arsenic in a wastewater matrix to 106.4% for selenium in the seawater matrix. The accuracy was also determined via a proficiency test, resulting in acceptable Z-scores of 0.65 and 0.4 for arsenic and selenium, respectively.

3.3.6 Flow-Injection Analysis

Yaqoub et al.[136] have developed a simple and rapid flow-injection method for the determination of arsenic(V) based on luminol chemiluminescence detection. The molybdoarsenic heteropoly acid formed by arsenic and ammonium molybdate in the presence of vanadate in acidic conditions generated chemiluminescence emission via the oxidation of luminol. The limit of detection was 0.15 $\mu g L^{-1}$, with a sample throughout of 7120 $L h^{-1}$. A linear calibration graph was obtained over the range 0.15–7.5 $\mu g L^{-1}$, with a relative standard deviation ($n=4$) in the range of 0.8–2.5%. Interfering cations were removed by passing the sample through an inline iminodiacetate chelating column, and phosphate (at 0.6 $mg L^{-1}$) was removed offline by a magnesium-induced co-precipitation method. The method was applied to freshwater samples and the results obtained were in reasonable agreement with the results obtained using hydride-generation atomic-absorption spectrometry as the reference method.

Arsenic speciation studies have been carried out by several workers.[137–139]

Gian and Tong[140] have developed an inductively coupled plasma mass-spectrometric method for the determination of arsenic and selenium. After a first phase of optimization, where the torch alignment, flow of gases, and ion optic adjustments were calibrated, the method was validated successfully. The limits of detection, linearity, working range, sensitivity, interferences, precision, and accuracy were studied with three procedures for the method: groundwaters were spiked at four different matrices at three levels. The limits of detection obtained were 0.2 and 0.8 $\mu g L^{-1}$ for arsenic and selenium, respectively. Percentages of linearity obtained were 99.2% for arsenic and 99.8% for selenium. All recovery values were 95.3% for arsenic and 104.6% for selenium. The accuracy was also determined via proficiency testing, resulting in acceptable Z-scores of 0.65 and 0.4 for arsenic and selenium, respectively. The complete method allowed analysis of water samples according to European Directive 200/60 EC.

Solanki et al.[141] have described a surface plasma resonant-based DNA biosensor for the detection of As(III). This sensor was fabricated using a self-assembled monolayer of β-merceptoethanol (MCE) deposited on a gold substrate. The double-stranded calf thymus deoxyribonucleic acid (dsCT-DNA) was immobilized onto an MCE/Au electrode using *N*-ethyl-*N*-*N'*-(3-diemethylaminopropyl) carbodiimide and *N*-hydroxysuccinimide chemistry. This dsCT-DNA/MCE/Au bioelectrode was characterized using cyclic voltammetry, scanning electron microscopy, and X-ray photoelectron spectrometry, respectively. The detection limit of arsenic trioxide was $0.01 \times 10^{-6} g mL^{-1}$.

Amberger et al.[142] described a continuous-flow thin-layer electrolysis cell with platinum cathode in combination with a microwave plasma torch operated with argon as the working gas for the optical-emission spectrometric determination of arsenic with the hydride technique. Under the optimized conditions, the limit of detection (3σ) in the case of As(I) 228.1-nm emission line was 81 $ng mL^{-1}$. The influence of the transition metals Cu(II), Fe(III), and Ni(II), of the hydride-forming elements Sb(III), Se(V), and Sn(II) and of sodium on the determination of arsenic was studied. Cu(II) was found to be the strongest interferent, as in the presence of 100 $\mu g mL^{-1}$ of Cu(II) the signal for 3 $\mu g mL^{-1}$ of arsenic was reduced to 4% of the signal without interferents. Sn(II) and Sb(III) were found to yield an increase of the signal for arsenic. L-cysteine and KI/ascorbic acid (1:1) at a concentration of 2% were found effective to reduce the interferences of Cu(II), Fe(III), and Ni(II). For a solution containing 3 $\mu g mL^{-1}$ of arsenic and 100 $\mu g mL^{-1}$ of Ni(II), it was shown that in the presence of L-cysteine or KI/ascorbic acid, the signal for arsenic was 99% and 94% of the one without interference, whereas it was only 43% without masking reagents. Arsenic could be determined in water samples from a copper refinery. It was found that the amount of arsenic determined with hydrogen-generation microwave plasma-torch optical-emission spectrometry agrees with values determined by FAAS ICP-OES at the 0.02 and 1.6 $\mu g mL^{-1}$ level respectively.

3.3.7 Phosphorimetry

Liu et al.[143] have described a new catalytic solid substrate room-temperature phosphorimetry method for the determination of trace arsenic(V). It is based on the fact that fullerenol emitted strong and stable room-temperature phosphorescence on nitric acid cellulose membrane substrate. The method could oxidize fullerenol to cause the quenching of room-temperature phosphorescence. As(V) could catalyze hydrogen peroxide to fullerenol and decrease the room-temperature phosphorescence signal of fullerenol sharply. After adding tween-80 in the system, its Δlp enhanced 7.7 times compared with the without-tween-80 levels. Under the optimum conditions, the linear dynamic range of this method was 2.3×10^{-17} g mL^{-1}.

Phosphorescence spectre of fullerenol-tween-80, fullerenol-hydrogen peroxide, and fullerenol-hydrogen peroxide-As(V) showed that fullerenol will emit weak room-temperature phosphorimetry on a nitric acid-cellulose membrane at 478 nm with a reagent blank intensity of Ip = 55.6.

Tween-80 can enhance the room-temperature phosphorescence signal of fullerenol with a blue shift of (max for 9.3 nm, curve 5.5′). The reason for this might be that tween-80 could react with fullerenol and then forms a micelle complex, making fullerenol molecules in the micelle orderly.

In the presence of hydrogen peroxide, the room-temperature phosphorescence signal of fullerenol was quenched sharply and remains almost unchanged, which might be explained by the fact that hydrogen peroxide oxidizes fullerenol to form a nonphosphorescent component. After arsenic is added, the room-temperature phosphorescence of fullerenol was quenched more sharply.

Results showed that As(V) catalyzed hydrogen peroxide so that it oxidizes fullerenol Consequently, 470/637 nm was chosen as the working wavelength to determine traces of As(V).

3.3.8 Voltammetry

Pallier et al.[144] adapted voltammetry to quantify As(III) in model water samples with a low mineral content and compared results to those obtained by hydride-generation atomic-absorption spectrometry method. The specificity is based upon a study of the influence of iron salts on the reported arsenic III concentration. A square-wave cathodic-stripping voltammetry method at a hanging mercury drop electrode (SW-CSV–HMDE) was optimized and compared with the hydride-generation atomic-absorption method. The influence of the presence of Fe(II), Fe(III), and As(V) was evaluated. No interferences from As(V) were observed; however, for both methods, the As(III) signal decreased by 10% in the presence of iron. The standard addition method was thus required to efficiently quantify As(III) in a complex matrix and the SW-CSV (limit of quantification = 0.5 μg L^{-1}) method was preferred to the hydride generation atomic absorption spectrometry technique to determine low As(III) levels ([As(III)] ≤ 2 μg L^{-1}).

The deposition time is an important factor in cathode voltammetry.

3.3.9 Miscellaneous Techniques

Other techniques for the determination of inorganic arsenic in natural waters are reviewed briefly in Table 3.1.

TABLE 3.1 Miscellaneous Techniques for the Determination of Inorganic Arsenic in Natural Waters

Form of Arsenic	Type of Sample	Techniques	Detection Limit	References
As; also Se, Sn, Sb	Water	Gas chromatography	0.001 ppb	158
As03, As04, methyl arsenate	Water	High-performance liquid chromatography with hydride-generation atomic absorption	2 ng	159
As	Water	Absorption spectrometry detection, high-performance liquid-chromatography with platinum disc detector	–	160
Methyl arsenic species	Water	High-performance liquid chromatography, separated species reduced by hydride generation and determined by inductively coupled plasma atomic absorption spectrometry	–	161
As(III), As(V)		Ion exclusion chromatography	0.012 μm	162
As(III), As(V)	Water	Natural radioanalytical method	–	163
As(III) As(V)	Freshwater	Differential-pulse anodic stripping voltammetry using rotating gold electrode	0.2 μg L^{-1}	164
As		X-ray fluorescence spectroscopy	0.2 μg	165,166
As, Sb	Water	Co-precipitation with $Zr(OH)^4$ then determination by X-ray	0.8 μg As	167
			6.1 μg Se	
As(III)	Water	Neutron activation analysis of pyrrolidiene carbodithioate	–	168
As(III), As(V)	Water	Selective hydride generation coupled with neutron activation analysis; numerous elements do not interfere; yields above 97%	–	169

Ali and Kimar Jain[157] reviewed the speciation techniques of arsenic. The principal advanced techniques discussed are gas chromatography, reversed-phase liquid chromatography, ion chromatography, and capillary electrophoresis. Some other techniques are also mentioned. The extraction procedures of arsenic species from unknown samples are also discussed. Arsenic speciation is summarized in tabular form and optimizing parameters are also discussed.

3.4 ANTIMONY

3.4.1 Spectrophotometry

Abu Hilal and Riley[170] have described a spectrophotometric procedure for the determination of antimony in river water. After a preliminary oxidative digestion, the element is quantitatively co-precipitated at pH 5.0 with hydrous zirconium oxide. The precipitate is dissolved in acid, and after reduction with titanium(III) chloride, antimony is oxidized to antimony(V) with sodium nitrite. The ion-pair of the Sb C^l_b ion with crystal violet is extracted with benzene and its absorbance is measured at 610nm. The detection limit is 0.005 $\mu g L^{-1}$; relative standard deviations are 0.5% for spiked water (0.5 $\mu g L^{-1}$). A wide range of anions and cations cause no interference at levels many times those in river waters.

An indirect spectrophotometric determination of antimony(III) in geothermal waters has been described.[171] The method involved oxidation of the pentavalent form with potassium dichromate and diphenylcarbazide. Optimal conditions were established for the procedure and for the elimination or reduction of interferences from arsenic and vanadium. Determinations in the 0.05–5.0 $mg L^{-1}$ antimony range were rapidly and easily performed.

3.4.2 Atomic Absorption Spectrometry

Stauffer[172] has described a procedure for determining down to 6 $\mu g L^{-1}$ total antimony in geothermal waters in which antimony is oxidized to the pentavalent state with sodium nitrite, and a methyl isobutyl ketone extract containing quinquivalent antimony is centrifuged to remove silica, followed by use of a flame volatilization technique. The detection limit is equivalent to 6 $\mu g L^{-1}$ in the original sample, with a coefficient of variation of 4% at a level of 250 μg antimony L^{-1}. Bertine and Lee[173] have described hydride generation techniques for determining antimony(V) and antimony(III).

Xu and Fang[174] combined online ion-exchange separation with flow-injection hydride-generation atomic-absorption spectrophotometry for the determination of antimony. The antimony was preconcentrated on a microcolumn with CPG-8Q and then eluted with hydrochloric acid into the hydride generator. The detection limit was 1.5 $ng L^{-1}$ with a sampling frequency of 50/h. The relative standard deviation was 1.0% at a concentration of 0.5 μg of SbL^{-1}, and the recoveries were 97–102%.

Xu and Wan-Dxa[175] co-precipitated both Sb(III) and Sb(V) with ammonium pyrrolidinedithiocarbamate and then extracted into methyl isobutyl ketone prior to determination by atomic absorption spectrometry. The detection limit was 2 $\mu g L^{-1}$.

La Lalle Guntinas et al.[176,177] immobilized Sb(III) with fructose-6-phosphate in a glass microcolumn prior to elution with lactic acid and determination by electrothermal atomic absorption spectrophotometry.

3.4.3 Stripping Voltammetry

Capodaglio et al.[178] used cathodic stripping voltammetry for the determination of antimony in natural waters. The antimony was preconcentrated by adsorptive deposition with catechol. The deposition potential was −1.0V, in order to eliminate interference from uranium. The limit of detection was approximately 0.2 nM antimony.

3.4.4 Miscellaneous Techniques

Neutronivation water on analysis has been applied to the determination of SbIII in water.[179] Sharma and Patel[180] have discussed the speciation of antimony in natural waters, and Calle Guntinas[181] reviewed the speciation of antimony in river waters.

3.5 BARIUM

3.5.1 Atomic Absorption Spectrometry

Rollenberg and Curtius[182] determined traces of barium in natural waters including sea water using a standard addition–atomic absorption spectrometric technique. For saline waters, the barium is first separated from interfering ions by ion-exchange chromatography with 0.1 M EDTA solution.

3.5.2 Graphite Furnace Atomic-Absorption Spectrometry

Sun[183] determined barium in natural waters by utilising a tantalum-lined graphite furnace and atomic absorption spectrophotometry. Sensitivity was improved by 20-fold over the standard graphite tube. A technique for the preparation and operation of the tantalum-lined tubes is described.

3.5.3 Miscellaneous Techniques

Ferrus and Torrades[184] have investigated the limit of detection in the determination of barium in water by the gravimetric barium-sulfate-precipitation procedure.

3.6 BERYLLIUM

3.6.1 Gas Chromatography

Beryllium has been determined[185] in natural waters. To the sample was added 0.1 M EDTA, 1.0 M sodium acetate, and 1,1,1-trifluoro-2-4-pentanedione. Following liquid–liquid extraction with benzene using detailed handling procedures, the organic phase was mixed with 1.0 M sodium hydroxide (de-emulsifier), washed several times with distilled water, and the resultant beryllium 1,1,1-trifluoro-2,4-pentanedione complex was analyzed by gas chromatography with electron capture detection. The detection limit was approximately 2.0 pM.

Tao et al.[186] describe a method combining gas chromatography and inductively coupled plasma emission spectrometry. Beryllium is extracted from a natural water sample with acetylacetone into chloroform and concentrated by evaporation.

The beryllium acetylacetonate is separated in a gas chromatograph and injected into the helium plasma emission spectrometer. The detection limit is 10 pg in a 30-mL water sample and the standard deviation was 4.1% at 10 ng of beryllium.

3.6.2 Miscellaneous Techniques

Other methods that have been used to determine beryllium in natural waters include atomic absorption spectrometry[187] with a detection limit of 4–8 $\mu g\,L^{-1}$, spectrofluorimetry (sub ppb),[188] catalytic polarography (40 $\mu g\,L^{-1}$),[189] and photothermal spectrometry (0.1 ppm).[190]

3.7 BISMUTH

3.7.1 Atomic Absorption Spectrometry

Arai et al.[191] complexed bismuth(III) 2,4,6-tri-2-pyridyl with 1,3,5-triazine perchlorate followed by extraction with nitrobenzene. The complex was determined by flame atomic absorption spectrometry. The detection limit was 5 $\mu g\,L^{-1}$, and the relative standard deviation was 2.3% for a concentration of 50 μg of Bi L^{-1}. Shijo et al.[192] completed bismuth with dithiocarbamate and then extracted into xylene, and back-extracted into nitric acid. Bismuth was then determined by electrothermal atomic-absorption spectrophotometry detection with a detection limit of 0.27 $ng\,L^{-1}$.

3.7.2 Hydride-Generation Atomic-Absorption Spectrometry

Lee[193] has given details of a procedure for the determination of traces of bismuth in freshwater, sea water, and marine samples. It involves the reduction of bismuth to bismuthine with sodium borohydride, stripping with helium gas, collection in a modified carbon rod atomizer, and detection by atomic absorption spectrometry.

The absolute detection limit of this method is 3 pg of bismuth. The precision of the method is 2.2% for 150 pg and 6.7% for 25 pg bismuth. Less than 0.15 $ng\,L^{-1}$ bismuth was found in U.S. lake water samples by this method.

3.7.3 Inductively Coupled Plasma Atomic-Emission Spectrometry

Nakahara et al.[194] describe a method for the continuous reduction of bismuth in natural waters with sodium borohydride followed by introduction of the bismuthine into an inductively coupled plasma atomic emission spectrometer.

The method has a detection limit of 0.35 μg L^{-1} of bismuth. The relative standard deviation was 2.8% at 2 μg L^{-1} and 1.3% of 200 μg L^{-1}.

Froediro et al.[195] developed an inductively coupled plasma mass spectrometry method for determining bismuth in natural waters.

3.7.4 Anodic Stripping Voltammetry

Mal'kov and Fedoseeva[196] have described a procedure based on anodic stripping voltammetry from a mercury-graphite electrode for the determination of nanogram amounts of bismuth in natural waters. In this procedure, bismuth is extracted from the water sample into chloroform as the diethyldithiocarbamate in the presence of sulfosalicylic acid, EDTA, and aqueous ammonia and re-extracted with aqueous hydrochloric acid.

This aqueous extract is washed with chloroform and mixed with water, and then filtered, boiled, cooled, and treated with 25% aqueous ammonia. Deposition is carried out on a mercury-graphite electrode[197] with passage of nitrogen through the solution. The potential, initially at −1.0 V versus the SCE for 1–2 s, is reduced to −0.6 V for the deposition of bismuth as an amalgam. Anodic stripping in the unstirred solution is carried out at −0.35 to −0.1 V, with the peak for bismuth occurring at 0.2 V.

3.8 BORON

3.8.1 Spectrophotometry

Various chromogenic reagents have been used in the determination of boron including methylene blue,[198] 4,6-di-tertbutyl-3-methyloxy catechol,[199] curcumin,[200] and H-resorcinol.[201]

In a relatively new method, Balogh et al.[202] used the reaction of tetrafluoroborate and Victoria blue 4R (VB4R) reagent to develop a new, simple, rapid, and sensitive method for the spectrophotometric determination of boron. The method is based on the reaction of boric acid with fluoride, which forms the tetrafluoroborate anion, and followed by the extraction of Victoria blue 4R into benzene and subsequent spectrophotometric detection. The molar absorptivity of the investigated complex is 9.6×10^4 L mol^{-1} cm^{-1} at 610 nm. The absorbance of the colored extracts obeys Beer's law in the range 0.03–0.55 mg L^{-1} of B(III). The limit of detection calculated from a blank test ($n = 10$; $p = 0.95$) based on 3 s is 0.02 mg L^{-1} of B(III). Under appropriate extraction conditions, the majority of metals (excluding probably tantalum and some others) did not form extractable fluoride complexes with Victoria blue 4R. Therefore, the presence of small quantities of metals should not interfere with the determination of boron in the presence of a sufficient surplus of fluoride. Exceptions, however, are metal ions such as Sn(V), Ti(IV), Hg(II), and Ti(I), which are strongly hydrolyzed under experimental conditions. Some anions formed complexes with the cation of Victoria blue 4R and are easily extractable using benzene. Examples of such ions are 1-, SCN, bromide, chlorate, iodate, and perrhenate. These anions strongly interfere with boron determination and therefore must first be extracted with Victoria blue 4R into benzene for determination. The Cl^-, CH_3COO^- and other anions are extracted in very small amounts or not at all and do not interfere with boron determination.

3.8.2 Miscellaneous Techniques

Other methods that have been used to determine boron in natural waters include atomic absorption (L-D 20 μg L^{-1}[203], inductively coupled plasma-emission spectrometry,[204,205] ion-exclusion chromatography,[206] BF4 selective electrodes,[207] oscillopolography,[208] and isotope dilution mass spectrometry[208–212] (LD 2 μg-B L^{-1}).

3.9 CADMIUM

3.9.1 Spectrophotometric and Conductiometric Titration

Lukionets and Kulish[213] used high-frequency conductiometric and spectrophotometric titration and complexane(III) to determine cadmium in natural waters. Spectrophotometric titration curves were obtained using complexane(III) with methylthiourea blue. The titration was conducted in buffer solutions of hexamethylenetetramine. The spectrophotometric titration method was most accurate in the concentration range 0.0–10 mM cadmium. High-frequency titration had high sensitivity and accuracy in the range 4.4–880 μM. High-frequency titration (neutral and weakly acidic media) and spectrophotometric titration of cadmium with methylthiourea blue (pH 6) were not affected by calcium or magnesium. With eriochrome black T (pH 10), the sum of cadmium, calcium, and magnesium was determined. In the presence of pyridylazoresorcinol, separate titration of cadmium and magnesium was possible; the presence of calcium interfered with cadmium determination with this indicator.

Chen et al.[214] have shown that cadmium inhibitation of the oxidation of arsenazo-I by hydrogen peroxide is the basis of a spectrophotometric method for the determination of cadmium. The method has a detection limit of 2.7×10^{-3} μg L^{-1}.

3.9.2 Spectrofluorometry

Lorserna et al.[215] described a method for the spectrofluorimetric determination of cadmium that is sensitive (detection limit 11 μg L^{-1}), rapid, and selective (only zinc seriously interferes). The method is based on chelation of cadmium with benzyl-2-pyridylketone 2-pyridylhdrazone as pH 11–13 and measurement of the fluorescence intensity in the dark at 555 nm, with excitation at 469 nm. The fluorescence development is instantaneous and remains stable for 1 h. Fluorescence intensity diminishes by 50% on raising the temperature from 2 to 60 °C. For measurements of 337 and 45 μg L^{-1}, cadmium(II), relative errors of 1.8% and 4.0%, and relative standard deviations of 2.6% and 5.3%, respectively, were obtained.

Kabasakalis and Tsitouridou[216] have described a fluorescence spectroscopic technique for the determination of cadmium in natural waters.

3.9.3 Atomic Absorption Spectrometry

Various workers have discussed the application of atomic absorption spectrometry to the determination of low levels of cadmium in natural waters in amounts down to 0.01 $\mu g L^{-1}$.[217–234]

Hasan and Kumar[227] studied the interference due to calcium and magnesium in the flameless atomic absorption spectrometric determination of cadmium in groundwater. A matrix-modifying reagent consisting of ammonium nitrate and orthophosphoric acid was used to suppress this interference. The modified samples are atomized in a graphite furnace and a cleaning burn after each analysis was accomplished by eliminating the drying and charring steps and atomizing at 2700 °C for 15 s.

The volatility of the analyte is decreased and that of the matrix is increased by the modification technique, almost completely removing interference. The accuracy of the method ranged between 96.0 and 102.3%, the coefficient of variation was 7.7% at 1.3 $\mu g L^{-1}$, and the limit of detection was 0.1 $\mu g L^{-1}$.

The analytical Quality Control Committee of the Water Research Centre UK[228,229,231] has carried out a study of the accuracy of the atomic absorption spectrometric determination of less than 1 $\mu g L^{-1}$ dissolved cadmium in river waters. The requirement for participating laboratories in this study was the maximal acceptable standard deviation would be 50% of the determined concentration or 0.025 $\mu g L^{-1}$, whichever was the greater. This aim was not achieved, and the results were adversely affected by both random and systematic errors, which could not be correlated within the time limits involved. Further inspection of results showed that they would meet a criterion according to which the error should not exceed 0.5 $\mu g L^{-1}$ or 20% of the determined concentration, but this was judged to be inadequate for the assessment of cadmium loads entering the sea from rivers and for monitoring compliance with an EC Directive concerning the discharge of cadmium to river waters.

Flame atomic absorption spectrophotometry has been used in a method for determining cadmium in natural waters described by Okutani and Arai.[230] The cadmium is complexed with 2,4,6-tri-2-pyridyl-1,3,5-trianzine in the presence of iodide ion, which is then extracted into nitrobenzene. The method was effective in removing interferences by a large number of ions found in water. The detection limit was 0.06 $\mu g L^{-1}$ and the relative standard deviation was 8.7% at the 1.0 μg of cadmium level.

Sun and Suo[233] have described a sensitive and accurate method for the determination of ultra-trace levels of cadmium in environmental waters by atomic absorption spectrometry with in situ trapping of hydride in an iridium-coated graphite tube. Filik et al.[232] used glyoxal-bis(2-hydroxy-anil) in the micelle-mediated extraction of cadmium from water samples prior to determination by atomic absorption spectrometry.

Hu[234] has proposed a co-precipitation that does not require collection of the precipitate for the determination of trace lead and cadmium in water with flame atomic absorption spectrometry. Lead and cadmium are preconcentrated by using cobalt(II) and ammonium pyrrolidine dithiocarbamate as co-precipitant and a known amount of cobalt as an internal standard. Since lead, cadmium, and cobalt were all distributed in the homogeneous precipitate, the concentration ratio of lead to cobalt, and cadmium to cobalt, remained unchanged in any part of the precipitate. The amount of lead and cadmium in the original sample solution can be calculated from the ratio of the absorbance values of lead and cadmium to cobalt in the final sample solution that is measured by flame atomic absorption spectrometry, and from the known amount of the lead and cadmium in the standard series solutions, respectively. The optimum pH range for quantitative co-precipitation of lead and cadmium is from 3.0 to 4.5. The 16 diverse ions tested gave no significant interferences in the lead and cadmium determination; under optimized conditions, lead ranging from 0 to 40 μg and cadmium ranging from 0 to 8 μg were quantatively co-precipitated with cobalt II ammonium pyrrolidine dithiocarbamate from 100 mL sample solution (pH ~3.5). This co-precipitation technique coupled with flame atomic absorption spectrometry was applied to the determination of lead and cadmium in water samples with satisfactory results, with recoveries in the range of 94.0–108%, and a relative standard deviation of <6.0%.

3.9.4 Graphite Furnace Atomic-Absorption Spectrometry

Lum and Callaghan[235] determined cadmium directly in natural waters by graphite furnace electrothermal atomic-absorption spectrometry without matrix modification. A direct injection method was used, which relied on the background correction capability of the polarized Zeeman effect combined with an L'vov platform. Results were obtained on various samples of natural water. The limit of detection was less than $2\,ng\,L^{-1}$.

More recently, Baysal et al.[236] have discussed the use of solid-phase extraction and direct injection of a co-polymer sorbent as a slurry into the graphite furnace prior to determination of cadmium by electrothermal atomic-absorption spectrometry.

3.9.5 Polarography

Stewart and Smart[237] give details of a procedure developed to determine cadmium in natural waters by differential-pulse anodic-stripping voltammetry using a rotating membrane-covered mercury film electrode constructed by placing a dialysis membrane over a glassy carbon rotating disc electrode and plating a thin mercury film on the electrode through the membrane. The pH of the solution should not exceed 6.0. The response was linear from $4.0\times10^{-9}\,M$ Cd^{2+} to $1.07\times10^{-5}\,M$ Cd^{2+} with a standard deviation of $\pm9.30\times10^{-10}\,M$ Cd^{2+}

for a 1.73×10^{-3} M Cd^{2+} solution (RSD ±11.1%), and a standard deviation of 6.44×10^{-9} M Cd^{2+} for a 1.78×10^{-7} M Cd^{2+} solution (RSD ±3.64%). The limit of detection was estimated to be 8.6×10^{-10} M.

Kemula and Zawadowska[238] developed a hanging mercury drop electrode and applied it to the determination of cadmium. The electrode produces a mercury drop of the required size. The electrode was tested in water with a reproducibility of results of about 1%. The electrode is easy to handle and steady even in the negative potential range.

Cadmium has been found to occur in Lake Michigan water at concentrations of 30–40 μg L^{-1}. Mass spectrometric isotope dilution analysis and atomic absorption spectrometry following electrodisposition were the techniques used in this study.[239]

Pundo-Botello et al.[240] used differential pulse polarography to determine cadmium in natural waters (see Section 3.63 for further details).

3.9.6 Miscellaneous Techniques

Other techniques that have been used to determine cadmium in environmental waters include differential electroanalysis,[230] neutron activation analysis,[231] speciation studies.[232] inductively coupled plasma atomic-emission spectrometry,[233] anodic stripping voltammetry,[234] and flow-injection chemiluminescence.[235]

Javar and Hoek[241] evaluated the removal of cadmium ions from water by nanoparticle-enhanced ultra-filtration using polymer and Zeolite nanoparticles. This evaluation considered nanoparticle physical–chemical properties, metal-binding kinetics, capacity and reversibility, and ultra-filtration separation for Linde type A Zeolite nanocrystals, poly(acrylic acid), alginic acid, and carboxyl-functionalized dendrimers in simple, laboratory-prepared ionic solutions. The three synthetic materials exhibited fast binding and strong affinity for cadmium, with good regeneration capabilities. Only the Zeolite nanoparticles were completely rejected by the ultra-filtration membranes tested. Overall, colloidal zeolites performed similar to conventional metal-binding polymers, but were more easily recovered by using relatively loose filtration membrane (i.e., lower energy consumption). Further, the superhydrophilic colloidal zeolites caused relatively little flux decline even in the presence of divalent cations, which caused dense, highly impermeable polymer gels to form over the membranes. These results suggested zeolite nanoparticles may compete with polymeric materials in low-pressure hybrid filtration processes designed to remove toxic metals from water.

3.10 CAESIUM

3.10.1 Atomic Absorption Spectrometry

Frigieri et al.[242] have described a method for the determination of caesium in river waters which involves preliminary chromatographic separation, and ammonium hexacyancobalt ferrate, on a strong cation-exchange resin followed by electrothermal atomic absorption spectrometry. The procedure is convenient, versatile, and reliable, although decomposition products from the exchange, namely iron and cobalt, can cause interference.

Recoveries ranged from 103% at the 10-μg caesium level to 106% at the 100-μg caesium level.

Other methods that have been used to determine caesium include flame emission spectrometry,[243] spectrophotometry,[244] flame photometry,[245] isotope dilution analysis,[246] and ion-exchange chromatography.[247]

3.11 CALCIUM AND MAGNESIUM

3.11.1 Titration Method

Jackson et al.[248,249] carried out consecutive amperometric titrations of calcium and magnesium in natural waters. Calcium is titrated first with ethylene glycol-bis(beta-aminoethyl ether)-*N,N,N′,N′*-tetraacetic acid and then magnesium is titrated with ethylenediaminetetra-acetic acid. The end-point for each titration is determined by amperometric detection of the excess chelate at a dropping mercury electrode.

3.11.2 Spectrophotometry

Qui[250] has described a sequential spectrophotometric determination of calcium and magnesium in river and well waters by complexation with beryllon(II) (2,8-hydroxy-3,6-disulfo-1-naphthylazo)-1,8-dihydroxynaphthalene-3,6-disulfonic acid) at pH 11.8. The total is measured and the calcium complex is then destroyed by addition of lead ethylene glycol-bis(2-amino-ethylether)-tetraacetic acid (EGTA), solution.

Song and Wu[251] presented a photometric method using chlorophosphonazo-pB for the determination of calcium in natural waters. Calcium in the typical range for waters gave results comparable to the volumetric EDTA method. The complex has an absorption maximum at 660 nm in the range of 0–20 mg L^{-1} and follows Beer's law.

3.11.3 Flow-Injection Analysis

Kempster et al.[252] determined calcium in natural water by flow-injection analysis–inductively-coupled plasma-emission spectrometry.

Tecator Ltd, in a series of Application Notes,[253–256] describes flow-injection analysis methods for the determination of calcium in natural waters in the concentration ranges 0.2–5, 1–20, and 5–100 mg L^{-1}. The detection limit of the first of these methods is 0.05 mg L^{-1}.

In these methods the sample is injected into a carrier stream of distilled water and mixed with 8-hydroxyquinoline and *o*-cresol-phthalein. The resulting colored solution is evaluated spectrophotometrically at 570 nm.

3.11.4 Atomic Absorption Spectrometry

A British Standards Institution method[257] gives details of a method for the determination of dissolved calcium and magnesium by flame atomic absorption spectrometry for use with waters containing up to $50\,mg\,L^{-1}$ calcium and up to $5\,mg\,L^{-1}$ magnesium. Absorbances were measured at 422.7 and 285.2 nm for calcium and magnesium, respectively. Lower limits of detection using the prescribed operating conditions were $3\,mg\,L^{-1}$ calcium and $0.9\,mg\,L^{-1}$ magnesium.

3.11.5 Ion-Selective Electrodes

Hulanicki et al.[258] have studied the effects of cationic, anionic, and nonionic detergents on the potentials of calcium selective electrodes. Electrodes with solid silver contacts were less sensitive to interference by surfactants than were electrodes with an internal reference solution.

Li et al.[259] developed a calcium-selective electrode by coating a copper wire with a mixture of calcium bis-[bis(*p*-isooctylphenyl)] phosphate, dioctyl phenyl phosphate, and PVC in cyclohexane. The electrode gave a Nernstian response in the range of 2×10^{-5} to 10^{-1} mol of $Ca^{2+}L^{-1}$, and has a detection limit of 8×10^{-6} mol of $Ca^{2+}L^{-1}$. The electrode has a low internal resistance, high selectivity, and good stability in natural waters.

3.11.6 Size Exclusion Chromatography

Gardner et al.[260] recognized that because component separations on size exclusion columns with distilled water are affected by chemical physical interactions as well as component molecular size, distilled water size exclusion chromatography should also fractionate dissolved metal forms. They interfaced distilled water size exclusion chromatography with inductively coupled argon plasma detection to fractionate and detect dissolved forms of magnesium and calcium in lake and river waters. Inductively coupled argon plasma detection is ideally suited to this application because it provides continuous metal monitoring for aqueous samples and accepts sample flow rates appropriate for distilled water size exclusion chromatography.

When Grand River or Lake Michigan filtrates were analyzed by distilled water size-exclusion chromatography–inductively coupled argon plasma, three peaks resulted for both magnesium and calcium. Although retention times varied, similar separations were obtained with both the TSK 2000 sw and TSK 3000 sw columns. These results suggest occurrence of at least three forms of each metal in these natural waters.

3.11.7 High-Performance Liquid Chromatography

Liquid chromatography was used by Rho and Choi[261] for the simultaneous determination of calcium and magnesium. A column of Zipax SCX was used for the separation. Linear calibration curves were obtained in the range of 1×10^{-4} to 5×10^{-4} M and the correlation of the calibration curves was in the range of 0.9952–0.9996.

3.11.8 Ion Chromatography

Jones[262] carried out a simultaneous separation of nonorganic cations (e.g., calcium, potassium, and sodium) and also anions (e.g., nitrate, thiosulfate, cyanide, and thiocyanate) by ion chromatography using a single column coated with weak/strong charged three welterionic bile sale micelles.

3.11.9 Capillary Zone Electrophoresis

Wang et al.[263] developed a method based on capillary zone electrophoresis with UV detection at 214 nm for the simultaneous determination of phosphate and calcium in waters, where Ca^{2+} reacted with 2,6-pyridinedicarboxylic acid in the electrolyte to form an anion complex. Subsequently, calcium and phosphate were separated by using an electrolyte containing 10 mM 2,6-dipyridinecarboxylic and calcium and 0.75 mM tetradecyltrimethylammonium bromide at pH 7.0. The results showed that reasonable resolution with low interference from other ions in the water was achieved. Linear calibration curves were obtained in the concentration range of 0.01–0.5 mM with hydrodynamic injection. The detection limits (S/N = 3) were 5 μm for phosphate and 2 μm for calcium. The separation of calcium and phosphate occurred within 6 min with small relative standard deviation of peak areas (<5%). The method was successfully applied to the determination of phosphate and Ca^{2+} in river water samples.

3.12 CERIUM

See under Lanthanides, Section 3.24.

3.13 CHROMIUM

The toxicity of chromium depends on its oxidation state, chromium(VI) being significantly more toxic than chromium(III). Hence, oxidation-state-specific determinations of chromium are of particular interest. Element-specific techniques, such as atomic absorption spectrometry, require a preliminary chemical separation of chromium(VI) from chromium(III) for the selective determination of chromium(VI). The separation is generally achieved by liquid–liquid extraction or ion exchange, requiring additional sample preparation prior to the actual determination.

Amperometric (electrochemical) determination of chromium(VI) inherently discriminates against chromium(III) without preliminary chemical separation. However, amperometric techniques are not element-specific, and other species that are reduced at the potential used for reduction of chromium(VI) interfere with its determination. One particularly significant interference in environmental samples is iron(III). Polymer-modified electrodes or liquid chromatographic separation have been used to eliminate the interference of iron(III) in amperometric determinations of chromium(VI).

3.13.1 Spectrophotometry

Various chromogenic reagents have been applied in the determination of low levels of chromium in natural waters. These include 1,5-diphenylcarbazide,[167,264,265] beryllon,[266] and iodide hexadecylpyridinium couplex.[267,268]

A spectrophotometric method[264] for determination of total chromium in natural waters with high mineral contents (e.g., alkaline earths) is based on digestion of the sample at 80 °C with alkaline hydrogen peroxide followed by decomposition of excess hydrogen peroxide with trichloroacetic acid and addition of ethanolic diphenylcarbazide. The chromium-diphenyl carbazide complex is evaluated at 540–550 nm.

Pettine et al.[265] studied hydrogen peroxide interference in the determination of chromium(VI) by the diphenylcarbazide method. Hexavalent chromium is reduced to the trivalent state under acidic conditions. The extent of the interference depended on the concentration of hydrogen peroxide and the period between addition of acid and diphenylcarbazide. The interferences can be avoided by adding a larger amount of reagent before the acid. These findings suggested that storage of samples under acidic conditions could result in a decrease in the concentration of hexavalent chromium.

Piying et al.[267] determined chromium(VI) in natural waters by reaction with iodide ion, followed by measurement of the tridide ion formed by spectrophotometry. Down to 0.2 $\mu g\,cm^{-3}$, chromium could be determined.

Zaitoun[269] described a simple, rapid, and sensitive method for spectrophotometric determination of chromium(VI) based on the absorbance of its complex with 1,4,8,11-tetraazacyclotetradecane (cyclam). The complex showed absorptivity of $1.5\times10^4\,LmoL^{-1}\,cm^{-1}$ at 379 nm. Under optimum experimental conditions, a pH of 4.5 and $1.960\times10^3\,mg\,L^{-1}$ cyclam were selected, and all measurements were performed 10 min after mixing. Major cations and anions did not show any interference; Beer's law was applicable in the concentration range 2–20 $mg\,L^{-1}$ with a detection limit of 0.001 $mg\,L^{-1}$. The standard deviation in the determination is ±0.5 $mg\,L^{-1}$ for a 15.0 $mg\,L^{-1}$ solution (n = 7). The described method provides a simple and reliable means of determination of Cr(VI) in real samples.

3.13.2 Flow-Injection Analysis

Workers at Tecator Ltd[270] have described a flow injection analysis method for the determination of total chromium(III) and chromium(VI) in the 1–10 $mg\,L^{-1}$ concentration range in natural waters. In this method, chromium(III) is oxidized to chromium(VI) by cerium(VI), and the sum of chromium(III) and chromium(VI) is determined by the 1,5-diphenylcarbazide spectrophotometric method by evaluation at 540 au absorption maximum. Riz et al.[232] determined both chromium(III) and chromium(IV) in natural waters by flow-injection analysis in a simultaneous or sequential mode. Chromium(VI) was reacted with 1,5-dipehnylcarbazide in one carrier stream and chromium(III) was oxidized to chromium(VI) by cerium(IV) in a separate stream. The method was applied to chromium in the range of 0.2–10 mg of chromium L^{-1}.

Al and Xing[271] complexed Cr(VI) in water with diphenylcarbohydrazide in the presence of sodium dodecyl sulfate, which in turn is extracted into chloroform. The complex is measured by spectrophotometry at 550 nm. They report recoveries of 96.7–115% and a relative standard deviation of 1.1% for measurement of Cr at the 2-μg level. The detection limit was 1 μg of Cr/I.

3.13.3 Atomic Absorption Spectrometry

Abdallah et al.[272] described a method for the determination of chromium(III) and have studied the interfering effects of various ions on the determination of chromium. Although the sensitivity for chromium by atomic absorption spectrometry is greatest in a fuel-rich flame, interferences are also greater.

The object of the work carried out by Abdallah et al.[272] was to investigate the feasibility of using a continuous titration technique for studying the interfering effects of foreign species on the atomic absorption signal of chromium and thus establishing the possibility of using a simple method for eliminating such interferences. The action of boron in an air-acetylene flame constitutes a promising universal flame buffer. Its action in the flame, when present in excess, is to interact with the matrix components to form relatively stable, unreactive species, leaving the analyte atoms free. Boric acid is a powerful releasing agent and acts very effectively on oxidizing species such as nitrate or chlorate and possible reducing species such as nitrite.

Muzzucotelli et al.[273] have described a rapid electrothermal atomic absorption method for the selective determination of chromium(VI) in waters. The sample is added directly to a liquid exchanger solubilized in methyl isobutyl ketone and hydrochloric acid. Analysis was carried out by atomic absorption spectroscopy.

No interference was observed in this procedure by 1000-fold molar excesses of iron, aluminum, calcium, and magnesium. It was pointed out that analyses must be carried out immediately after sample collection to avoid the reducing effects due to the presence of organic substances present in groundwater samples.

Magnesium nitrate was used by He[274] as a matrix modifier for the determination of chromium in water. A methylisobutylketone solution of ammonium pyrollide carbodithioate was used to extract chromium(VI) from natural water prior to determination by electrothermal atomic absorption spectrophotometry. Total chromium was determined after oxidation in a separate aliquot and chromium(III) was calculated by difference. The detection limit is 0.7 μg of chromium L^{-1} and the relative standard deviation is in the range of 7.6–24.7%.

Electrothermal atomic absorption spectrometry has been used to determine down to 30 ppt of chromium in natural waters.[275]

Catalytic Cr(VI) oxidation of fuchsine acid by hydrogen peroxide followed by absorption spectrophotometry has been applied.[276]

Alc and Sarab[277] have described a simple and sensitive method for speciation and preconcentration of trace amounts of Cr(III) and Cr(IV) in water samples using Amberlite CG-50. The Cr(III) content of the sample solutions was sorbed on the resin at pH of 5.5, desorbed by 3% hydrogen peroxide at pH of 9.5, and then determined by flame atomic absorption spectrometry. After reduction of Cr(VI) to Cr(III) by potassium sulfate at pH 3, the total chromium was determined and the Cr(VI) was obtained by the difference. The optimum conditions for reduction and sorption/desorption processes were investigated on several experimental parameters, including pH of sample and eluent, contact time, the resin quantity, type of the reductant eluent, and their concentrations. No considerable interferences have been observed due to the presence of a number of anions and cations, which may be found in the natural water samples. The dynamic range for the determination of both Cr(III) and Cr(VI) was found to be 2.0×10^{-7} to 2.0×10^{-6} M. The relative standard deviation (RSD) and detection limit (DL) were 2.56% and 8.00×10^{-8} M, respectively. This method has been successfully applied for the determination of the chromium species in various natural water samples. The recoveries for the spiked amounts of chromium species to the water samples were found to be more than 95% at the 95% confidence level, which satisfactorily confirmed the reliability of the method.

3.13.4 Inductivity Coupled Plasma Atomic-Emission Spectrometry

He et al.[278] used a surface ion-impregnated silica gel as a selective absorbent coupled with inductively coupled plasma atomic spectrometry for the determination of Cr(III) and total chromium in natural water samples. The Cr(III)-imprinted and nonimprinted absorbent were prepared by an easy one-step reaction with a surface imprinting technique. The maximum static absorption capacities for Cr(III) were 11.12 and 3.81 mg g^{-1}, respectively. The relative selectivity factors (a_r) for Cr(III)/Co(II), Cr(III)/Au(III), Cr(III)/Ni(II), Cr(III)/Cu(II), Cr(III)/Zn(II), and Cr(III)/Cr(VI) were 377, 2.14, 15.4, 27.7, 26.4, and 31.9, respectively. Under the optimal conditions, Cr(III) can be absorbed quantitatively, but Cr(IV) was not retained. Total chromium was obtained after reducing Cr(VI) to Cr(III) with hydroxylammonium chloride. The detection limit (3a) for Cr(III) was 0.11 ng mL^{-1}. The relative standard deviation was 1.2%. The method was validated by analyzing certified reference materials and successfully applied to the determination and speciation of chromium in natural water samples with satisfactory results.

3.13.5 Inductively Coupled Plasma Mass Spectrometry

Brydy et al.[279] determined down to 100 pg chromium(III) and 200 pg chromium(VI) in natural waters by inductively coupled plasma mass spectrometry after online concentration and separation by high-performance liquid chromatography.

3.13.6 Electrochemical Methods

3.13.6.1 Voltammetry

Ruan and Wang et al.[280] used a 1.5th-order differential voltammetric method for the determination of chromium in natural waters. The method used ethylenedianine-1 N sodium nitrite electrolytic solution and has a catalytic wave at -1.89 V and a linearity range of 2×10^{-11} to 8×10^{-9} g of chromium mL^{-1}.

Wang et al.[281] used an ammonium chloride–ammonium hydroxide supporting electrolyte containing cupferron to obtain an absorptive wave of the chromium(VI)–cupferron complex by linear sweep voltammetry. The reduction peak potential of the wave is at -1.55 V and the derivative peak height is 1×10^{-9} to 9×10^{-8} M. The detection limit for chromium in natural waters is 6×10^{-10} M.

Yang and Tang[282] determined chromium in waters by cathodic stripping voltammetry. The chromium was preconcentrated as Hg_2CrO_4 at 0.45 V (vs SCE) and determined from the stripping peak at 0.35 V. The stripping peak height is linearly proportional to Cr(VI) in the concentration range of 3×10^{-8} to 8×10^{-7} M.

3.13.6.2 Amperometry

Pratt and Koch[283] have reported a procedure using phosphoric acid as the supporting electrolyte for the trace-level amperometric determination of chromium(VI) at gold or palladium electrodes. This procedure suppresses the interference from iron(III) since the complex species of $Fe–(PO_4)_3^{6-}$ and $Fe(HPO_4)_3^{3-}$ are formed and are not reduced at the potential used for the amperometric detection.

3.13.6.3 Polarography

Su et al.[284] determined chromium(VI) in natural waters by using a polarographic catalytic wave method. Recoveries in the range of 97–102% for samples containing 10–80 g L^{-1} chromium were measured. The detection limit is 2 μg of chromium L^{-1}.

3.13.6.4 Capillary Isotachoelectrophoresis

Zelensky et al.[285] used capillary isotachophoresis in the trace determination of chromium(VI) in water samples at low (μg L^{-1}) concentrations. The separation unit in the analyzer employed a photometric detection at 405 nm. Losses of chromium(VI) due to adsorption on the walls of the glassware was prevented by the addition of sulfate (0.0001 M) to the sample solutions. At lower pH, addition of napthalene-1,3,6-trisulfonate prevented adsorption on to the walls of the separation unit. With these precautions detection limits were in range of 4–5 $\mu g L^{-1}$ for a 30-L sample.

3.13.7 High-Performance Liquid Chromatography

High-performance liquid chromatography coupled with time-resolved quenched phosphorescence detection has been used for the determination of chromium in natural waters by Baumann et al.[286] Paired-ion reversed-phased high-performance liquid chromatography is used for separation, followed by detection of the time-resolved phosphorescence signal of biacetyl at 515 nm. The detection limit is 1.4×10^{-7} M, and the method has a linear calibration curve over three orders of magnitude.

Posta et al.[287] coupled an atomic absorption spectrometric detector to high-performance liquid chromatography in the online determination of chromium(III) and chromium(VI) in amounts down to 30 $\mu g L^{-1}$ in natural waters.

Powell et al.[288] used high-performance liquid chromatography, direct injection nebulization, and inductively coupled plasma mass spectrometry to determine 30–180 ppt levels of chromium(III), chromium(VI), and total chromium in natural waters. The determination of chromium is also discussed in Section 3.65.

3.13.8 Miscellaneous Techniques

Other techniques that have been used in studies of inorganic chromium in water include spectrofluorimetry,[289–291] chemiluminescence,[292] gas chromatography,[293] thermal ionization isotope dilution mass spectrometry,[294,295] cationic ion exchange film coupled with attenuated total reflectance spectroscopy,[295] X-ray fluorescence spectroscopy,[296] and inhibition-based enzymes biosensor.[297]

Various speciation studies have been carried out on chromium in natural waters.[298–304]

3.14 COBALT

3.14.1 Spectrophotometry

Abbasi and Ahmed[305] have described a procedure for the microdetermination of cobalt in natural waters. 5-sulfo-4-methylsalicylic acid is used as the chromogenic reagent. Cobalt is determined in the presence of hydrogen peroxide. Natural chelating agents such as fulvic acid interfere in the procedure. It was found that water with a biochemical oxygen demand of greater than 10 had a retarding effect on the color development. Organic interferences can be eliminated on digestion of the sample with nitric and perchloric acids.

Gharehboghi et al.[306] have described a new, simple, and rapid dispersive liquid–liquid microextraction method to preconcentrate trace levels of cobalt as a prior step to its determination by spectrophotometric detection. In this method, a small amount of chloroform as the extraction solvent was dissolved in pure ethanol as the disperser solvent, and then the binary solution was rapidly injected by a syringe into the water sample containing cobalt ions complexed by 1-(2-pyridylazo)-2-naphthol (PAN). This forms a cloudy solution. The cloudy state was the result of chloroform fine droplets formation, which has been dispersed in the bulk aqueous sample. Therefore, the Co–PAN complex was extracted into the fine chloroform droplets. After centrifugation (2 min at 5000 rpm), these droplets were sedimented at the bottom of a conical test tube (about 100 μL), and then the whole of the complex-enriched extracted phase was determined by spectrophotometry at 577 nm. Complex formation and extraction are usually affected by some parameters, such as the types and volumes of extraction solvent and disperser solvent, salt effect, pH, and the concentration of chelating agent, which are optimized in this method. Under optimum conditions, the enhancement factor (as the ratio of slope of preconcentrated sample to that obtained without preconcentration) of 125 was obtained from 50 mL of water sample, the limit of detection of the method was 0.5 $\mu g\,L^{-1}$, and the relative standard deviation (RSD), n=5) for 50 $\mu g\,L^{-1}$ of cobalt was 2.5%. The method was applied to the determination of cobalt in tap and river water samples.

3.14.2 Chemiluminescence

Boyle et al.[307] and Rodionova et al.[308] employed luminol for chemiluminescence of cobalt in natural waters after cation exchange liquid chromatography. Cobalt can be determined directly in 500-μL samples with a detection limit of 20 $pmol\,kg^{-1}$.

3.14.3 Graphite Furnace Atomic-Absorption Spectrometry

Graphite-furnace Zeeman atomic-absorption spectrometry has been used to determine low $\mu g\,L^{-1}$ levels of cobalt in lake water. The method required small sample volumes, minimal sample pretreatment, and preparation and chelation or solvent extraction procedures. Analysis of a water sample of known cobalt content yielded a result ($5.4 \pm 0.2\,\mu g$ cobalt L^{-1}) that compared favorably with the reported mean of $5.2 \pm 1.2\,\mu g\,L^{-1}$ obtained from pooling results obtained by other techniques.

Koizumi et al.[309] also applied this technique to the determination of $\mu g\,L^{-1}$ levels of cobalt in surface waters. Ophel and Judd[310] also determined levels of cobalt in freshwater lake samples by flame atomic absorption spectrometry. This method, however, requires preconcentrating 4 L of water sample and subsequent chelation and solvent extraction after removal of iron. Detection limits between 10[311] and 0.4 $ng\,L^{-1}$[312,313] have been reported in the determination of cobalt in natural waters.

Souza and Tarley[314] have described a method based on sorbent separation and enrichment for the determination of cobalt in water by graphite furnace atomic absorption spectrometry.

3.14.4 Polarography

Hao et al.[315] used differential pulse adsorption stripping voltammetry for the determination of cobalt in natural waters. Dimethylglyoxime complexes of cobalt in the presence of triethanolamine and were used in the determination. A detection limit of approximately 3 ppt was reported.

3.14.5 X-ray Fluorescence Spectroscopy

X-ray fluorescence spectrometry without preconcentration was used to determined cobalt in natural waters by Okashita and Tanaka.[316] The detection limit was as low as 0.03 μg of cobalt for samples with small amounts of dissolved iron.

3.14.6 Gas Chromatography

Schaller and Neeb[317] chelated cobalt in natural waters with di(trifluoroethyl) dithiocarbamate and measured the complex by capillary gas chromatography. Detection is by electron capture and a detection limit of 0.2 $\mu g L^{-1}$ is reported.

3.14.7 Optical Sensing Film

Arvard et al.[318] have proposed an optical sensing film for sensitive determination of cobalt (Co) (II) ion in aqueous solutions. The cobalt-sensing membrane was prepared by incorporating N^5-(2,4-dinitro-phenyl)-N^1,N^1-diethyl-penta-1,3-diene-1,5-diamine as ionophone in the plasticized PVC membrane containing *o*-nitrophenyl octyl ether as plasticizer. The membrane responds to cobalt ion by changing color irreversibly from yellow to green in acetate buffer solution at pH=5.5. This film displays a linear range of 0.028–29.68 $\mu g mL^{-1}$ with a limit of detection 0.012 $\mu g mL^{-1}$. Moreover, upon the introduction of a negatively charged lipophilic additive (oleic acid) into the membrane, the optode displayed enhanced sensitivity. In addition, satisfactory analytical sensing characteristics for determining the Co(II) ion were obtained in terms of selectivity, stability, and reproducibility. The response time of the optode was about 15–25 min, depending on the concentration of Co(II) ions. The optode membrane has been applied to determine Co(II) in various water samples and copper-free alloys.

3.15 COPPER

3.15.1 Spectrophotometry

Moffett et al.[319] described a sensitive procedure for the selective determination of monovalent copper in the presence of divalent copper based on spectrophotometric measurement of its complex with bathocuproine. Ethylenediamine was a suitable masking ligand, which reacted rapidly with divalent copper to form complexes that were inert to reduction and did not interfere with the determination of the cuprous ion. The effect of variable such as pH and ionic strength were investigated.

Themelis and Vasilikiotis[320] used a catalytic method based on the copper-catalyzed oxidation of chromotrophic acid by hydrogen peroxide for the determination of copper in natural waters. The reaction is followed spectrophotometrically by measuring the rate of change of absorbance at 430 nm. The method is applicable in the range of 12–190 $\mu g\,L^{-1}$ and the precision and accuracy were within 2%.

Itoh et al.[321] determined copper in freshwater by a spectrophotometric technique involving complexation with a α,β,γ,δ-tetrakis(4-*N*-trimethylaminipheny) porphine. The relatively slow reaction was accelerated by the addition of sodium L-ascorbate. A linear calibration curve was obtained for copper in the range of 0–10 $\mu g\,L^{-1}$. The relative standard deviation was 1.3%.

Other chromogenic reagents that have been employed in the determination of copper include 4,7-diphenyl-2,9-dimethyl-1,10-phenanthrolinedisulfonate (LD 0.08 $\mu g\,L^{-1}$),[322] acid fuchsin LD 0.53 μg,[323] methyl red (LD 0.4 $\mu g\,L^{-1}$),[324] chrome azurol ≤(0.06 $\mu g\,L^{-1}$),[325] rhodamine B (LD 0.03 $\mu g\,L^{-1}$),[326] amino black 10 B (LD 0.06 $\mu g\,L^{-1}$),[327] and 2-amino methyl pyridine grafted silica.[328]

3.15.2 Chemiluminescence

Yamada and Suzuki[329] applied a flow-injection system to the determination of copper in natural waters. The chemiluminescence reagents β-nitrostyrene/sodium hydroxide/hexadecyltrimethylammonium bromide sensitized with fluorescein allowed 0.1–10 ng of copper(II) to be determined at a high sampling rate.

Flow-injection chemiluminescence based on the reaction of luminol, potassium cyanide, and divalent copper has been described.[330] The detection limit was 0.01 μg of $Cu\,L^{-1}$, and the linear operating range was 0.4–80 $\mu g\,L^{-1}$. Copper(II) catalyzed the oxidation of eosin Y by hydrogen peroxide in an ammonia/ammonium chloride medium, which allows the fluorescent determination of copper. The detection limit was 0.2 μg of $Cu\,L^{-1}$. Reaction of Cu(II) with 5-[(4-acetaminiophenyl) aso]-8-amino-quinoline forms a fluorescent complex which is excited at 334 nm, and emission is measured at 414 nm.[331] The detection limit was 1 $\mu g\,L^{-1}$, and the fluorescence was linear in the range of 4–36 $\mu g\,L^{-1}$.

3.15.3 Atomic Absorption Spectrometry

Ejaz et al.[332] extracted and preconcentrated copper from water by the use of 4-(5-nonyl)pyridine as an extractant for copper from aqueous 0.1-M thiocyanate solutions before its determination by atomic absorption spectrometry. The procedure provides efficient extraction in a single step from neutral or acidic solutions. The results show that 1 μg copper can be extracted into 1 mL of the organic phase from 500 mL of natural water.

Silva and Valcarcel[333] discuss a procedure for the determination of traces of copper in solution by atomic absorption spectrometry, based on formation of a complex with naphthoquinone thiosemicarbazone. The complex is extracted into isobutyl methyl ketone, and the absorbance is measured at 324.8 nm.

Copper can be determined in the presence of a 5000-fold excess of many of a great number of diverse ions including most of those normally encountered in river waters.

Water samples from Japanese rivers were treated by solvent extraction, passed through an anion-exchange resin, and then atomic absorption spectrometry was used to measure the concentrations of reactive copper and organic copper present.[334]

Sweileh et al.[335] utilized ion exchange/atomic absorption spectrophotometry to study the complexation of copper by organic ligands in natural waters. The method was found to be more sensitive than the ion-selective electrode methods and was used to determine copper(II) in lakes.

Nishoika et al.[336] coupled copper in natural waters with diethyldithiocarbamate, prior to determination by electrothermal atomic absorption spectrophotometry. The method concentrates copper 50-fold from water samples and enables the determination of copper at concentrations ≥2.4 µg of copper L^{-1} without significant interferences.

Bradshaw et al.[312] described a procedure for determination of copper in natural waters over the range of 0.0001–0.01 ng L^{-1}. It involved complexation with 1-pyrrolidinecarbodithioate, extraction into isobutylmethylketone, and determination by atom-trapping atomic absorption spectrometry.

Ueda and Yamazaki[337] describe a method for the determination of copper in freshwater in which the copper is co-precipitated with hafnium tetrahydroxide ($Hf(OH)_4$) prior to determination by electrothermal atomic-absorption spectrophotometry. The calibration curve was linear in the range of 4–400 µg of copper L^{-1}. Many elements were examined for interferences in the determination of copper and no significant interferences were found.

Zhang et al.[338] electrodeposited copper on a tungsten wire prior to determination by electrothermal atomic absorption. The detection limit was 0.012 µg L^{-1}, and the coefficient of variation was 3.0% at copper concentrations of 0.2 µg L^{-1}. Flow injection coupled with microwave plasma torch atomic-emission spectrometry has been used for the determination of copper.[339] It was found that flow injection was superior to continuous injection, allowing for the preconcentration of the sample, and the detection limit was lowered by a factor of 25 over that of continuous sample injection. The detection limit was 0.16 µg of Cu L^{-1}, and the relative standard deviation was 3.2% at a copper concentration of 500 µg L^{-1}. Copper preconcentration by ammonium pyrrolidine dithiocarbamate on an activated carbon minicolumn prior to analysis by flame atomic-absorption spectrophotometry has been described.[340] The preconcentrated copper was desorbed by methyl isobutyl ketone, which was carried by water flow directly into the aspirator. Preconcentration factors between 35 and 100 were achieved with a relative standard deviation of 1.8–3.5%.

3.15.4 Ion-Selective Electrodes

Hulanicki et al.[341] determined copper in natural water by means of a chalcocite ion-selective electrode. Gulens et al.[342] used a copper ion-selective electrode in studies of the hydrolysis of copper(II).

The copper ion-selective electrode has been used to study copper behavior in weakly basic and hydrocarbonate solutions in a concentration range simulating natural systems.[343] Accurate pH measurements were simultaneously made with cupric ion activity. Examples are given of applications of the technique to Italian rivers.

Distributions of $CuOH^+$, $Cu(OH)^2$, $CuCO_3(aq)$, and $Cu(CO_2)^{2-}$ were deduced measuring Cu^{2+} and OH^2 CO_3^{2-}. $CuCO_3(aq)$ predominates at natural pH and alkalinity levels.

Copper(II) has been determined in natural waters by a method based on its catalysis of the reaction of persulfate and iodide ions that was monitored by an iodide-selective electrode, as described by Tong and Yang.[344] A fixed reaction time was used prior to the measurement of the potential with the ion-selective electrode. The detection limit was 2.5 μg of copper L^{-1}.

An electrode constructed by incorporating cupron in a carbon paste mixture has been described by Peng et al.[345] The copper was deposited from an NH_3/NH_4Cl buffer solution. The detection limit was 0.33 μg L^{-1} with a 10-min preconcentration. The linear working range was 0.64–64 μg of Cu L^{-1}. A potentiometric method for measuring the catalytic effect of copper on the reaction of hexachlorantimonate(V) with hydroxylamine in an acetic acid/acetate buffer has been described.[346] The reaction progress was monitored with a Sb(V) ion-selective electrode. The working range was 5–510 μg of Cu L^{-1} with a coefficient of variation of 3.5%.

Singh et al.[347] investigated the use of membranes using 2-[{(2-hydroxyphenyl) imino}methyl]-phenol (L1) and 2-[{(3-hydroxyphenyl)imino}methyl]-phenol (L2), as Cu^{2+} ion-selective sensors. The effect of various plasticizers, namely, dibutyl phthalate, dibutyl sebacate, benzyl acetate, and *o*-nitrophenyloctyl ether; and anion excluders, oleic acid and sodium tetraphenylborate, were studied and improved performance was observed in several instances. Optimum performance was observed with membranes of 2-[{2-hydroxyphenylimino}methyl]-phenol:dibutal sebacate:oleic acid:PVC in the ratio of 6:54:10:30 (w/w, %). The sensor works satisfactorily in the concentration range of 3.2×10^{-8} to 1.0×10^{-1} mol L^{-1} with a Nernstian slope of 29.5 ± 0.5 mV decade^{-1} of Cu^{2+}. The detection limit of the proposed sensor is 2.0×10^{-8} mol L^{-1} (1.27 ng mL^{-1}). Wide pH range (3.0–8.5), fast response time (7 s), sufficient (up to 25% v/v) non-aqueous tolerance, and adequate shelf-life (3 months) indicate the utility of this sensor. The potentiometric selectivity coefficients as determined by the matched potential method indicate selective response for Cu^{2+} ions over various interfering ions; therefore the sensor could be successfully used for the determination of copper in river water.

Brazil et al.[348] used 2-aminomethylpyridine-anchored silica gel as a sorbent in a simple spectrophotometric flow system for preconcentration of Cu^{2+} in natural water samples, using sodium diethyldithiocarbamate as chromogenic agent (460 nm). The system was optimized using a full factorial design 2^5 to determine better analytical conditions to determine copper in the natural water samples such as those from river, tap, stream, spring, well, waste, and synthetic brackish water, and a water reference material (NIST-1640). The best conditions were 180 s loading; 30 s elution; 30 s regeneration of the column; loading flow rate 6.6 mL min^{-1}; buffer solution for the preconcentration and regeneration of the column-acetate buffer 5.75; elution flow rate 1.6 mL min^{-1}; and eluent composition 0.20 mol L^{-1} nitric acid. Under these conditions, the preconcentration factor obtained was 77, and the detection limit achieved was 3.0 ng mL^{-1}. The recovery of spiked water samples ranged from 95.2% to 104.7%.

3.15.5 Miscellaneous Techniques

Other techniques that have been employed in the determination of copper in natural waters include anodic scanning voltammetry,[349,350] LD 0.05 ng L^{-1} polarography,[351–355] size exclusion chromatography,[353] and high-performance liquid chromatography.[356]

Various separation studies have been conducted on complex copper ligands in water.[245,357–362]

Bigalke et al.[363] studied the bioavailability, mobility, and toxicity of copper on copper speciation in solution. In natural systems like soils, sediments, lakes and river waters, organo-Cu complexes are the dominating species. Organo-complexation of copper may cause a fraction of stable copper isotopes. The knowledge of copper isotope fractionation during sorption on humic acid may help to better understand copper isotope fractionsion in natural environments, and this facilitates the use of copper-stable isotope ratios ($\delta^{65}Cu$) as tracers of the fate of copper in the environment. Bigalke et al.[363] therefore studied copper isotope fractionation during complexation with insolubilized acid (IHA) as a surrogate of humic acid in soil organic matter with the help of sorption experiments at pH 2–7. The workers used NICA-Donnan chemical speciation modelling to describe copper binding in IHA and to estimate the influence of copper binding to different functional groups in copper isotope fractionation. The observed overall copper isotope fractionation at equilibrium between the solution and immobilized humic acid was $\Delta^{65}Cu_{IHA\text{-}solution} = 0.26 \pm 0.11‰$ (2 SD). Modelled fractionations of copper isotopes for low-affinity sites and high-affinity sites were identical with $\Delta^{65}Cu$ LAS/HAS solutions = 0.27; pH did not influence copper isotope fractionation in the investigated pH range.

Copper has been identified as a pollutant of concern by the U.S. Environmental Protection Agency (EPA) because of its widespread occurrence and toxic impact in the environment. Choyyck et al.[364] evaluated three nanoporous sorbents containing chelating diamine functionalities for Cu^{2+} adsorption from natural water: ethylenediamine functionalized self-assembled monolayers on mesoprous supports (EDA-SAMMS), ethylenediamine functionalized activated carbon (AC-CH_2-EDA), and 1,10-phenanthroline functionalized mesoporous carbon (phen-FMC). The pH dependence of Cu^{2+} sorption, Cu^{2+} sorption capacities, rates, and selectivity of the sorbents were determined and compared with those of commercial sorbents (Chelex-100 ion-exchange resin and Darco KB-B activated carbon). All three chelating diamine sorbents showed excellent Cu^{2+} removal (~95–99%) from river water and sea water over the pH range of 6.0–8.0. EDA-SAMMS and AC-CH_2-EDA demonstrated rapid Cu^{2+} sorption kinetics (minutes) and good sorption capacities (26 and 17 mg Cu g^{-1} sorbent, respectively) in seawater, whereas phen-FMC had excellent selectively for Cu^{2+} over other metal ions (e.g., Ca^{2+}, Fe, Ni^{2+}, and Zn^{2+}) and was able to achieve Cu below the EPA-recommended levels for river and sea waters.

Wetlands are reactive zones of the landscape that can sequester metals released by industrial and agricultural activities. Copper-stable isotope ratios ($\delta^{65}Cu$) have recently been used as tracers of transport and transformation processes in polluted environments. Babecsanyi et al.[365] used copper-stable isotopes to trace the behavior of copper in a stormwater wetland receiving runoff from a vineyard catchment (Alsace, France). The copper and stable isotope ratios were determined in the dissolved phase, suspended particulate matter, wetland sediments, and vegetation. The wetland retained >68% of the dissolved copper and >92% of the suspended particulate matter-bound Cu which represented, 84.4% of the total copper in the runoff. The dissolved copper became depleted in ^{65}Cu when passing through the wetland ($\Delta^{65}Cu_{inlet-outlet}$ from 0.3‰ to 0.33‰), which reflects copper adsorption to aluminum minerals and organic matter. The $\delta^{65}Cu$ values varied in the wetland sediments (0.04 ± 0.10‰), which stored >96% of the total copper mass with the wetland. During high-flow conditions, the copper flowing out of the wetland became isotopically lighter, indicating the mobilization of reduced Cu(I) species from the sediments and Cu reduction within the sediments. These results demonstrate that the copper-stable isotope ratios may help trace copper behavior in redox-dynamic environments such as wetlands.

3.16 DYSPROSIUM

See under Lanthanides, Section 3.24.

3.17 GADOLINIUM

See under Lanthanides, Section 3.24.

3.18 GALLIUM

3.18.1 Spectrofluorometry

Tenteno et al.[366] have described a fluorescence spectroscopic method for the determination of down to 10 ppb of aluminum and gallium in natural waters.

3.18.2 Neutron Activation Analysis

Honda et al.[367] have described a two-step procedure involving solvent extraction as the dithiocarbamate and neutron activation analysis for the determination of down to 10^{-3} μg L^{-1} gallium in natural waters.

Indium and gallium have been determined[368] in natural waters by a two-stage procedure involving chloroform extraction of the diethyldithiocarbamates and neutron activation analysis. The detection limit is 100 μg L^{-1} for each element.

3.18.3 Voltammetry

Trace levels of gallium have been determined in natural waters by linear-sweep voltammetry after adsorptive preconcentration of Ga-Solochrome violet RS chelate in a method described by Wang and Zadell.[369] The chelate was adsorbed on a hanging mercury drop electrode for a 2 min preconcentration time. The detection limit was 0.08 μg L^{-1} of gallium.

3.19 GERMANIUM

The earliest methods for the determination of germanium in natural waters involved the concentration of germanium from large water samples by co-precipitation and extraction procedures and its spectrophotometric measurement as the phenylfluorone complex.[370,371] At the concentrations typical of natural waters, these methods are working close to their limits of detection and require time-consuming enrichment steps.

Due to its tendency to form very stable oxide species, germanium shows relatively poor sensitivity in flame atomic absorption methods.[372] The high temperatures and relatively long residence times available in graphite tube atomizers made significant improvement in sensitivity possible; Johnson et al.[372] report an absolute limit of detection of 0.3 ng of germanium obtained with a graphite tube atomizer of their own construction. The restriction to sample volumes in the microliter range, however, results in concentration limits of detection of about 15 μg L^{-1}, several orders of magnitude above those characteristic of natural waters.

The reduction of germanium in solution to the volatile germane (GeH_4, bp −88.5 °C) by sodium borohydride and the subsequent detection of the gaseous germane by atomic absorption was first used by Pollock and West,[373] who achieved a relatively high limit of detection (about 0.5 μg germanium) by injecting the gas into a standard atomic absorption flame. Similar limits of detection are achieved with an externally heated silica-tube atomizing furnace.[374] Braman and Tompkins[375] combined the borohydride techniques with a dc-discharge atomic-emission detector and achieved a detection limit of 0.4 ng of germanium.

3.19.1 Graphite Furnace Atomic-Absorption Spectrometry

Zheng and Zhang[376] have investigated factors influencing the atomization of germanium in graphite furnace atomic-absorption spectrometry. The presence of oxidizing agents or alkalis and modification of the graphite surface all have an effect. It was shown that reduction of germanium(IV) to germanium(II) can be suppressed by the addition of perchloric acid or nitric acid or by using tungsten in the furnace or a zirconium-coated graphite tube.

Graphite furnace atomic-absorption spectrometry has been used to determine germanium in water. The method had a linear range of 0–200 ng of Ge mL^{-1} and a limit of detection of 1.69×10^{-9} g of germanium.[377] A method has been reported for determining germanium in water using flow-injection hydride generation followed by trapping and electrothermal atomization in a Pd-coated graphite tube. The detection limit was 0.004 μg of Ge L^{-1} using a 4.5-mL sample of germanium.[378]

3.19.2 Hydride Absorption Spectrometry

Andrae and Frohlich[379] applied this technique to the determination of down to 140 pg (0.56 μg L^{-1} for a 250-mL sample) of germanium in natural waters.

Germanium is determined in aqueous matrix at the part-per-trillion level by a combination of hydride generation, graphite furnace atomization, and atomic absorption detection. The germanium is reduced by sodium borohydride to germane (GeH_4), stripped from solution by a helium gas stream, and collected in a liquid-nitrogen-cooled trap. It is released by rapid heating of the trap and enters a modified graphite furnace, which is synchronized to reach the analysis temperature of 2600 °C before arrival of the germane peak. The atomic absorption peak is recorded and electronically integrated. The dynamic range of the method spans three orders of magnitude. The precision of the determination is 8% when peak absorbance is used; by peak integration in the nanogram range, the precision is 4%.

4.1 ng L^{-1} of germanium were found in the water of an American river by this method. High concentrations of many metals are known to cause negative interferences in hydride generation systems. In most natural waters, the concentrations of these elements are many orders of magnitude below these causing interference. Nevertheless, it is recommended that recovery checks are conducted, especially when a new type of water is being analyzed.

3.19.3 Hydride-Generation Inductively Coupled Plasma Mass Spectrometry

Jin et al.[380] have applied hydride generation inductively coupled plasma mass spectrometry to the determination of down to 0.08 pg of inorganic germanium in natural waters. In this method, inorganic and methylated germanium species were determined at below parts-per-trillion levels by a combination of hydride generation and inductively coupled argon plasma mass spectrometry. The germanium species in solution were reduced to the corresponding hydrides by sodium tetrahydroborate, transferred with a helium gas stream, and trapped in a liquid nitrogen cooled U-trap. The hydrides were evaporated and introduced into the ICP torch, and the ion count at $m/z=74$ was monitored. The reduction efficiencies for methylated germanium species in a malic acid matrix were more than 97%. The absolute detection limits were 0.08 pg of germanium for inorganic germanium, 0.1 pg of germanium for monomethyl-germanium and dimethyl-germanium, and 0.09 pg of germanium for trimethyl-germanium. The dynamic ranges of the detection span four orders of magnitude. The proposed method was applied to natural waters and wastewaters, and germanium, monomethyl-germanium, and dimethyl-germanium were detected in all of the samples studied.

3.19.4 Anodic Stripping Voltammetry

Choi et al.[381] determined germanium[382] in water with a detection limit of 210 μg L^{-1} by anodic striping polarography using a hanging Hg drop electrode. Methods for determining germanium using linear sweep voltammetry and a spectrophotometric method using the Ge–phenylfluorone complex were compared. The linear range of the voltammetric method was 2.5–80 μg of Ge L^{-1}, and that of the spectrophotometric method was 10–300 μg of Ge L^{-1}.[381]

3.19.5 Speciation

Two organogermanium species, monomethyl germanium and dimethyl germanium, have been identified in natural waters by methods based on the hydride generation technique,[383,384] and it should be noted that monomethyl germanium is the major germanium species in sea water. However, information on organogermanium species in the environment is still limited because of the very low concentrations of the element.

Although the hydride generation technique has improved the detection limit of germanium in various atomic spectrometric methods, the absolute detection limits so far reported are still at ng to sub-ng levels, and few studies have been done for methylated germanium species. Typically, detection limits of 75–150 pg of germanium have been achieved for inorganic germanium, monomethyl germanium, dimethyl germanium, and trimethyl germanium by hydride generation = graphite furnace atomic absorption spectrometry.

Many σ-bonded organometallic compounds are stable in aqueous solution and involve two types of bonds between the central metal atom and the ligands. The first is the metal-carbon (M-C) bond, which is relatively nonpolar and kinetically inert. The second is the more polar and labile M-X bond, of which X is a donor atom such as oxygen, nitrogen, or the halides.[385]

Studies have been carried out of the aqueous solution chemistry of organo-geranium compounds that include highly inert Ge-C bonds. Halide complex formation of the methylated and inorganic germanium have been investigated and separation of the germanium compound by liquid–liquid extraction of the halide complexes developed.[385–388]

Germanium compounds dissolve in water mainly as nonionic tetrahedral hydroxide ($(CH_3)_nGe(OH)_{4-n}$; n=0, 1, 2, and 3).[381,389–392] In hydrochloric, hydrobromic, and hydriodic acids, halide complexes ($(CH_3)_nGeX_{4-n}$; X=Cl^-, Br, and I^-) are formed and extracted into carbon tetrachloride. As the number of methyl groups increases, the stability constant of the halide complexes increases, and the germanium species are extracted in a lower concentration range of hydrohalogenic acid. This is attributed to the inductive effect of the methyl group, which weakens the Lewis acidity of the germanium atom and reduces the stability of the hydroxide complexes.

Sazaki et al.[393] have studied liquid–liquid extraction of methylated and inorganic germanium ($(CH_3)_nGe(OH)_{4-n}$; n = 0, 0, 2, and 3) in aqueous solution (pH 1–12) with organic ligands to develop a separation method for germanium compounds. Ligands containing a negatively charged oxygen donor were proved to be the most powerful extractants for germanium compounds. Using benzoic acid, trimethyl germanium is extracted into carbon tetrachloride, while monomethyl germanium, dimethyl germanium, and inorganic germanium are not extracted into the organic phase. The extracted species is trimethyl germanium-benzoic acid, of which the benzoate ion is monodentate. Catechol and mandelic acid produce monoanionic complexes of germanium and monomethylgermanium ($[(CH_3)_nGe{-}(OH)_{1-n}L_2]^-$; n=0 and 1), of which the coordination number about the central germanium atom is 5 or 6. These compounds are extracted into the nitrobenzene phase accompanied by tetrabutylammonium as a counter cation. Dimethyl germanium and trimethyl germanium are not extracted as the catecholate and mandelate complexes, because they have low affinity for higher coordinated states. This study demonstrates for the first time that germanium compounds can be separated on the basis of their stereochemistry in solution.

3.20 GOLD

3.20.1 Atomic Absorption Spectrometry

Feingerg and Bowyer[394] described a method for the determination of gold in natural water that involves evaporation of the sample, placing it in a solution of hydrobromic acid-bromine, extraction with methyl-isobutylketone, and determination by electrothermal atomization in an atomic absorption spectrometer. The limit of detection was 0.001 μg gold in L^{-1}. Good results were obtained in studies conducted to assess precision, recovery, and interference. This method is precise, effective, and free of interferences. The relative standard deviation of 15.9–18.3% is well within the limits of precision for the nanogram range. The method recovers gold at an average of 93%.

Schvova et al.[395] determined gold and silver by electrothermal atomic absorption spectrophotometry after extraction with polyorgs X1-N. The detection limits for gold and silver are 0.02 and 0.005 μg L^{-1}, respectively. Hall et al.[396] preconcentrated gold from waters with activated charcoal, followed by determination using electrothermal atomic absorption spectrophotometry. The pH of the water sample is adjusted and intimate mixing of the activated charcoal with the sample is required. A detection limit of 0.5 ng L^{-1} was obtained for a 1-L sample, and a relative standard deviation of about ±10% at the 5 ng L^{-1} concentration level was obtained.

3.20.2 Inductively Coupled Plasma Mass Spectrometry

Faulkner and Edmond[397] and Gomez and McLeod[398] determined gold at the femtomolar level (10^{-15} M L^{-1}) in natural waters by flow-injection inductively coupled plasma quadruple-mass spectrometry. The technique involves preconcentrations by anion exchange of gold as a cyanide complex $[Au(CN)_2]^{1-}$ using 195 gold radiotracer with a half-life of 183 days to monitor recoveries. Samples are then introduced by flow injection into an inductively coupled plasma quadruple-mass spectrometer for analysis. The method has a relative precision of 15% at the 100-fM level.

3.20.3 Direct Potentiometry

Ol'khovich[399] has described a direct potentiometric method for the determination of the concentration of gold(III) in natural water. Two identical platinum electrodes were used, one immersed in a reference electrolyte and the other in a solution containing the reference electrolyte and gold connected by an electrolytic bridge of sodium nitrate or potassium chloride. The potential difference between the electrodes was measured after evaporation of an aliquot with aqua-regia and hydrochloric acid (twice) and leaching with sodium chloride. The difference from a calibration provided a measure of gold concentration with detection limits of 0.4 mg L^{-1}. Data obtained by this method agreed with those obtained by atomic absorption.

3.20.4 Neutron Activation Analysis

Asamov et al.[400] have described a two-step procedure involving anion-exchange chromatography and neutron activation analysis for the determination of total gold in natural water.

Turaev et al.[401] also used neutron activation for the simultaneous determination of gold and silver in natural waters. Evaporated residues were dissolved in aqua regia and then extracted with potassium *O,O*-diisopropyldithiophosphate. The detection limit was 0.2 μg L^{-1} for gold and 0.0011 μg L^{-1} for silver.

3.21 HAFNIUM

Boswell and Elderfield[402] described an isotope-dilution mass-spectrometric method for the determination of hafnium. Boswell and Elderfield[402] have applied polarography to the determination of hafnium in natural waters in amounts down to 0.5 $\mu g L^{-1}$, and various workers used atomic absorption spectrometry to determine hafnium.[403–405]

3.22 INDIUM

3.22.1 Spectrofluorometry Method

Fluorescence of the indium complex of 4,4′-oxalyl-bis(hydrazonomethyl) diresorcinol has been described in a method by Pastor et al.[406] The complex was excited at 425 nm and detected at 475 nm. The detection limit was 2.6 μg of indium L^{-1}.

3.22.2 Atomic Absorption Spectrometry

A method for the extraction of indium from waters with 8-metcaptoquinoline in an amyl acetate solution has been described.[407] The indium was then determined by atomic absorption spectrophotometry with a detection limit of 4 μg of In L^{-1}. Gallium phosphate has been used for the co-precipitation of indium from water.[403] The indium was determined by electrothermal atomic-absorption spectrophotometry with a detection limit of 0.3 $\mu g L^{-1}$, using a 500-mL sample. Acetylacetone was used to complex indium, which was then absorbed on activated carbon.[404] The activated carbon containing the indium was dispersed in a glycerine solution and then determined by electrothermal atomic-absorption spectrophotometry. The detection limit was 0.025 μg of In L^{-1}, and the relative standard deviation was 4–5% for samples with a concentration of 10 $\mu g L^{-1}$.

Ueda and Matsui[405] co-precipitated indium(III) in natural waters with hafnium tetrahydroxide ($Hf(OH)_4$) prior to determination by electrothermal atomic absorption spectrophotometry. The calibration curve is linear for indium in the range of 8–160 $\mu g L^{-1}$, and the detection limit was 0.5 $\mu g L^{-1}$.

3.22.3 Neutron Activation Analysis

Neutron activation analysis[408] and polarography[409] have also been applied for the determination of indium in water.

3.23 IRON

3.23.1 Spectrophotometry

Gibbs[410] described a simple method for the rapid determination of ferrous iron in natural waters. This method relies on the formation of a magenta-colored chromogen with ferrozine. It is capable of analyzing samples with an iron content of 5 $mg L^{-1}$ with high precision. Reduction of ferric iron, which does not react with ferrozine, to ferrous iron with hydroxylamine hydrochloride enables a distinction to be made between ferrous and ferric acid.

Copper, cobalt, nickel, cyanide, and nitrite interfere in concentrations over 500 μg L^{-1}. Where eutrophic lakes stratify in midsummer, the anoxic bottom water can contain high concentrations of dissolved hydrogen sulfide. This was found to be a major interfering agent in analysis using ferrozine. Hydrogen sulfide can be removed by acidifying the solution with hydrochloric acid and bubbling nitrogen gas through it; standing the acidification sample overnight also eliminates this interference.

The method employing the ferrozine chromogen[403] has been modified to allow for the presence of humic acid in the samples and could in fact be used to determine both iron and humic acid. This method has been applied to river and stream samples. Two absorbance measurements are required, one on an untreated sample aliquot and the other on an aliquot treated to enhance iron.

Macaldy et al.[411] have shown that, under certain, conditions (low pH value and presence of trivalent iron) there is a small positive interference in the determination of ferrous iron by the bathophenanthroline spectrophotometric method; this is attributed to the formation of a colored complex in the presence of Fe OH^{2+}. The workers suggest that this observation casts doubt on reports by other investigators that measureable amounts of ferrous iron exist in the oxygenated surface layers of lake water. Of the *N*-sulfoalkyl derivatives of 2-(2-thiazolylazo-5-amino phenol),[412] 2-(4-methyl-2-thiazolylazo)-5-(*N*-sulfopropyl) amino phenol showed high selectivity with the iron(II) complex, having a characteristic absorption maximum at 745 nm mg L^{-1}. The recommended procedure for the determination of less than 1.6 mg L^{-1} of iron involved the sequential addition of ascorbic acid (0.1%) and the chromogen (0.1%) and acetate buffer (1.0 M) solutions to an optimal pH of 5.5. The method was relatively interference-free and has a limit detection of 0.01 mg L^{-1} for humic acid and 0.04 μM for iron.

Nigo et al.[413] have described a method for determining iron(II) and iron(III) in natural waters based on ion-exchange calorimetry using 1,10-phenanthroline as color reagent for the former, and citrate as the masking reagent for the latter. Total iron is determined after the reduction of iron(III) and iron(II) with hydroxylamine. Most common foreign ions at 100 times the concentration of cobalt did not interfere, although zinc did interfere. Copper at the 0.05-μg L^{-1} level did interfere. The method was applied to the determination of iron in hot spring water from Ijiri Spar (Fukuoka, Japan).

The two forms of iron could be determined at L^{-1} levels by this method. Detection limits obtained ranged from 6 μL^{-1} using a 200-mL sample to 0.9 μg L^{-1} using a 1 L sample.

Pakalns and Farrar[414] investigated the effect of surfactants on the determination of soluble iron in natural waters. Cationic anionic and nonionic detergents, and also sodium tripolyphosphates, pyrophosphates, and nitriloacetic acid were included in the study. The chromogenic reagents investigated were 1:10 phenanthroline[415] and tripyridine.[415]

The tripyridyl method is superior to the other two methods for the determination of iron in the presence of up to 1000 mg L^{-1} of various surfactants but not for up to $100\,mg\,L^{-1}$ of nonionic detergents. The phenanthroline method can be used to determine iron in the presence of up to $1000\,mg\,L^{-1}$ of cationic, anionic, and nonanionic detergents, but sodium tripolyphosphates interfered above $2\,mg\,L^{-1}$.

Iron available for uptake by phytoplankton is usually assumed to be present as soluble or complexed iron, which is separated from particulate forms by filtration. Experiments have been carried out by Box[416] to study the effect of different reaction conditions on the spectrophotometric determination of iron, using a number of compounds capable of complexing divalent iron. It was found that there were changes in the absorbance of the iron complex in the presence of acetate buffer both with and without a reduction agent, and the apparent concentration of ferrous iron increased with time. However, acidification of the sample with dilute hydrochloric acid for at least 1 h before addition of the reducing agent and the complexing agent resulted in a stable iron concentration (the acid extractable fraction of the total filterable iron). Box[416] discussed the results obtained with this method for two lakes in the Lake District, England.

Other spectrophotometric methods for the determination of iron in natural waters are reviewed in Table 3.2.

TABLE 3.2 Spectrophotometric Methods for the Determination of Iron in Natural Water

Chromogenic Reagent	Form of Iron	Absorption Maximum (nm)	Detection Limit	References
Stilbexon (4,4′-bis[bis(carboxy-methyl)amino])	Fe(II)	–	–	
Silbene-2,2′-disulfuric acid		–	$0.1\,mg\,L^{-1}$	417,418
1,10-phenanthroline	Fe(II)	–	$0.2\,mg\,L^{-1}$	419,420
2,2′-bipyridyl	Fe(II)	515–518		421
Pyrogallol red and zephiramine	Fe(II)	–	–	420
2,4,6-tripyridyl 1,2,5-triazine	Fe(II)	595	–	421
Ferric thiocyanete	Fe(III)	–	$2\,mg\,L^{-1}$	422
N,N-dimethyl-*p*-phentylene diamine/hydrogen peroxide	Fe(III)		–	423

TABLE 3.2 Spectrophotometric Methods for the Determination of Iron in Natural Water—cont'd

Chromogenic Reagent	Form of Iron	Absorption Maximum (nm)	Detection Limit	References
Tiron	Fe(II), Fe(III)	–	–	424
Sodium bathophenanthroline disulfonate	Fe(II), Fe(III)	–	–	425
Chromotropic acid	Fe	440	–	426
Indigocarmin	Fe(II)	540–640		427–429
Iron catalyzed oxidation of 4-amino antipyrene with *N*,*N*-dimethyl aniline 3-(2-pyridyl)-5,6-bis(4-phenylsulfonic) acid	Total	555	–	430
	Fe(II)	8.8 $\mu g L^{-1}$	–	431
Di-2-pyridyl ketone	Fe(II), Fe(III)	–	–	432
Oxidative decolorization of methylene green with hydrogen peroxide	Fe(II) or Fe(III)	654	0.2 $\mu g L^{-1}$	433
4,7-diphenyl-1,10-phenanthroline disulfonic acid	Fe(II)	541	–	434
O-phenanthroline-dipped sensor	Fe(II)	–	–	435
Chromasurol sensor	Fe(III)	–	10 $\mu g L^{-1}$	436

3.23.2 Flow-Injection Analysis

Burguerra and Burguerra[437] used flow-injection analysis followed by atomic absorption spectrometry for the determination of iron(II) and total iron. They give details of equipment for the determination of divalent iron by measuring the absorbance of its complex with 1,10-phenanthroline at 510 nm, followed by determination of total iron by atomic absorption spectrometry at 248.2 nm. Linear calibration ranges were 0.1–35 and 0.1–10 $mg L^{-1}$ for iron(II) and total iron, respectively.

Mortatti et al.[438] give details of a procedure for the determination of total iron in natural waters by flow-injection analysis, using 1,10-phenanthroline and spectrophotometric measurement at 512 nm.

Flow-injection analysis using 2-nitroso-5-(*N*-propyl-*N*-sulfopropylamino) phenol for the determination of iron in natural waters is reported by Ohno and Sakai.[439] The method has a recovery of 100 ± 1% and does not have significant interference from other transition-metal ions. Iron is determined by spectrophotometry at 753 nm in the 4–100 $\mu g\,L^{-1}$ range of concentration.

3.23.3 Chemiluminescence

Rehman et al.[440] have described a flow-injection chemiluminescence method for the determination of total iron in freshwater samples. The enhanced chemiluminescence emission was caused by the iron(II) from the neutralization of hydrochloric acid and sodium hydroxide without the use of any chemiluminescence reagent. The calibration graph was linear in the concentration range of 2.8–560 $\mu g\,L^{-1}$ (r2 = 0.9983, n = 8), with relative standard deviation of (RSD; n = 4) in the range of 0.8–2.6%. The limit of detection (S/N = 3) was 0.56 $\mu g\,L^{-1}$ with injection throughout of 180 $L\,h^{-1}$. The effect of common anions and cations were studied over their environmentally relevant concentrations in freshwater. The method was successfully applied to determine total iron in freshwater samples. Iron(III) was reduced to iron(II) by using hydroxylammonium chloride. The method was compared with a spectrophotometric method and there was no significant difference between the two methods at the 95% confidence level (*t*-test). Analysis of river water (certified reference material SLRS-4) for iron(III), after reduction of iron(III) with hydroxylammonium chloride, gave good results (2.17 ± 0.22 μM compared with the certificate value of 1.85 ± 0.1 μM).

3.23.4 Miscellaneous Techniques

Other techniques that have been used in studies of iron concentration in natural waters are spectrofluorimetry,[441] direct current spectrometry,[442] high-performance liquid chromatography,[425,443,444] electrophoresis,[426] ultraviolet spectrometry,[427] electrothermal vaporization/inductivity coupled plasma atomic-emission spectrometry,[445] resonance ionization isotope-dilution mass spectrometry,[446] electron spin-resonance spectrometric titration,[447,448] fluorescence quenching,[449] kinetic methods,[450] bioluminescence,[451] laser-induced breakdown spectroscopy,[452] spectrofluorimetry,[453] DC current plasma spectrometry,[453] and cyclic voltammetry.[454,455]

3.24 LANTHANIDES

Various workers have reviewed methods for the determination of lanthanides in water by different techniques including inductively coupled plasma mass spectrometry,[456] ion-exchange chromatography,[457] isotope-dilution inductively coupled plasma mass spectrometry,[458–464] induced fluorescence (ppb levels),[465,466] ion chromatography,[458] and kinetic phosphorimetry.[467]

3.24.1 Cerium

3.24.1.1 Ion Chromatography

Rubin and Heberling[468] have reviewed the applications of ion chromatography to the analysis of elements, including the lanthanides in water.

3.24.1.2 Spectrophotometric Method

Abassi and Ahmed[469] have described a procedure for the determination of cerium in natural waters. 5-sulfo-4-methyl salicylic acid is used as the chromogenic reagent.

3.24.1.3 Spectrofluorometry Methods

Xiao[470] described a direct fluorometric method for the determination of cerium in natural waters. The solution is irradiated with UV radiation at 256 nm and the detection of fluorescence is at 358 nm. The detection limit is 40 $\mu g L^{-1}$ and the relative standard deviation is 10%.

Kubitz et al.[465] used a laser-induced fluorescence method[466] to determine ppb levels of cerium in natural waters.

3.24.1.4 Inductively Coupled Plasma Mass Spectrometry

Inductively coupled plasma mass spectrometry has been applied to the determination of 14 lanthanides in amounts down to 1 ppb in natural waters.[458–463]

3.24.1.5 Neutron Activation Analysis

Neutron activation analysis has been employed in the determination in cerium in natural waters.[367,471]

3.24.2 Dysprosium

3.24.2.1 Fluorescence Spectroscopy

Panehrahi et al. and Moulin et al.[466] used inductively coupled plasma mass spectrometry to determine 14 lanthanides in natural water in amounts down to 0.1 ppt.[459]

3.24.2.2 Ion-Exchange Chromatography

Ion-exchange chromatography has been employed to separate lanthanides as a group from more common metals.[472]

3.24.3 Europium

3.24.3.1 Fluorescence Spectroscopy

Fluorescence spectroscopy has been used to determine the EDTA or DTPA complexes of europium in amounts down to 45 $\mu g L^{-1}$.[473]

3.24.3.2 *Inductively Coupled Plasma Mass Spectrometry*

Inductively coupled plasma mass spectrometry has been used to determine 14 lanthanides in natural waters.[458–463]

3.24.3.3 *Neutron Activation Analysis*

Neutron activation analysis has been used to determine europium in natural waters.[367,471]

3.24.3.4 *Ion-Exchange Chromatography*

Ion-exchange chromatography has been employed to separate lanthanides as a group from more common metals.[472]

3.24.4 Gadolinium

3.24.4.1 *Inductively Coupled Plasma Mass Spectrometry*

Hall et al.[458] and Aggarwal et al.[459] used inductively coupled plasma mass spectrometry to determine 14 lanthanides in amounts down to 0.1 ppb.

3.24.4.2 *Ion-Exchange Chromatography*

Ion-exchange chromatography has been used to separate lanthanides as a group from more common elements.[472]

3.24.4.3 *Ion Chromatography*

Rubin and Heberling[468] reviewed the application of ion chromatography to the determination of 14 lanthanides in water.

3.24.5 Holmium

Inductively coupled plasma mass spectrometry,[458–463] ion-exchange chromatography,[472] and ion chromatography[468] have been applied to the determination of holmium in water.

3.24.6 Lanthanium

3.24.6.1 *Inductively Coupled Plasma Mass Spectrometry*

Inductively coupled plasma mass spectrometry,[458–463] ion-exchange chromatography,[472] and ion chromatography[468] have been applied to the determination of lanthanides in water.

Fan and Fang[474] preconcentrated ion curates lanthanium by online flow injection prior to determination by inductively coupled plasma atomic-emission spectrometry. A detection limit of 0.7 $\mu g L^{-1}$ was achieved.

3.24.6.2 *Neutron Activation Analysis*

Neutron activation analysis has been used to determine lanthanium in water.[367,471]

3.24.7 Lutecium

3.24.7.1 Inductively Coupled Plasma Mass Spectrometry

Inductively coupled plasma mass spectrometry,[458–463] ion-exchange chromatography,[472] and ion chromatography[468] have been applied to the determination of lutecium.

3.24.8 Neodymium

Inductively coupled plasma mass spectrometry,[458–463] ion-exchange chromatography,[472] and ion chromatography[468] have been applied to the determination of neodynium in water.

3.24.9 Praeseodynium

Inductively coupled plasma mass spectrometry,[458–463] ion-exchange chromatography,[472] and ion chromatography[468] have been applied to the determination of praeseodynium in water.

3.24.10 Promethium

Inductively coupled plasma mass spectrometry,[458–463] ion-exchange chromatography,[472] and ion chromatography[468] have been applied to the determination of promethium in water.

3.24.11 Samarium

Inductively coupled plasma mass spectrometry,[458–463] ion-exchange chromatography,[472] and ion chromatography[468] have been applied to the determination of samarium in water.

3.24.11.1 Neutron Activation Analysis

Neutron activation analysis has been used to determine samarium in water.

3.24.12 Terbium

Inductively coupled plasma mass spectrometry,[458–463] ion-exchange chromatography,[472] and ion chromatography[468] have been applied to the determination of terbium in water.

Anubaker et al.[473] determined terbium as its EDTA and DTPA complexes by fluorimetry in amounts down to $95\,\mu g\,L^{-1}$.

3.24.13 Thulium

Inductively coupled plasma mass spectrometry,[458–463] ion-exchange chromatography,[472] and ion chromatography[468] have been used to determine thulium in water.

3.24.14 Ytterbium

Inductively coupled plasma mass spectrometry,[458–463] ion-exchange chromatography,[472] ion chromatography,[468] and neutron activation analysis[367,456,457,464,466–468,471–473,475] have been used to determine ytterbium in water.

3.25 LEAD

3.25.1 Spectrophotometric Method

Various workers have studied the determination of lead in natural water by atomic absorption spectrometry.[474,476–479] Detection limits down to 5 μg L^{-1} have been achieved.[476,477] Breueggemeyer and Caruso[477] determined lead as a tetramethyl derivative in amounts down to 5 μg L^{-1} with the upper working range limit of 200 μg L^{-1}. Sinemus et al.[478] improved sensitivity in lead determinations by using an electron-discharge lamp to utilize the more sensitive resonance frequency at 217 nm in place of the 283.3-nm line usually employed. Improved detection limits and higher signal-to-noise ratio are obtained.

3.25.2 Spectrofluorimetry

Cheam et al.[480,481] used laser excited atomic fluorescence spectroscopy to study the distribution of lead in the concentration range of 4–25 ppb in the Great Lakes of the United States.

Cheam et al. used laser-excited fluorescence for the determination of lead in freshwater. Recoveries were 100 ± 10%, the detection limit was 0.4 ng L^{-1}, and the relative standard deviation was 4.9% for samples with a concentration of 10 ng L^{-1} lead. Another laser-excited atomic fluorescence method has been described with a reported detection limit of 1 ng of Pb L^{-1} for water samples.[482]

3.25.3 Atomic Absorption Spectrometry

Kumar et al.[483] have described a matrix modification using a mixture of ammonium nitrate and diammonium hydrogen phosphate for the determination of lead in natural waters, prior to electrothermal atomic absorption spectrophotometry. The detection limit was 1 μg of lead L^{-1} and the sensitivity was 0.4 μg L^{-1}.

Chen et al.[484] have described an atomic absorption spectrometric method for the determination of down to 0.2 μg L^{-1} lead in natural waters.

Zhang et al.[485] have described an online preconcentration of lead by a flow-injection system prior to flame atomic absorption spectrophotometry. The analyte was deposited on an alumina microcolumn and subsequently eluted with nitric acid. The limit of detection was 0.36 μg L^{-1} and the relative standard deviations at 40 and 4 μg L^{-1} levels in natural waters were 1.4% and 12%, respectively.

Martinez-Jimenez et al.[486] employed a continuous precipitation and filtration flow system coupled with an atomic absorption spectrophotometer for the preconcentration and determination of lead. Lead(II) forms a precipitate with ammonia that is retained on a stainless steel filter, then redissolved with nitric acid. The method was proposed for the determination of lead in natural waters in the range of 1.2–1500 $\mu g L^{-1}$. The relative standard deviation was 3.6%.

Lead was preconcentrated with a fibrous aluminum microcolumn and then eluted into a flow-injection flame atomic-absorption spectrometer.[477] The detection limit was 0.7 μg of Pb L^{-1} and the relative standard deviation was 4.9% for a sample at 5 $\mu g L^{-1}$. Lead was adsorbed on a sulfonated dithizone-loaded resin, followed by hydride generation, prior to the flame absorption determination.[478] The detection limit was 0.025 $\mu g L^{-1}$ and the response was for a range of 5–200 $\mu g L^{-1}$.

Hasseini and Hassan-Abodi[487] have discussed the use of flotation separation and electrothermal atomic absorption spectrometry in the determination of low levels of lead in water.

Other studies on the determination of lead in natural waters have been exported by Carbrara et al.,[488] Granadillo et al.,[489] Ohta and Suzuki,[490] and Vandegans et al.[491]

Expanded polystyrene (EPS) foam waste (white pollutant) has been utilized for the synthesis of novel chelating resin, i.e., EPS-N=N-α-benzoin oxime (EPS-N=N-Box). This synthesized resin was characterized by FT-IR spectroscopy, element analysis, and thermogravimetric analysis. Siyal et al.[492] have described a selective method for the preconcentration of Pb(II) ions on EPS-N=N-Box resin packed in a minicolumn. The sorbed Pb(II) ions were eluted with 5.0 mL of 2.0 mol L^{-1} hydrochloric acid and determined by microsample injection system coupled flame atomic-absorption spectrometry (MIS-FAAS). The average recovery of Pb(II) ions achieved was 95.5% at optimum parameters such as pH 7, resin amount 400 mg, and flow rates 1.0 mL min^{-1} (of eluent) and 3.0 mL min^{-1} (of sample solution). The total saturation capacity of the resin, limit of detection, and limit of quantification of Pb(II) ions were found to be 30 mg g^{-1}, 0.033 $\mu g L^{-1}$, and 0.107 $\mu g L^{-1}$, respectively, with a preconcentration factor of 300. The accuracy, selectivity, and validation of the method was checked by analysis of sea water (BCR-403), wastewater (BCR-715), and Tibet soil (NCS DC-78302) as certified reference materials (CRMs). This method was applied successfully for the trace determination of Pb(II) ions in aqueous samples.

3.25.4 Graphite Furnace Atomic-Absorption Spectrometry

Bertenshaw et al.[479] studied methods of reducing matrix interference in the determination of lead in river water, potable water, and sewage and trade effluents by graphite furnace atomic-absorption spectroscopy with electrothermal atomization and lanthanum pretreatment.

The amounts of lanthanum and nitric acid employed were optimized such that the technique is applicable to a wide range of samples. The technique was found to be satisfactory for samples containing up to 1150 mg L^{-1} of chloride, 1420 mg L^{-1} of sulfate, 760 mg L^{-1} of sodium, and 1530 mg L^{-1} total hardness (as calcium carbonate). The optimum pretreatment conditions for samples was 1% v/v nitric acid and 0.05 m/v of lanthanum (as lanthanum chloride), which completely overcome suppressive interferences in the determination of lead and gave a furnace tube lifetime of approximately 600 firings.

3.25.5 Inductively Coupled Plasma Mass Spectrometry

Wang et al.[493] describe a lead hydride generation system for use in total and isotopic analysis of lead in natural waters by inductively coupled plasma mass spectrometry. The limit of detection for lead was restricted to 0.01–0.05 $\mu g L^{-1}$ by reagent blanks, significantly higher than when "ultra-clean" techniques are used. The lead-hydride generation is interfered with by iron and copper, and this was overcome by the addition of sulfosalicylic acid and sodium cyanide, dissolved in sodium tetrahydroborate(III). The authors found good agreement for the inductively coupled plasma mass spectrometry method and the certified value for SRM 1643a.

Javanbaktit et al.[494] used two functional nanoporous silica gels containing dipyridyl subunit (SiL_1 and SiL_2) as selective solid-phase extraction materials for separation preconcentration and determination of trace levels of Pb(II) ions by inductively coupled plasma optical-emission spectroscopy. The experimental parameters including pH, amounts and type of sorbent, sample volume, eluent type, and interfering ions on the recovery of the target analytes were investigated, and the optimal experimental conditions were established. Under the optimized operating conditions with SiL_2 as sorbent, an enrichment factor of 300 was obtained. The detection limit based on three times standard deviations of the blanks was 15 $ng L^{-1}$. The proposed method was applied to the determination of lead in natural and wastewater samples with satisfactory results (recoveries greater than 96.5%, RSDs lower than 5.0%).

3.25.6 Voltammetry

Voltammetry has been used to determine very low concentrations of lead in natural water.[394,495–501] Apte and Badke[495] have studied the estimation of lead in natural water by anodic scanning voltammetry using a graphite electrode. It is reproducible and sensitive and needs only 5 mL of test solution. The curve of lead estimation is linear in the range of 100–600 $\mu g L^{-1}$. The cadmium wave occurs at −0.78 V and the copper wave at −0225 V. The lead wave is at −0.515 V. No appreciable interference of cadmium occurs when it is present at 100 time the concentration of lead. However, 30 $mg L^{-1}$ copper suppressed the stripping peak to 82% and 40 $mg L^{-1}$ copper addition suppressed the peak to 73%.

Benes et al.[496] showed that anodic stripping voltammetry can be used to distinguish between the different complexed forms of lead. Problems may be encountered in samples containing large amounts of organic impurities. Stripping voltammetry is one of the very few methods that in principle permit differentiation between free ionic forms of metals and their complexed forms, on the basis of shifts in the deposition and peak potentials on complexation, and are simultaneously sufficiently sensitive for trace analysis. Under normal experimental conditions, only hydrated metal ions and weak metal complexes are deposited and stripped within the potential range available. Therefore, by carrying out stripping determinations in untreated water samples and in the same samples after decomposing the complexes, e.g., by mineralization or acidification with mineral acid, the strongly complexed fraction of the total metal content can be estimated.

Goa et al.[502] described a method for the determination of Pb^{2+} in water samples with bismuth film electrodes based on magneto-voltammetry. In the presence of an 0.6 T external field, square-wave voltammetry of Pb^{2+} was performed with bismuth film electrodes. A high concentration of Fe^{3+} was added to the analytes to generate a large current during the preconcentration step. A Lorentz force from the flux of net current through the magnetic field resulted in convection. Then, more Pb^{2+} deposited onto the electrode and larger stripping peak currents were observed. Bismuth film electrodes that were prepared by simultaneous depositing the bismuth and the Pb^{2+} on an electrode offered a mercury-free environment for this determination. This method exhibits a high sensitivity of 4.61 $\mu A\,\mu M^{-1}$ for Pb^{2+} over the 1×10^{-8} to 1×10^{-6} M range. A detection limit as low as 8.5×10^{-10} M was obtained with only 1-min preconcentration.

3.25.7 Miscellaneous Techniques

Other techniques that have been applied to the determination of inorganic lead in natural waters include ion-selective electrodes,[503,504] X-ray fluorescence spectrometry,[505,506] liquid chromatography-[507] and metastable transfer-emission spectrometry.[508,509]

3.26 LITHIUM

3.26.1 Spectrophotometry

Morgen and Vlazov[510] described an extraction-spectrophotometric procedure for the determination of lithium in natural water. The sample is evaporated to dryness and the residue extracted with acetone. The acetone extract is dried and the residue dissolved in water and lithium determined by formation of the chromophore with nitroanthranilazo in dimethylformamide medium. The detection limit of this method is about 50 $\mu g\,L^{-1}$ lithium. Numerous anions and cations do not interfere in this procedure, but no information is given on what ions do interfere.

3.26.2 Atomic Absorption Spectrometry

Chen et al.[511] report a unique method of standard addition for the determination of lithium and other alkali-metal ions in natural waters. The authors describe an inverted Y-shaped tube for the simultaneous aspiration of sample and standard into an atomic absorption spectrophotometer.

3.26.3 Inversion Voltammetry

Khakhanina et al.[512] determined micro amounts of lithium in natural waters by inversion chromatography and flame photometry. The supporting electrolyte was 0.02 M tetrabutylammonium iodide in dimethylformamide. The detection limit for lithium was 0.1 mM in the presence of a 100-fold to 1000-fold excess of sodium and potassium.

3.26.4 Mass Spectrometry

Chan[513] determined total lithium and lithium isotope using thermal ionization mass spectrometry. A chemical procedure for the quantitative separation of lithium from natural waters is described in which lithium tetraborate is precipitated for analysis and the Li_2BO^{2+} ion is used for the determination.

3.26.5 Neutron Activation Analysis

Yang et al.[514] determined lithium by neutron activation analysis.

Itoh et al.[515] applied neutron activation analysis for the determination of down to 3 ppm of lithium in natural waters.

Chao and Tseng[516] determined sub-ppb levels of lithium in natural waters using neutron activation analysis.

3.26.6 Ion Chromatography

Hoshika et al.[517] determined lithium in freshwaters by ion chromatography. A linear calibration curve based on peak areas was obtained for lithium in the range of 2–3000 $\mu g L^{-1}$. The detection limit was 1 $\mu g L^{-1}$ and the relative standard deviation was 4.4%.

3.27 LUTECIUM

See under Lanthanides, Section 3.24.

3.28 MAGNESIUM

3.28.1 Spectrophotometry

Qui et al.[518] determined magnesium by a photometric method using the nitrophosphonazo complex of magnesium at 584 nm. Beer's law is obeyed in the range of 0–1.25 $mg L^{-1}$ of magnesium L^{-1} in natural waters.

3.28.2 Flow-Injection Analysis

Forteza et al.[519] used a flow-injection spectrophotometric technique for the determination of magnesium in natural waters. They report a sampling of 60 samples L^{-1} if in the range of 0.5–8 mg L^{-1}. A magnesium sample is injected in an nitriloacetic acid flowing solution, which suppresses most interferences, including those from calcium in a ratio of 30 Ca:1 Mg.

3.28.3 Amperometry

Downrad et al.[520] have described an amperometric method for determining down to 6 mg L^{-1} of magnesium in natural waters.

3.28.4 Miscellaneous Techniques

Various other techniques for the determination of magnesium in natural waters are discussed in Section 3.65. These include atomic absorption spectrometry, inductively coupled plasma atomic-emission spectrometry, inductivity couples plasma mass spectrometry, ion-exchange chromatography, ion chromatography, ɩ-particle-induced X-ray emission spectrometry, emission spectrometry, neutron activation analysis, prompt gamma neutron activation analysis, and high-performance liquid chromatography.

3.29 MANGANESE

3.29.1 Spectrophotometry

Various chromogenic reagents have been employed for the determination of manganese in water. These include *o*-dianisidine (absorption maximum 445 nm),[521] *o*-tolidine or 3,3′-dimethylnaphthidine (440 nm),[522] alizarin red S with hydrogen peroxide,[523] and leuco crystal violet (4,4′,4″-metylidynetris-(*N*,*N*-dimethylaniline) (591 nm).[524] The *o*-dianisidene method[521] is capable of determining manganese down to 0.3 μg L^{-1} and is relatively free from interference effects by other ions likely to be present in natural waters. The detection limit of the leuco crystal violet method[524] is claimed to be 0.1 μg of manganese(IV), and this can be improved to 0.02 μg manganese(IV) by extracting the crystal violet into 1:1 isobutylalcohol-benzene. There is negligible interference from manganese(II) in this method.

Li and Li[525] evaluated a spectrophotometric method for manganese by using Eriochrome Black T to complex manganese(II) in natural waters. Beer's law is followed for manganese in the range of 0–12 μg of manganese L^{-1}.

Salinas et al.[424] used a kinetic spectrometric method for the determination of manganese in natural waters. The manganese(II) catalyzes the oxidation of salicylaldehyde guanylhydrazone by hydrogen peroxide. The calibration curve is linear in the range of 8–80 μg L^{-1} with a relative error of 1%.

Wang et al.[526] reported a kinetic spectrophotometric determination of manganese in natural waters. The method is based on the catalytic effect of manganese on the oxidation of malachite green by potassium periodate. The method was applied to manganese concentrations in the range of 0.4–5.0 μg L^{-1}.

3.29.2 Spectrofluorometry

Morgan et al.[527] carried out a kinetic determination of manganese by its attenuation of the fluorescence of the beryllium-morin complex. The sample is treated with diethanolamine and a reagent comprising beryllium sulfate and morin. After 20 min, the reaction is stopped by the addition of EDTA and the fluorescence measured at 525 nm (excitation at 436 nm). Down to 5 $\mu g L^{-1}$ manganese can be determined by this procedure.

Other nonfluorescence kinetic methods for the determination of manganese include the use of osmium and EDTA,[528] and iridium nitriloacetic acid and 1,2-diaminocyclohexane-*N,N,N′,N′*-tetraacetic acid.[529]

Zhang et al.[530] have described a fluorescence spectroscopic method for determining down to 18 ppt of manganese in natural waters.

Catalytic kinetic fluorometry has been used to determine manganese in natural waters.[311] The catalytic effect of manganese on the reaction of acid chrome blue K by UV radiation increased the fluorescence. The detection limit was 2 μg of $Mn L^{-1}$, and the calibration was linear in the range of 1–120 $\mu g L^{-1}$. A similar method based on the catalytic oxidation of rhodamine 6G with potassium periodate by nitriloacetic acid has been described.[531] The detection limit was 0.018 μg of $Mn L^{-1}$, and the linear working range was 0.04–1.00 $\mu g L^{-1}$.

3.29.3 Continuous-Flow Analysis

Hydes[532] has described a continuous-flow method for the determination of manganese in natural waters containing iron. Interference from up to 100 M iron could be removed by addition of EDTA after formation of the manganese/formaldoxime complex. The extent of formation and destruction of the complexes of iron and manganese with formaldoxime depended on the pH value of the solution and on the period between addition of the reagent and measurement of absorbance.

3.29.4 Atomic Absorption Spectrometry

A limited amount of work has been carried out using atomic absorption spectrometry.[533–541]

Kumar et al.[533,542] used a matrix modification reagent consisting of a mixture of ascorbic acid and ammonium nitrite prior to the determination of manganese by electrothermal atomic absorption spectrometry. This method eliminates interference by alkali metals, alkaline earth metals, and iron. The detection limit for manganese is 0.5 $\mu g L^{-1}$. Shijo et al.[534] employed diethyldithiocarbamate for the chelation of manganese prior to solvent extraction by 1-chlorotoluene. The manganese was subsequently determined by electrothermal atomic absorption spectrometry. Complexation of manganese by nicotinohydroxamine acid and trioctylmethyl ammonium cation, followed by extraction with methylisobutyl ketone, was used by Abbasi.[535] The manganese was determined by atomic absorption spectrometry with a detection limit of 0.1 ppb. Separation of soluble manganese into two fractions has been reported by Corsini et al.[536] A sequential preconcentration on XAD-7, followed by chelex 100, left manganese on the XAD-7 column. The second fraction consisted of at least one form of unidentified manganese.

Manganese has been preconcentrated by complexation with thenoyltrifluoroacetone and dibenzo-18-crown-6, followed by determination by atomic absorption spectrometry.[537] A tungsten atomizer has been used for the electrothermal atomic absorption determination of manganese.[538] Matrix interferences were reduced with ascorbic acid. The detection limit in water was 1.2 pg, which corresponded to 0.12 μg of Mn L^{-1}. Manganese has been preconcentrated on a flow-through electrochemical microcell with deposition of manganese in reticulated vitreous carbon.[539] The manganese was determined by electrothermal atomic absorption spectrometry with a detection limit of 8.7 pg of manganese.

3.29.5 Miscellaneous Techniques

Gine et al.[543] have described a semiautomatic flow-injection analysis system for the determination of manganese in natural waters. Potentiometric stripping analysis has been used to determine manganese concentrations in the range of 2 nM to 30 μM. Interference resulting from the interaction of manganese and copper in the mercury electrode can be overcome by adding zinc or gallium.

Maggi et al.[544] used radiotracers to study the distribution of manganese[545] and zinc in the ultrafiltrate fraction of freshwater. Chiswell and Makhtar[545] applied electron spin resonance spectroscopy to a study of speciation of manganese in fresh waters.

A method for natural water analysis based on the catalytic effect of manganese on pyrogallol red discoloration by hydrogen peroxide has been reported by Cheng.[546] The sensitivity is 0.018 μg of manganese L^{-1}, with a range of 0–40 μg of manganese L^{-1}, and a relative standard deviation of 2.8–3.6%.

Beinrohr et al.[547] have described a galvanic stripping method for the determination of down to 5 ppt manganese in natural waters.

3.30 MERCURY

3.30.1 Atomic Absorption Spectrometry

Atomic absorption spectrometry using a cold vapor-generation technique has generally been used for the determination of low concentrations of mercury in solution. The detection limits of this technique are between 1 and 0.05 ng. Analysis of mercury in natural waters at sub-nanogram-per-liter levels with cold vapor atomic absorption spectrometry requires large sample volumes (nearly 1 L per analysis). Detection limits better than 10 pg have only been achieved with some novel instrumentation, e.g., dc-mode operation of the light source and double-beam compensation of vacuum ultraviolet spectrometry.

While plasma emission spectrometry using various types of plasma has high sensitivity for mercury comparable to or better than that of cold vapor atomic absorption spectrometry generally, detection limits better than 10 pg have been reported. These reports are classified with the plasma source: dc discharge plasma microwave-induced plasma, and low-pressure ring-discharge plasma.

A convenient sample volume for natural water analysis is less than about 100 mL per analysis, when ease of sample handling and the necessity for repeat analyses are taken into consideration. This implies that the necessary detection limit for analysis of sub-nanogram-per-liter levels of mercury is better than 10 pg. Plasma emission spectrometry thus offers a more useful approach for the analysis of sub-nanogram-per-liter levels of mercury in natural water samples than does cold-vapor atomic-absorption spectrometry.

Gold trap techniques are being used to improve sensitivity in the determination of mercury in natural water. In one such technique, the mercury is reduced to its elemental state using stannous chloride and then swept with a current of air onto a gold-treated graphite furnace tube. The tube is then inserted into the carbon rod of a graphite furnace atomic-absorption spectrometer and analyzed. Detection limits of 10 $\mu g L^{-1}$ have been obtained by this procedure.

Atomic absorption spectroscopy has found extensive use in the determination of mercury in river waters.[548–553] Inter-laboratory tests carried out on this method indicated that only 30% of UK laboratories were achieving accuracy targets of a total error not exceeding 20% of the standard concentration or 0.1 $\mu g L^{-1}$, whichever was greatest. Thompson and Godden[551] claim to have considerably improved the detection limit of mercury determination using an improved mercury fluorescence detector system.

Lutze[553] devised a sensitive method using cold-vapor absorption spectroscopy for determining mercury in river waters in amounts down to 0.1 $\mu g L^{-1}$ (10-mL sample) or even lower (0.1 $\mu g L^{-1}$ quoted). This method eliminates gas-purge dilution of mercury vapor during partitioning and aeration in the cold vapor generation for atomic absorption spectrophotometry determinations. A dual-bubbler system of aeration apparatus was used for the mercury vapor generation and compared with other techniques. The sensitivity of this method is superior to those of earlier methods.

BITC[554] has described two procedures based on cold-vapor atomic-absorption spectroscopy for the determination of mercury in river water and other waters. Both methods give reliable results at concentrations as low as 0.2 μg mercury L^{-1}. Inter-laboratory comparisons with 22 participants show that there were no significant differences at the 0.75 $\mu g L^{-1}$ level. The repeatability variation coefficient was 3.8–10.9% and that for reproducibility was 7.2–29.4% for the two methods.

Pinstock and Umland[555] have used a cold-vapor atomic absorption technique to measure different forms of mercury (mercurous, mercuric, elemental mercury) at the $\mu g L^{-1}$ level in natural water. Mercury at the nanogram per liter level in natural water has been determined by atomic emission spectrometry.[555–557]

Pratt and Elrick[558] studied the interference effects of selenium on the determination of mercury by cold-vapor atomic-absorption spectrometry. An automated online digestion of mercury with potassium permanganate was used for the oxidation of selenium(IV) to selenium(VI) prior to the determination of mercury.

Harsanyi et al.[559] enhanced the sensitivity for the determination of mercury in natural waters by preconcentrating the mercury in a 4-L sample by reduction to metallic mercury with stannous chloride. The solution is then aspirated into 50 mL of sulfuric acid-potassium permanganate solution; then, after further reduction with stannous chloride, the mercury was determined by atomic absorption spectrometry.

Joensun[560] has described an apparatus by means of which 1–300 ng of mercury can be determined in a variety of samples without pretreatment. The sample (10–500 mg) is placed in a stainless steel boat with 10–100 mg of sodium nitrate and heated in a furnace at 700° for 1–2 min. The mercury vapor is swept by a stream of nitrogen over silver maintained at 400–500 °C to remove chlorides, and is amalgamated on gold maintained at 100–200 °C. After 3 min, the gold is heated at 400 °C and the mercury vapor released is carried by the nitrogen through a cooling coil and dust filter to a second gold amalgamator. The mercury is determined by connecting the amalgamator to a measuring chamber that can be evacuated, and heating the amalgam to 400 °C. The atomic absorption of mercury vapor in the chamber is measured at 253.7 nm.

Cold-vapor atomic-absorption spectrophotometry was used in the method for mercury determination described by Temmerman et al.[561] in which mercury in the natural water samples is extracted by reduction with stannous ion, preconcentrated by amalgamation with gold, prior to introduction into the light path for measurement. A detection limit of 1 ng L^{-1} was obtained when samples as large as 1 L were used.

Boehnke[562] determined mercury in natural waters by flameless atomic absorption spectrometry, following sodium borohydride reduction of the ionic and organic forms of mercury. The detection limit for a 10-mL sample volume is 2 μg L^{-1} for the hollow cathode lamp and 1 μg L^{-1} for the electrodeless discharge lamp.

Aoki et al.[563] have proposed a method for the continuous-flow determination of mercury in natural water samples. The mercury is reduced with sodium borohydride followed by separation of the mercury by diffusion through a microporous PTFE membrane with a determination by flameless atomic absorption spectrometry. The detection limit was 0.007 ppb and the relative standard deviation was 1.7% at the 5 μg L^{-1} level.

Churchwell et al.[564] evaluated the U.S. EPA Methods 7470 and 7471 for the cold-vapor atomic-absorption spectrophotometric determination of mercury. They found that the recirculating cold vapor methods are not sufficiently flexible to permit special quality-control measures, have an inadequate detectability for low-level mercury concentrations, and are plagued by spectral interferences by organic vapors. They suggest sample digestion is carried out to remove organic interferences, in reconfiguration of the glassware, and amalgamation prior to introduction into the atomic absorption spectrophotometric.

Korenaga et al.[565] eliminated interference in the cold-vapor atomic-absorption spectrophotometric determination of mercury by alkaline tin(II) reduction instead of the conventional acidic reduction. Aqueous samples were digested with potassium persulfate to decompose organic-bound mercury, prior to alkaline stannous chloride addition. The detection limits in natural waters were 0.5 μg L^{-1} and the precision was 3%.

Gill and Fitzgerald[566] isolated mercury from natural water samples by reduction with stannous chloride combined with collection and two-stage concentration onto gold, in a flameless atomic absorption spectrophotometric method. The detection limit was 0.21 pM with an analytical precision of approximately 10% for 500 mL samples in the 2–20 pM range.

Welz and Schuber-Jacobs[567] describe a method where sodium borohydride was found equivalent or superior to stannous chloride as a reducing agent for the cold-vapor atomic-absorption spectrometric determination of mercury in natural waters. The mercury vapor is washed with sodium hydroxide and dried with magnesium perchlorate prior to amalgamation with gold. Detection limits were 15 ng L^{-1} for a 10-mL sample and 3 ng L^{-1} for a 50-mL sample. The calibration curve was linear to 40 ng of mercury, and the relative standard deviation was less than 2%.

An automated flow-injection analysis method for the digestion of organic mercury forms and reduction to elemental mercury in natural waters is described by Birnie[568] Mercury is removed by aeration and determination by flameless atomic absorption spectrophotometry, with sampling rates approaching 20 samples L^{-1}. The detection limit was 2 μg L^{-1} of mercury.

Numerous other workers have discussed atomic absorption spectrometric methods for the determination of inorganic mercury in natural waters.[569–576]

3.30.2 Inductivity Coupled Plasma Atomic-Emission Spectrometry

Nojiri et al.[577] have described a method for the determination of sub-ng L^{-1} levels of mercury in lake water using atmospheric pressure helium microwave-induced plasma-emission spectrometry. In this method, mercury vapor was generated from water samples by reduction and purging and was collected with a gold amalgamated trap. The mercury vapor, removed by heating the trap, was introduced into a helium microwave-induced emission spectrometer. The atomic emission line of 253.7 nm was used for the determination of mercury. The detection limit, defined as three times the standard deviation of the blank operations, was 0.5 pg in 50 mL of water sample, corresponding to 0.01 ng L^{-1}.

Anderson et al.[578] have described an inductively coupled plasma atomic emission spectrometric method for the determination of down to 2 μg L^{-1} of mercury in natural waters.

Tong et al.[579] have also discussed the application of inductively coupled plasma atomic emission spectrometry to the determination of mercury in natural waters and compared it to cold-vapor atomic-absorption spectrometric methods.

Borgnon and Cadet[580] describe a system for generating mercury vapor and the hydrides of mercury, selenium, arsenic, antimony, and bismuth for the simultaneous determination by inductively coupled plasma emission spectrometry.

Camunan-Aguilar et al.[581] coupled argon and helium inductively coupled plasma for the determination of mercury in natural waters and found helium produced the best detection at 10 ng L^{-1} mercury.

Krull et al.[582] have described a procedure for the determination of inorganic and organomercury compounds using high-performance liquid chromatography with an inductively coupled plasma emission spectrometric detector with cold vapor generation. In this method, postcolumn cold vapor generation was used to obtain improved detection limits. The replacement of the conventional polypropylene spray chamber of the inductively coupled plasma by an all-glass chamber is described. A comparison of band broadening indicates that the glass chamber is useful when a severe memory effect is observed with the polypropylene spray chamber. Detection limits ranged from 32 to 62 μg L^{-1} of mercury, based on a signal-to-noise ratio of 2:1. This represents a three to four order of magnitude enhancement over detection limits obtained without cold vapor generation. The approach is linear over three orders of magnitude.

The postcolumn reaction system has been described by Bushee et al.[583] An aqueous solution of 0.5% m/v sodium tetrahydroboration(III) in 0.25 M sodium hydroxide solution and a 1.2 M solution of hydrochloric acid served as the two reagents.

3.30.3 Inductively Coupled Plasma Mass Spectrometry

Smith[584] described an isotope-dilution inductively coupled plasma mass spectrometric method for the determination of mercury in amounts down to 6 ppt in natural waters. In the method, the sample was spiked with 201 Hg. Natural concentrations of mercury in water samples require preconcentrations on gold traps and subsequent electrothermal heating and purging of the traps with argon directly into the ICPMS torch. The detection limit was 0.2 ng mercury L^{-1} using a 200-mL sample.

3.30.4 Anodic Stripping Voltammetry

Overall, the key issue in making an electrochemical analytical technique one of the conventional methods for detecting mercury in the category associated with high sensitivity, such as cold-vapor atomic-absorption spectrometry, cold-vapor atomic-fluorescence spectrometry, and neutron activation analysis, is the availability of a suitable working electrode.

Many studies have been carried out on the choice of electrode materials for determining mercury anodic stripping voltammetry.

Jyh-Myng and Mu-Jye[585] have described a square-wave voltammetric stripping analysis of mercury(II) at a poly(4-vinylpyridine)/gold film electrode (PVP/GFE). In this method, mercury is preconcentrated as the anionic forms in the chloride medium, onto the modified electrode by the ion-exchange effect of the poly(4-vinylpyridine). The high solubility of mercury in gold also helps to increase the preconcentration effect. The preparation of the PVP/GFE is performed by first spin-coating a solution of the poly(4-vinylpyridine) polymer onto the electrode surface. Subsequently, gold is plated onto the electrode. Various factors influencing the determination of mercury(II) were thoroughly investigated in this study. In comparison with the conventional gold film electrode, this modified electrode showed improved resistance to interferences from surface-active compounds and common ions, especially for copper(II), which is generally considered as a major interference in the determination of mercury(II) on gold film electrodes. The PVP/GFE also showed increased sensitivity and better mechanical stability of the gold film when used in conjunction with the square-wave voltammetric method. In addition, detection can be achieved without deoxygenation, and the electrode can be easily renewed. The analytical utility of the PVP/GFE was demonstrated by application to various water samples.

A flow-potentiometric and constant current stripping analysis for mercury(II) has been described by Huang et al.[586] Gold, platinum, and carbon fibers with a diameter of 10 μm were mounted in PVC tubes and used as flow sensors in the determination of mercury. The detection limit for mercury in natural waters after 10 min of electrolysis was $45\,ng\,L^{-1}$.

Zhang and Wang[587] used a glassy carbon electrode modified with bipyridyl/ethanol solution for the determination of mercury by semi-differential anodic stripping voltammetry.

Hosseini et al.[588] used a PVC membrane electrode for detecting Hg(II) ions based on a new cone-shaped calix[4]arene (L) as a suitable ionophore. The sensor exhibits a linear dynamic in the range of 1.0×10^{-1} M, with a Nernstian slope of $29.4 \pm 0.4\,mV\,decade^{-1}$ and a detection limit of 4.0×10^{-7} M. The response time is quick (less than 10 s), it can be used in the pH range of 1.5–4, and the electrode response and selectivity remained almost unchanged for about 2 months. The sensor revealed comparatively good selectivity with respect to most alkali, alkaline, earth, and some transition and heavy metal ions. It was successfully employed as an indicator electrode in the potentiometric titration of Hg^{2+} ions with potassium iodide, and the direct determination of mercury content and water samples.

In order to examine the calix[4]arene suitability as an ion carrier for the Hg^{2+} ion, Hosseini et al.[588] constructed several PVC ion-selective electrodes for the wide variety of cations, including alkali, alkaline earth, and transition metal ions. The slope of the potential responses is much lower than expected for mono-, di-, and trivalent metal ions.

3.30.5 Spectrophotometry Method

Gharchbaghi et al.[589] applied a new, simple, and rapid dispersive liquid–liquid microextraction based on ionic liquid preconcentrate trace levels of mercury as a prior step to its determination by spectrophotometric detection. In this method, a small amount of an ionic liquid (1-hexyl-3-methyllimmidazolium bis(trifluormethylsulfonyl)imide) as the extraction solvent was dissolved in acetone as the disperser solvent and the binary solution was then rapidly injected by a syringe into the water sample containing Hg cations that were complexed by 4,4′-bis(dimethylamino)thiobenzophenone in the presence of sodium dodecyl sulfate as the anti-sticking agent. Thereby, a cloudy solution was formed and the Hg-4,4′-bis(dimenthylamino)thiobenzophenone complex was extracted into the ionic liquid droplets. After centrifuging, the droplets of extractant were settled at the bottom of a conical test tube and the extracted phase was determined by spectrophotometry at 575 nm. Usually some parameters affect the complex formation and extraction, such as the type and volumes of extraction and dispenser solvents, type and concentration of anti-sticking agent, salt concentration, pH, and concentration of chelating agent, which have also been optimized for this method. Under optimum conditions, the enrichment factor of 18.8 was obtained from 10 mL of water. The detection limit of the method was found to be 3.9 $\mu g L^{-1}$ and the relative standard deviation (n=5) for 50 $\mu g L^{-1}$ of mercury was 1.7%.

Theranlaz and Thomas[590] determined down to 0.4 ppt of mercury in natural waters by a method based on indirect spectrophotometry.

3.30.6 Miscellaneous Techniques

Other procedures that have been employed in studies of mercury in natural waters include reverse-phase liquid chromatography,[591] metastable transfer emission spectroscopy,[592] spectrophotometry,[545,593–596] gas chromatography,[597] flow-injection analysis,[598] kinetic photometry gas,[599] piezocrystal detection,[600–602] photochroism,[603,604] and photo-acoustic spectroscopy.[605]

Silica gel modified with diaminothiourea as a selective solid-phase extractant[606] and polyaniline and polyaniline-methylene blue-coated screen-printed carbon dietrodes[607] have been used in the determination of inorganic mercury in natural waters.

Liu et al.[608] described a new solid substrate room-temperature phosphorimetry method for the determination of trace mercury. It is based on the fact that in the acidic medium of sulfuric acid, an ionic associate will be formed between $[HgI_4]^{2-}$ and rhodamine 6G ($[(Rhod.6G)_2]^{2+} \cdot [HgI_4]^{2-}$). The number of Rhod.6G molecules in the associate is higher than that in $Rhod.6G^+ \cdot I^-$, and the Hg^{2+} will induce a heavy atom perturbation effect. Both aspects can cause a sharp increase in photophorescent intensity in a solid substrate. The linear range of this method is 0.80–160.0 fg $spot^{-1}$ (corresponding concentration, 0.002–0.4 ng mL^{-1}; sample volume, 0.4 $\mu L\ spot^{-1}$) with a limit of detection of 0.15 fg $spot^{-1}$ (corresponding concentration, 3.8×10^{-13} g mL^{-1}). The regression equation for the working curve is $\Delta I_p = 114.0 + 1.956\ {}^{m}H^{2+}(fg\ spot^{-1})$

(r=0.9991, n=6). This method has been successfully applied to the determination of trace mercury(II) in human hair, cigarettes, and water. The reaction mechanism for the formation of ionic associate is also discussed.

3.31 MOLYBDENUM

3.31.1 Spectrophotometry

A catalytic spectrophotometric method for determining trace molybdenum in natural waters was described by Li[609] Molybdenum(VI) is reduced to molybdenum(III) by potassium borohydride and determined indirectly in the presence of iron(III) and phenanthroline. Molybdenum(III) reduced iron(III), forming the ferric(II)-phenanthroline complex. The detection limit was 0.5 $\mu g\,L^{-1}$ molybdenum and the relative standard deviation was 8.3%.

Ion-exchange spectrophotometry was used for the determination of molybdenum in a method described by Capitan et al.[610] The molybdenum in natural water reacts with thiocyanate ion in the presence of stannous chloride, the reaction product is adsorbed on Dowex 1-8X-8 anion exchange, and the resin-phase absorbance is measured at 467 and 800 nm.

Zheng et al.[611] determined molybdenum in natural waters by catalytic spectrophotometry using reduction of azure(I) with hydrazine hydrochloride. The method can be applied to molybdenum in a concentration range of 1–1000 $\mu g\,L^{-1}$. The detection limit was 0.6 μg of molybdenum L^{-1}.

3.31.2 Spectrofluorometry Method

Lu et al.[612] described a method for determining molybdenum in which this element is reduced to Mo(III), followed by reaction with luminal.[612] The florescent response was linear for molybdenum in the range of 1–40 $\mu g\,L^{-1}$, and the relative standard deviation was less than 2%. In a further method,[613] molybdenum was reacted with carminic acid prior to spectrofluorimetry.[614]

3.31.3 Flow-Injection Analysis

Molybdenum-catalyzed oxidation of thiazine red R by hydrogen peroxide has been used as a basis for determination of molybdenum.[615] The sensitivity was 0.64 μg of Mo L^{-1}. Stopped-flow injection spectrophotometry has been used for the determination of Mo.[616] Potassium thiocyanate/malachite green, with PVC and Triton X-100 as enhancing agents, was used for the termination. Recoveries were in the range of 103–108% for Mo in the range of 40–200 $\mu g\,L^{-1}$.

3.31.4 Atomic Absorption Spectrometry

Molybdenum has been determined in natural waters after preconcentration in Sephadex G 25 gel[617] at pH 3.5. Ethylenediamine tetraacetic acid desorbs molybdenum from the gel. To the evaporated sample containing 0–5–50 µg of molybdenum, 2 mL of lanthanum(III) solution is added and molybdenum determined by atomic absorption spectrometry. This method is suitable for the determination of molybdenum at the levels normally encountered in river water (0.2–0.6 µg L^{-1}).

Emerick[618] used calcium chloride for matrix modification to eliminate sulfate interference in the electrothermal atomic absorption determination of molybdenum in natural water. A 0.5% calcium chloride ($2H_2O$) (w/v) solution is added to the sample in a volume equal to that of the sample.

Kuroda et al.[619] used anion exchange to preconcentrate molybdenum prior to the electrothermal atomic absorption spectrophotometry. Molybdenum was adsorbed on a column of Bio-Rad AG 1 in the Cl^- form, which was subsequently eluted with ammoniacal ammonium chloride solution. The method gave a relative standard deviation ≤8% at a molybdenum level of 10 µg L^{-1}.

Palladium-containing matrix modifiers have been used in the electrothermal atomic-absorption spectrophotometric determination of molybdenum.[620]

3.31.5 Inductively Coupled Plasma Atomic-Emission Spectrometry

Online preconcentration by ion exchange, followed by flow-injection inductively coupled plasma atomic-emission spectrometry, has been used for the determination of molybdenum.[621] A microcolumn containing Dowex 1 resin in the Cl^- form was used to absorb molybdenum, followed by elution with ammonium chloride/ammonium citrate. Enrichment factors of 50, with recoveries of 90–110%, and a relative standard deviation of 5.3% for samples at the 0.2 mg of Mo L^{-1} level, were achieved. A similar method for molybdenum using anion exchange has been described.[622] Molybdenum(VI) and hydrazoic acid form a complex anion, which was separated on an AG 1-X8 anion exchanger, which was then eluted with 1 M NH_4Cl/NH_4OH. The molybdenum was determined online by inductively coupled plasma atomic emission spectrometry.

3.31.6 Electrochemical Methods

Willie et al.[623] used linear sweep voltammetry for the determination of molybdenum. The molybdenum was absorbed as an Eriochrome Blue Black R complex on a static mercury drop electrode. The method was reported to have a limit of detection of 0.50 µg L^{-1}.

Polarography has been used to determined down to 0.1 µg of molybdenum as it forms an Mo(VI)–β-mercaptopropionic acid–phenylarsenium chloride couplex.[624]

Yu and Li[625] reported a catalytic kinetic ion-selective electrode method for molybdenum in natural waters. The molybdenum(VI) was determined by its catalytic effect on the oxidation of iodide ions in an acidic hydrogen peroxide solution, the iodide being measured by the ion-selective electrode. Interferences were noted for iron, vanadium, tungsten, copper, and chromium, which can be partially masked by EDTA. The detection limit was 0.01 $\mu g L^{-1}$.

3.32 NEODYNIUM

See under Lanthanides, Section 3.24.

3.33 NICKEL

3.33.1 Atomic Absorption Spectrometry

Sun and Suo[626] determined ultra-trace amounts of nickel in environmental samples by atomic absorption spectrometry with in situ trapping of volatile nickel species in an iridium-palladium coated 8-graphite furnace.

Saraji et al.[627] used an ion-imprinted silica sorbent prepared using a sol-gel process for the selective extraction of Ni(II) ions from water samples. Bis(dibenzoylmethanto)nickel(II) complex was used as template, phenyltrimethoxysilane and 3-aminopropytriethoxysilane as functional monomers, and tetraethylorthosilicate as reticulating agent. The material was packed in a solid-phase extraction column. The effect of sampling volume, elution conditions, sample pH, and sample flow rate on the extraction of Ni ions from water samples were studied. The relative selectivity coefficients of imprinted sorbent for NI(II)/Co(II), Ni(II)/Cu(II), and Ni(II)/Cd(II) were 23.7, 30.3, and 24.4 times greater than nonimprinted sorbent, respectively. The relative standard deviation of the eight replicate determinations of Ni(II) was 4.2%. The detection limit was 0.9 $\mu g L^{-1}$ using flame atomic absorption spectrometry. This method was successfully applied to the determination of trace nickel in water samples.

3.33.2 Chemiluminescence Analysis

Lu et al.[628,629] used a chemiluminescence reaction of 2-thenoyltrifluoroacetone in alkaline hydrogen peroxide-sodium hypochlorite with Ni(II) as the basis for a method of determination of nickel in natural waters. The detection limit was 1.0 $\mu g L^{-1}$. The linear response range was 4–5000 μg of Ni L^{-1}.

Li and Yu[630] have described a flow injection on chemiluminescence method based on the reaction of nickel with anthracene green in hydrogen peroxide catalyzed by NiII. The detection limit was 0.11 $\mu g L^{-1}$.

3.33.3 Polarography

McCurdy[631] described a procedure for the differential pulse voltammetric determination of nickel. This method consists in the application of dc or differential pulse voltammetry after prior interfacial accumulation by an adsorption layer of dimethylglyoximate at the hanging mercury drop electrode. This is an example of the approach of substantial sensitization by chelate adsorption at stationary electrodes. In aqueous media determination limits of 1 ng L^{-1} are attainable for nickel(II) with good precision and accuracy.

3.33.4 Stripping Voltammetry

Wang and Zhang[632] determined nickel in natural waters by cathodic stripping voltammetry in the ppb concentration range with 90–105% recoveries and a relative standard deviation of ≤23%. The nickel is concentrated by anodic oxidation, followed by reaction with dimethylglyxoime forming a precipitate on a glassy electrode. The nickel peak appears at 1.1 V and the calibration curve is linear in the range of 1 mg L^{-1} to 1 μg L^{-1}.

Another electrochemical method for determining nickel in natural waters is reported by Sawamoto.[633] The nickel was determined by adsorptive stripping voltammetry of the 2,2′-bipyridine complex on the hanging mercury drop electrode. The calibration curve was linear at 1.0 μM concentration, the detection limit was 5 nM, and the relative standard deviation at 50 nM nickel was 5.5%.

Farisa et al.[634] complexed Ni(II) with hydroxynaphthol blue on a hanging mercury drop electrode, prior to determination by cathodic adsorption stripping. The detection limit was 0.1 μg L^{-1}. The linear response extended up to 25 μg of Ni L^{-1}.

Marques and Chierice[635] describe a method based on the adsorption of a nickel-phenyldithiocarbamate complex prior to the determination by adsorption stripping of voltammetry. The detection limit was 0.026 μg L^{-1} with 15 min of preconcentration.

3.33.5 Miscellaneous Techniques

Yoshimura et al.[636] have described the application of ion-exchange colorimetry with PAN to the determination of nickel at μg L^{-1} or less, in natural waters.

Wilson and DiNunzio[637] used Donnan dialysis to enrich nickel (and cobalt) in natural hard waters by factors of up to 20 at the 200-μg L^{-1} level. Recoveries were in excess of 99%.

Aliakbar and Jalali[638] have described a new polymer as a selective chelating agent for separation and reconcentration of nickel ions from water samples. This polymer was prepared by electropolymerization of 4-nitrophenol. Electrosynthesis was carried out on the lead cathode in aqueous sodium acetate solutions. The electrode-product is a dark-brown powder, insoluble in water but soluble in methanol, *N,N*-dimethylformamide, and tetrahydrofuran. The electrode-product was characterized by differential scanning calorimetry, gel permeation chromatography, FT-IR, ^{1}H-NMR, cyclic voltammetry, and UV-Vis spectrometry. A proper mechanism and structure of the prepared polymer was suggested. A few drops of methanolic solution of electrode-product formed a blue complex with nickel ions in an aqueous medium in the pH range of 6–10.

This new chelating reagent was used as a coating material on activated charcoal and applied for solid-phase extraction of trace amounts of nickel ions from natural water and wastewater. The effect of different parameters such as type of eluent, elution conditions, sample volume, sample flow rate, and mass of coating material were studied. In the presence of coexisting ions, no significant intereferences were observed. Under the optimal conditions, limits of detection and quanitifcation were 0.32 and 1 $\mu g L^{-1}$ Ni(II), respectively. The method was used for determination of Ni(II) in some lagoons south of the Caspian Sea. The validity of this method was confirmed by the comparison of the obtained results with the results of inductively coupled plasma atomic-emission spectrometry.

3.34 NIOBIUM

3.34.1 Spectrophotometry

Abbasi[639] described a spectrophotometric procedure for the determination of niobium in natural waters. A niobium complex for *N-p*-methoxyphenyl-2-furylacrylohydroxamic acid was extracted into chloroform. The complex has an absorbance maximum at 550 nm and gave a detection limit of 0.1 $\mu g L^{-1}$. The method was compared with the methyl isobutyl ketone extraction of niobium, which has a detection limit of 0.3 $\mu g L^{-1}$.

3.35 OSMIUM

3.35.1 Spectrophotometry

Ensafi and Rezaei[640] described a kinetic spectrophotometric method for determining down to 5 ppt of osmium in natural waters.

3.36 PALLADIUM

3.36.1 Spectrophotometry

Abbasi et al.[641] used *N-p*-methoxyphenyl-2-furylacrylohydroxamic acid and 5-(diethylamino)-2-(2-pyridylazo)phenol for the extraction of palladium from natural waters. The complex is measured at 560 nm by absorption spectrophotometry.

3.36.2 Atomic Absorption Spectrometry

Ma[642] evaluated several electrothermal atomizers for the atomic absorption spectrophotometric determination of palladium in natural waters. The normal graphite tube, pyrolytic graphite-coated tube, platform, and pyrolytic graphite tube lined with tungsten were evaluated. The sensitivity was highest with the pyrolytic graphite-coated tube, followed by the pyrolytic graphite-coated tube lined with tungsten.

Interferences by elements in sample waters were observed for all atomizers.

In a further method, palladium was immobilized on a chelate-forming resin containing the sulfonic acid derivative of dithizone.[643] The loaded resin was resuspended in water for electrothermal atomic-absorption spectrophotometry. The detection limit was 5 ng of Pd L^{-1} with a coefficient of variation of 3.94% for a 125-ng spike of Pd.

3.36.3 Stripping Voltammetry

Wang and Varmghese[644] employed absorptive stripping voltammetry to determine palladium in natural waters as the dimethylglyoxime complex. The palladium complex was concentrated on a hanging mercury drop electrode at 0.20 V. With a 10 min preconcentration time, the detection limit was 20 ng L^{-1}.

Abbasi and Wang[645] used *N-p*-methoxyphenyl-2-furylacrylohydroxamic acid and 5-(diethylamino)-2-(2-pyridylazo)phenol for the extraction of palladium from water samples. The complex is measured at 560 nm by absorption spectrophotometry. The detection limit is approximately 0.1 ppb.

3.36.4 Miscellaneous Techniques

Guo et al.[646] reported on a unique photoacoustic method for the determination of palladium in natural waters. The palladium(II) is extracted by dithizone and subsequently evaporated as an organic film.

The method offers a detection limit of 0.04 ppb of palladium(II) in a 100-mL water sample, a linear dynamic range of 2–200 ng palladium(II), and a relative standard deviation of 3.2% at the 100 ng of palladium(II) level.

3.37 PLUTONIUM

3.37.1 Miscellaneous Techniques

Selective adsorption of plutonium(VI) by silica gel has been used to separate plutonium(VI) from plutonium(V).[647] Only plutonium(V) was found in natural water samples. Plutonium(VI) is highly unstable in alkaline lake waters, reducing to plutonium(V).

Suutarinen et al.[648] have discussed the speciation of plutonium in river and lake waters.

Plutonium adsorption to and desorption from mineral phases plays a key role in controlling the environmental mobility of plutonium. Begg et al.[649] assessed whether the adsorption behavior of plutonium at concentrations used in typical laboratory studies ($\geq 10^{-10}$ [Pu] $\leq 10^{-6}$ M) are representative of adsorption behavior at concentrations measured in natural subsurface waters (generally $<10^{-12}$ M). Pu(V) sorption to Na-montmorillonite was examined over a wide range of intital plutonium concentrations (10^{-6}–10^{-16} M). Pu(V) adsorption after 30 days was linear over the wide range of concentrations studied, indicating the plutonium sorption behavior from laboratory studies at higher concentrations can be extrapolated to sorption behaviour at low, environmentally relevant concentrations. Pu(IV) sorption to montmorillonite was studied at intital concentrations of 10^{-16}–10^{-11} M and was much faster than Pu(V) sorption over the 30-day equilibration period. However, after one year of equilibration, the extent of Pu(V) adsorption was similar to that observed for Pu(IV) after 30 days. The continued uptake of Pu(V) is attributed to a slow, surface-mediated reduction of Pu(V) to Pu(IV). Comparison between rates of adsorption of Pu(V) to montmorillonite and a range of other minerals (hematite, goethite, magnetite, groutite, corundum, diaspore, and quartz) found that minerals containing significant Fe and Mn (hematite, goethite, magnetite, and groutite) adsorbed Pu(V) faster than those that did not, highlighting the potential importance of minerals with redox couples in increasing the rate of Pu(V) removal from solution.

3.38 POTASSIUM

3.38.1 Spectrophotometry

Motomizu et al.[650] have described a spectrophotometric determination of potassium based on solvent extraction of the complex formed with a crown ether and an anionic azo dye using flow-injection analysis. Three analogs of 18-crown-6 were examined as the crown ether, four analogs of the anionic azo dye, 4-(4-dialkylaminophenyl)azo-2,4-dichlorobenzene sulfonate and picrate were examined as counter anions and approximately 10 different solvent systems were investigated. Optimal conditions involved the use of 4-(4-diethyl-aminophenyl)azo-2,5-dichorobenzene sulfonate and benzo-18-crown-6 reagents with benzene-chlorobenzene (1:1 by volume) as the preferred solvent for extraction. The carrier stream was distilled water and the reagent stream contained the dye anion ether, EDTA dilithium salt, and lithium hydroxide. A PTFE porous membrane was used for separating the organic phase. The detection limit was 2×10^{-6} M and the relative standard deviation was 0.4%.

3.38.2 Atomic Absorption Spectrometry

Iwachido et al.[651] have reported a chromogenic crown ether method for the determination of potassium in natural waters. Iwachido et al.[651] used 4′-LN-(8-sulfo-1-napthyl)-4-aminophenylaxo/benzo-18-crown followed by adsorption spectrometry at 520 nm.

3.38.3 Potassium-Selective Electrodes

Ward[652] have evaluated different types of potassium ion selective electrodes, manufactured by different companies for suitability for application in monitoring or in situ chemical analysis systems. These were a glass membrane single electrode, a glass membrane combination electrode, and a liquid ion-exchange electrode. All three electrodes performed well in river water.

Hulanicki[653] discussed the interference by cationic anionic and nonionic surfactants on the potentials of potassium by ion-selective electrodes. Electrodes with a solid silver contact were less sensitive to interferences than were electrodes with an internal reference solution.

3.39 PRAESEODYNIUM

See under Lanthanides, Section 3.24.

3.40 PROMETHIUM

See under Lanthanides, Section 3.24.

3.41 RADIUM

Mcteos and coworkers[654] devised a sequential-injection preconcentration method for the determination of radium.

3.42 RHODIUM

Lazarev and Gerko[655] have described a spectrophotometric method for determining rhodium as an ion-phase with chrompyrozol and SnII. Measurements were made to 630 nm.

3.43 RUBIDIUM

3.43.1 Atomic Absorption Spectormetry

Atomic absorption spectrometry,[656] ion-exchange chromatography,[657] and l-particle induced X-ray emission spectrometry[656] have been applied to the determination of rubidium admixture with other elements.

3.44 RUTHENIUM

3.44.1 Spectrophotometry

Areceneau et al.[658] chelated ruthenium with disodium 5,5′-(3′(2-pyridyl)-1,2,4-trianzine-5,6-diyl)-bis(2-furansulfonate) (Ferene), followed by spectrophotometric determination at 520 nm. The method is applicable to the determination of ruthenium in the 0.03–4.0 $mg L^{-1}$ range in natural waters.

3.44.2 Miscellaneous Techniques

Vashal et al.[659] have described paper chromatographic and electrophoretic methods for the determination of 0.1-μg amounts of ruthenium in natural water. Ruthenium can be extracted from the zones on the chromatogram and determined by a kinetic method based on the formation of the blue color of the oxidation products of benzidine catalyzed by ruthenium.

A catalytic method for the determination of ruthenium in natural waters has been described by Yu and Li.[660] Recoveries were 95–105% and the sensitivity of the method was 5.4 $ng L^{-1}$. The relative standard deviation was ±5.8% at the 100-$ng L^{-1}$ level of ruthenium.

3.45 SAMARIUM

See under Lanthanides, Section 3.24.

3.46 SCANDIUM

3.46.1 Spectrophotometry

Scandium(III) is extracted with *N*-phenylbenzohydroxoamic acid into isoamyl alcohol and complexed with xylenol orange or morin in a method described by Agrawal and Nagar[661] for the analysis of natural waters. The xylenol orange complex is measured spectrophotometrically at 540 nm. The morin complex of scandium is determined fluorometrically in the range of 0–1.5 $mg L^{-1}$.

3.46.2 Neutron Activation Analysis

Neutron activation analysis has been used[662] to determine low levels of scandium in natural water. Ammonia, and if necessary a carrier, are added to the sample and the resulting precipitate filtered and dried prior to irradiation to produce 45 Se after 10–30 days depending on the content of trace elements. The activity at 1.12 MeV is measured in comparison with a standard. A detection limit of less than 0.4 $\mu g L^{-1}$ scandium was achieved.

3.47 SELENIUM

3.47.1 Spectrophotometry

Dessai and Paul[663] proposed a simultaneous colorimetric determination of selenium(VI) with sodium diethyldithiocarbamate. Other examples of spectrophotometric determinations are given in Table 3.3.

3.47.2 Spectrofluorometry

Chen and Liu[664] presented a fluorometric method for determining selenium in which Se(IV) is reduced hydrogen selenide and reacted with bis(8-hydroxyquinoline-5-sulfonic acid)–Pd solution. The selenium formed PdSe and the liberated 8-hydroxyquinoline-5-sulfonic acid coordinated with Al^3 to form a fluorescent chelate. The detection limit was 0.26 ng of Se mL^{-1}.

TABLE 3.3 Examples of Selenium Determination by Spectrophotometry

Reagent	Wavelength (nm)	Range of Determination Limit	References	Remarks
3,3′-Diamino-benzidine	420	0.1–10 mg L^{-1}	145,146	Toluene extraction, pH 6–7; interferences: Au, Br, Cr, Mo, Sn, V,W, Zr, and oxidation agents
2,3-Dianminaphthalene	380	0–4.00 mg L^{-1}	147	Toluene extraction, pH 1.5–2.5; interferences: Cu, hypochlorite, and reducing agents
1,2-Diaminobenzene	335	5.25 mg L^{-1}	148	Toluene extraction, pH 1.5–2.5; interferences; Fe(III), Sn(IV), and iodide ion, but Fe(III) can be masked with EDTA
4-Chlorodiaminobenzene	341	0.3–10 mg kg^{-1}	149	Toluene extraction at low pH
4-Nitrodiaminobenzene	350	3 mg kg^{-1}	150	Toluene extraction pH <2
–	–	0.8 mg L^{-1}	151	–
Rhodamine B and thiocyanate	606	0–120 μg L^{-1}	152	
N[1]-(2-hydroxy-propyl), 1,2-diamino-4-chloro-benzene	–	–	153	Extraction of complex with carbon tetrachloride
Phenyl hydrazine hydrochloride	–	ng L^{-1}	154	Catalytic method
K13 R03 and vitamin C	–	–	155	Catalytic method
Phenyl hydrazine and hydrogen peroxide	–	ng L^{-1}	156	Catalytic method

Yu et al.[665] presented a laser time-resolved fluorometric method for determining Se(IV) by reaction with *N*-(2-pyridyl)thioquinaldinamide. In a further method, a linear range of 0.02–10 ng selenium is reacted with 2,3-diaminonaphthylene separated by reversed-phase HPLC, and determined fluorometrically. The method has a detection limit of 0.2 ng of Se(IV).[666]

3.47.3 Atomic Absorption Spectrometry

Atomic absorption spectrometry in its classical form is not sensitive for determination of selenium. A lower delectability level of 200 $\mu g L^{-1}$ can be reached that can be extended down to 100 $\mu g L^{-1}$ by application of a deuterium lamp for background correction. The argon-hydrogen flame is often used for augmentation of sensitivity but it increased interferences, too.

Electrochemical preconcentration of selenium(IV) by reduction and deposition of elemental selenium on a platinum spiral was used by Lund and Bye[667] for air-acetylene flame atomic absorption spectroscopy. The electrolysis is done in the presence of hydrazine dihydrochloride to prevent the generation of chlorine, which would oxidize selenium(IV) to the nonreducible selenium(VI). A detection limit of 5 $\mu g L^{-1}$ was obtained for a 25-mL sample and a 5-min electrolysis time.

A slotted quartz tube positioned on top of a conventional atomic absorption burner has been used in a flame atomic-absorption spectrometric determination of selenium.[668] Cobo Fernandez et al.[669] reported a method for determination of selenium by flow-injection hydride-generation atomic-absorption spectrometry. An online reduction step was used to reduce Se(VI) to Se(IV).

Oernemark and Olin[670] preconcentrated selenium on Dowex 1X8 resin prior to determination by flow-injection hydride-generation atomic-absorption spectrometry. A flow-injection hydride-generation atomic-absorption spectrometric method with online preconcentration of selenium by co-precipitation with La-$(OH)_3$ gave a detection limit of 0.001 μg of Se L^{-1} using a 6.7-mL water sample.[671] An atomic fluorescence spectrometer for determination of the hydride-forming elements has been developed that achieved detection limits of 0.05 μg of Se L^{-1} and 0.10 μg of As L^{-1}. Guo et al.[672] reported a cold atomic fluorescence method for determination of selenium in water with a detection limit of 0.02 μg of Se L^{-1}.

The flameless technique in which a small sample aliquot is electrothermally atomized in a graphite furnace, is especially suitable for direct analysis of samples as it offers a high sensitivity (50 pg of selenium is the detection limit), but is not simple or free from interference or volatility losses. Treatment of the furnace or addition of various metals, such as nickel,[673–681] mercury,[682] molybdenum,[683] lanthanum,[675] copper,[682,684–687] mercury,[682] and chromium[687] diminishes the chemical interference in water analysis, allows use of higher ashing temperatures without losses, and results in a significantly enhanced sensitivity. Vickery and Buren[675] reported that added metal solutions counteract signal depression by interfering elements not only because they reduce the volatility of selenium but also because they modify the graphite furnace surface, leading to more efficient atom formation.

Surface treatment can replace the matrix modifier; use of metal-coated or pyrolytic graphite-coated curvettes[675] or graphite tubes lined with tantalum foil[688] yields optimal results.

Table 3.4 summarizes some published results. Pretreatment of the water samples and some slight modifications can give enhanced sensitivity.

3.47.4 Hydride-Generation Flameless Atomic-Absorption Techniques

The literature on hydride-generation flameless atomic absorption (HGAA) spectroscopy for environmental water analysis is summarized in Table 3.5.

It appears that with regard to sensitivity, the sample pretreatment step is of paramount importance. For most uncontaminated environmental waters, only the last three methods listed in Table 3.5 will be suitable.

TABLE 3.4 Determination of Selenium in Natural Waters by Flameless Atomic-Absorption Spectroscopy

Type of Water	Pretreatment	Detection Limit ($\mu g L^{-1}$)	References
Estuarine/ fresh	Addition of 0.2% Ni(II)	0.5	674
Fresh	Predigestion + addition of Ni(II) Chemical preconcentration	0.02	689
River/sea	APDC or DDTC/CCl_4 extraction	0.4	684
	DDTC/CCl_4 extraction + addition of Cu(II)	0.4	673
River/sea	APDC/MIBK extraction + addition of Cu(II)	0.3	685
River	APDC/MIBK extraction	0.3	690
Drinking/ source	Evaporation		
Rain	Piazselenol-toluene + addition of Ni(II)	0.02	676
River	Graphite furnace AAS	10^{-6}	691
River	Na diethyldithio carbamate extraction, graphite furnace AAS	0.4 ng	684,692

TABLE 3.5 Determination of Selenium in Environmental Waters by Hydride-Generation Flameless Atomic-Absorption Spectroscopy

Pretreatment	Reducing Agent	Flame Type	Detection Limit ($\mu g\,L^{-1}$)	References
None	$NaBH_4$	Ar/H_2	50	694
None	$NaBH_4$	N_2/H_2	5	695
HCl, H_2SO_4	$KI/SnCl_2/Zn$	Ar/H_2	2	696
Digestion with acid $KMnO_4$ solutions, reduction by HCl	$KI/SnCl_2/Zn$	N_2/H_2	2	697
None	$TiCl_3/Mg$ or $NaBH_4$	Air/C_2H_2	1.7	698
H_2SO_4, HNO_3, $K_2S_2O_8$	$NaBH_4$	Ar/H_2	1	699
Acid digestion	$NaBH_4$	N_2/H_2	0.6	700
None	$NaBH_4$	Ar/H_2	0.15–0.25	701
$K_2S_2O_8/HCl$	$KI/SnCl_2/Al$	Ar/H_2	0.1	692
HCl heating (1 h)	$NaBH_4$	Tube furnace	0.02	702
Absorption of hydrides on $HgCl_2$ or Ag/NO_3-impregnated filters, re-extraction with $HClO_4$	$NaBH_4$ or $KI/SnCl_2/Zn$	Ar/H_2	0.02	613
Co-precipitation with Fe(OH), flotation with air bubbles, scavenging, redissolution in HCl	$NaBH_4$	N_2/H_2	0.02	703

Nakahara et al.[693] used a nondispersive system to compare the zinc and sodium borohydride reduction systems. The best attainable detection limits for selenium were 0.43 and 0.2 ng, respectively. With the proposed method, 1 $\mu g\,L^{-1}$ levels of selenium can be determined accurately. The presence of several elements, including other hydride-forming elements, in 1000-fold ratio to selenium caused a negative interference. Tellurium gave a positive interference.

Rodan and Tallman[704] have pointed out that groundwaters containing appreciable levels of dissolved organic compounds often present difficulties in their analysis for selenium. Oxidative digestion of the sample usually removes organic interferences and permits the determination of total selenium. However, digestion also destroys the natural distribution of selenium between its common oxidation states, 4+ and 6+. These workers describe a procedure that permits speciation of inorganic selenium in groundwater samples. Many of these samples, even when spiked with appreciable amounts of selenium, totally suppress the release of selenium hydride. The method is based on separation of the selenium species from the organic interferents(s) by column chromatography at pH 1.6–1.8 on XAD-8 resin, which removes organic compounds while preserving the natural distribution of selenite and selenite. This is followed by hydride-generation–graphite furnace atomic-absorption spectrometric determination of selenium.

Krivan et al.[705] applied radiotracer error diagnostics in an investigation of the determination of selenium by flame hydride-generation atomic-absorption spectrometry. The method involved a preliminary decomposition of the sample with a mixture of sulfuric acid and hydrogen peroxide. Selenium(VI) was reduced to selenium(IV) with 5 M hydrochloric acid. Errors were caused by back-oxidation of selenium(IV) to selenium(VI) by the residual chlorine produced during reduction. Chlorine removal with nitrogen eliminated the back-oxidation.

3.47.5 Hydride-Generation Electrothermal (Flameless) Atomic-Absorption Spectroscopy

Hydride generation in combination with electrothermal atomic absorption spectroscopy is probably the most studied method for determination of selenium in environmental water, yet it is plagued by numerous interferences.[123,706–716] Meyer et al.[714] found that these depend very strongly on the concentration of hydrochloric acid in the sample solution. Vijan and Leung[715] utilized the capability of hydrochloric acid to form chloro-complexes with common interfering heavy metals, such as copper and nickel, to eliminate practically all of their suppressive effects.

Various automated hydride-generation flameless atomic absorption spectroscopy techniques have been described,[692,708,716] allowing 30–70 water analyses per hour. In the procedure of Pyen and Fishman,[716] organic selenium-containing compounds are first decomposed by hydrochloric acid-potassium persulfate digestion. The selenium(VI) produced, along with the inorganic selenium, is reduced to selenium(IV) with stannous chloride and potassium iodide and then to selenide with sodium borohydride. The hydrogen selenide is stripped from the solution with nitrogen and then decomposed in a tube furnace at 800 °C placed in the optical path of the electrothermal atomic absorption spectrometer.

Cutter[717] proposed a complex method for determination of selenite, selenite, dimethyl selenide, and dimethyl diselenide in natural waters. Volatile methyl species of selenium are removed from the sample by a stream of helium and inorganic forms of selenium are selectively reduced to the hydride with sodium borohydride and stripped. Both the organic selenides and the hydrogen selenide are trapped in a liquid-nitrogen trap. The methyl species are separated by gas–liquid chromatography and measured in the quartz-tube furnace. For selenium(IV) and (VI), the detection limit is 5 ng L^{-1} and the time of analysis is 15 and 30 min per sample, respectively.

Koelbl[718] compared flame atomic absorption spectrometry, graphite furnace atomic-absorption spectrometry, and inductively coupled plasma mass spectrometry for the determination of selenium in natural waters. Inductively coupled plasma mass spectrometry gave the best result, with a detection limit of 0.1 ng.

3.47.6 Inductively Coupled Plasma Atomic-Emission Spectrometry

Winge et al.[719] evaluated the accuracy, precision, and detection limits of inductively coupled argon plasma emission spectrometry. Pneumatic nebulization gives a detection limit of 30–40 µg L^{-1}. Only with ultrasonic nebulization is it possible to reach a limit of detection of 1 µg L^{-1} of selenium.

3.47.7 Inductively Coupled Plasma Mass Spectrometry

Ion-pairing reversed-phase or anion-exchange liquid chromatography followed by inductively coupled plasma mass spectrometry has been shown to yield sub-ppb detection limits for the determination of selenium in natural waters.[720]

Hydride generation ion inductively coupled plasma mass spectrometry has been used to determined selenium at levels of 10–100 ng L^{-1}.[135]

3.47.8 Stripping Methods

Howard et al.[721] determined selenium(IV) by differential-pulse polarography of the 4-chloro-*o*-phenylenediamine piazselenol, which gives a reduction peak at 0.11 V versus SCE at pH 2.5 in formate buffer. The detection limit was 0.4 µg L^{-1}. Interferences from chromium(VI), copper(II), molybdenum(VI), nickel(II), tin(II), tellurium(IV), and vanadium(V) were overcome by treatment of the sample with a Chelex-100 resin column.

Nguyen et al.[722] used differential-pulse anodic-stripping voltammetry for the determination of copper and lead.

Differential-pulse cathodic-stripping voltammetry at a hanging mercury drop electrode has been used for selenium determination, specifically in rain-water and snow, and pretreatment of selenium down to 10 ng L^{-1} in a 15-mL sample.

Henze[723] discussed the determination of selenium, arsenic, and tellurium at sub-μg L^{-1}.

Deldime and Hartman[724] discussed the electrochemical characteristics of the differential-pulse and cathodic-stripping polarography of selenium. The addition of copper and 2N hydrochloric acid as supporting electrolyte can decrease the interfering effect of certain trace metals. Denis et al.[725] used reduction of selenium to selenide by sodium borohydride, evolution of hydrogen selenide and its trapping in an alkaline cell, as a preparation step before differential-pulse cathode-stripping voltammetry. The detection limit was 1 μg L^{-1} with the precision of approximately 5% at concentrations above 4 μg L^{-1}. Breyer and Gilbert[726] determined selenium(IV) by formation of piazselenol with 3,3′-diaminobenzidine at pH 1.5 followed by differential-pulse voltammetry in borate-buffered electrolyte at pH 9, when the piazselenol gave reduction peaks at potentials of −0.64 V and −0.82 V against a saturated calomel electrode. The limit of detection was 0.10 μg L^{-1}. No interference was caused by up to 500-fold amounts of ions such as divalent copper and lead.

The determination of selenium by cathodic-stripping voltammetry has been discussed by several other workers.[727–730]

3.47.9 Neutron Activation Analysis

Because of its high sensitivity (10^{-8}–10^{-9} g of selenium), neutron activation analysis has been used for determination of selenium, especially in environmental waters. Thermal neutrons are most often used for the activation because radioactive selenium isotopes are forms in (n, γ) reactions. Only the 77 mSe and 75 Se radioisotopes are currently utilized in activation analysis.

The most important step in the neutron activation analysis of water is the enrichment of selenium. Methods of preconcentrating the element from the water include selective extraction, nonselective extraction, and freeze-drying; see Table 3.6. The irradiation can also be followed by radiochemical separation.

3.47.10 Gas Chromatography

Estimation of selenium by gas chromatography is based almost exclusively on measurement of the amount of piazselenol formed in the reaction of selenium(IV) with an appropriate reagent in acidic media. Piazselenols are easily extracted with organic solvents (most frequently toluene), in which they can be subsequently determined by spectrophotometric, fluorometric, or chromatographic methods. In gas chromatography, piazselenols are usually estimated with an electron capture detector due to its highest sensitivity and selectivity

TABLE 3.6 Determination of Selenium by Nondestructive Neutron-Activation Analysis without Chemical Preconcentration

Experimental Conditions	Type of Water	Detection Limit ($\mu g\,L^{-1}$)	References
10 mL Enclosed in quartz $2.6 \times 10^{12}\,n\,cm^{-1}\,s^{-1}$ (14 h), 30 days decay time	Subsurface waters	0.7	731–733
5 mL enclosed in quartz $1.4 \times 10^{13}\,n\,cm^{-2}\,s^{-1}$ (3 days), 17 decay/60 min counting	River water	0.34	734
250 mL in specially designed quartz bottles $10^{13}\,n\,cm^{-2}\,s^{-1}$ (20 h)	Sewage	–	735
5 L evaporation $10^{14}\,n\,cm^{-2}\,s^{-1}$	Rain water	–	736
Progressive evaporation $3 \times 10^{16}\,n\,cm^{-1}\,s^{-1}$	Rain water	–	737
250 mL, freeze-drying pressing into pellets $1.6 \times 10^{12}\,n\,cm^{-1}\,s^{-1}$ (32 h), 20 days decay time	Rain water	3.0	738
100 mL, freeze-drying $5 \times 10^{13}\,n\,cm^{-1}\,s^{-1}$ (4 h)	River water	–	739
200 mL, freeze-drying $10^{14}\,n\,cm^{-1}\,s^{-1}$ (1 day), 1 day decay time	River water	0.1	740
1,1 freeze-drying in the presence of 25 mg of ultra-carbon, $2 \times 10^{14}\,n\,cm^{-1}\,s^{-1}$ (35 h) 4 weeks decay time	River water	0.07	741
Preconcentrated activated carbon	River water	–	742–746

with respects to the compounds (Table 3.7). Apart from the superior sensitivity and selectivity, the gas chromatographic method allows for determination of selenium in natural waters based on 1,2-diamino-3,5-dibromobenzene with an extraction procedure that is specific for selenium(IV). Total selenium is determined by treatment of natural water with titanium trichloride and with a bromine-bromide redox buffer to convert selenide, elemental selenium, and selenite to selenious acid. After reaction, the 4,6-dibromopiazselenol formed from

TABLE 3.7 Examples of Selenium Determination in Various Materials by Gas-Liquid Chromatography

Materials	Detection limit $\mu g L^{-1}$	Reagent	References
Water	4×10^{-8} g	4-Chloro-Phenylenediamine	747
River water	5×10^{-10} g	2,3-Diamino-napthalene	748
River and sea water	2×10^{-9} g	1,2-Diamino-3,5-dibromobenzene or 4,6-dichloro-piazeslenol	749
Se(IV)	2	2,3-Diamino-piazselenol	750
Se(total)	2	Reduction + 2,3-diamino-piazselenol	750
Se(IV)	0.1	5-Nitropiazselenol	751
Se(total)	0.1	Reduction + 5-nitropiazselenol	751
Se(total)	0.05	Reduction + 5-chloropiazselenol	751
Se(IV)	0.01	2,3-Diamino-piazselenol	752
Se(IV)	0.01	4-Nitropiazselenol	753
Se(total)	0.01	Photo-oxidation + 4-nitropiazselenol	753
Se(IV)	0.002	4,6-Dibromopiazselenol	749
Se(-II, 0, IV)	0.002	Bromine oxidation + 4,6-dibromopiazselenol	754
Se(-II, 0, IV, VI)	0.002	Br_2/Br-buffer + 4,6-dibromopiazselenol	754
Se(IV)	0.008	5-Nitropiazselenol	754
Se(total)	0.0008	Photo-oxidation + 5-nitropiazselenol	755–757

as little as 1 ng of selenium can be extracted quantitatively into 1 mL of toluene from 500 mL of natural water; up to 2 ng L^{-1} of selenium(IV) and total selenium can be determined. The percentage of selenium(IV) in the total selenium in river water varies from 35% to 70%.

Uchida et al.[754] determined various senelnium species in river water and sea water by electron capture detection gas chromatography after reaction with 1,2-diamino-3,5-dibromobenzene. This reagent reacts with selenium IV to form 4,6-dibromopiazselenol, which is extracted into toluene. After Se(-11) and Se(0) has been reduced by a bromine-bromide solution to a selenium(IV) state, total selenium is determined by the same method. The limit of detection is 0.002 μg L^{-1}.

Flinn and Aue[758] proposed a photometric detector for selenium analysis that enables determination of 2×10^{-12} g selenium s^{-1}.

Johansson et al.[759] determined selenium in natural waters by derivativization followed by gas chromatography using an electron capture detector.

De La Calle Guntinas et al.[760] volatilized selenium from natural water samples by reaction with sodium tetraethylborate and measured the volatilized selenium by gas chromatography microwave-induced plasma-atomic spectrometry. The detection limit for a 5-mL sample was 8 ppt.

Gallus and Henmann[761] used gas chromatography coupled to an inductively coupled plasma mass spectrometric detector to determine down to 20 ppt of selenium in natural waters.

Sarkouhi et al.[762] have described a liquid-phase microextraction and gas chromatography method for the determination of selenium(IV) as its piazselenol complex in natural waters. The samples reacted with *o*-phenylenediamine in 0.1 M hydrochloride acid at 90 °C for 15 min, and then liquid-phase microextraction was performed. A microdrop of carbon tetrachloride was applied as the extracting solvent. After extraction, the microdrop was introduced directly into the injection port of gas chromatography for analysis. Several important extraction parameters, such as the type of organic solvent, sample and organic drop volumes, salt concentration, stirring rate, and exposure time were controlled and optimized. In this liquid phase, micro-extraction was achieved by suspending a 3-μL carbon tetrachloride drop from the tip of the microsyringe immersed in 12.5 mL of aqueous solution. Under optimized conditions, a dynamic linear range was obtained in the range of 20–1000 μg L^{-1}. The preconcentration factor and the limit of detection of selenium in this method were 91 and 0.9 μg L^{-1}, respectively. The relative standard deviations for the spiking levels of 50–100 μg L^{-1} in the real samples were in the range of 3.2–6.1%, and the relative errors were located in the range of 5.4–5%.

Sarkouhi et al.[762] studied the reaction time and formation of the complex for 1 h under different reaction temperatures in the range from 15 to 100 °C. The results showed stabilization at 90 °C after 15 min. The signal increases to an *o*-phenylene diamine-SeIV ratio of 10.1. At a ligand-SeVI ratio of 9:10 greater than 10.1, a constant signal was obtained, and therefore a ratio of 10.1 was adopted in the method. Also a reaction time of 20 min was adopted, as at this reaction time the amount of analyte extracted was constant.

3.47.11 High-Performance Liquid Chromatography

Nakagawa et al.[763] have described a method based on selenotrisulfide formation followed by high-performance liquid chromatography with fluorometric detection for the determination of selenium(IV). The method involves pre-column reaction of selenium(IV) with penicillamine (Pen) to produce stable selenotrisulfide (Pen-SSeS-Pen) and subsequent derivatization to a fluorophore by reaction with 7-fluoro-4-nitrobenzo-2,1,3-oxidazole. The fluorophore was separated by reversed-phase high-performance liquid chromatography and selenium content was determined by fluorometric detection. The calibration plots showed a linear relationship in the range of 10–2000 ppb of selenium(IV) with a detection limit of 5 $\mu g L^{-1}$ (signal-to-noise ratio (S/N) > 2).

3.48 SILVER

3.48.1 Spectrofluorometry

An extraction-spectrofluorometric method for the determination of silver in natural waters has been described by Que et al.[764] Benzothiacrown ether was highly selective for silver and served as the ligand in an ion-pair extraction. Eosin Y was the fluorescent anion. The method was applied to the determination of silver in the 2–10 $\mu g L^{-1}$ range.

Fluorometric quenching using eosin as the reagent has been used to determined down to 20 $\mu g L^{-1}$ of silver in lake water.[765]

3.48.2 Atomic Absorption Spectroscopy

Various workers have discussed the determination of silver in natural water.[700,766] Chau et al.[767] filtered 1-L samples containing 0.1–1.0 $\mu g L^{-1}$ silver through a membrane filter, adjusted to pH 1, then passed the eluate through an Amberlite AG 1-X8 anion-exchange resin column. Silver was eluted from the column with acetone-nitric acid-water (20:1:1), and acetone removed from the eluate by evaporation. After adjusting to pH 2.3, the solution was treated with ammonium pyrrolidine dithiocarbamate and the silver chelate extracted into methylisobutyl ketone. Evaluation at 328.1 nm gave the silver concentration of the original sample. Results are not affected by trace levels of iron, zinc, copper, and lead.

McHugh[766] has described a method for the determination of silver in natural waters using electrothermal atomization and a deuterium background corrector to help eliminate interference. Recovery ranged from 88% to 100%. The method has a detection limit of 0.02 $\mu g L^{-1}$ silver.

No serious interferences from various cations and anions were observed, except at the 100 mg L^{-1} level for calcium, iron, potassium, sodium, and chloride. The levels of recovery of silver for these elements were 48, 54, 55, and 68, respectively.

3.48.3 X-ray Fluorescence Spectroscopy

Guo et al.[768] determined silver in natural waters by X-ray fluorescence spectrometry after adsorption on cation exchange paper. The water sample is buffered to pH 4.5 with acetic acid-sodium acetate. The detection limit is approximately 2 μg of silver L^{-1}.

3.48.4 Differential-Pulse Anodic-Scanning Voltammetry

Wu et al.[769] and others[770–774] used differential pulse anodic stripping voltammetry to determined silver to a detection limit of 0.1 μg L^{-1}. Linear response was found over a range of 0.5–1000 μg of Ag L^{-1}. Anodic stripping voltammetry using Nafion 130-modified glassy C electrode has been used for the determination of silver. The recoveries were in the range of 96–102.5% for silver at the 0.1 μg L^{-1} level.

3.48.5 Miscellaneous Techniques

Silver-catalyzed oxidation of $K_4Fe(CN)_6$ by oxygen was the basis for a spectrophotometric method for silver. The detection limit was 0.5 μg of Ag L^{-1}, with recoveries of 9–103% for 20–40 μg of silver added. The relative standard deviation was 4.1% and a linear working range of 0–300 μg L^{-1} was observed.

Whitlow and Rice[775] studied the complexation of silver in river water. The capacity of a river to complex trace metals involved both a dissolved and a particular fraction. Studies were carried out to determine the dissolved component of the apparent complexation capacity for silver in samples of water. Silver ion activities detected by the silver sulfide ion selective electrode during potentiometric titration of the river water with silver nitrate were lower than the silver activities calculated using an inorganic equilibrium speciation model. The difference in silver activity was attributed to the presence of one or more constituents in the river water, possibly dissolved organic matter or collided material, which were capable of binding silver strongly. There was evidence that most of the dissolved silver was in the form of a silver chloride complex, with little free silver ion present.

A catalytic kinetic method for the determination of silver in natural waters has been described by Jiang et al.[776] The silver reacts in a system containing potassium persulfate α,α-bipyridine and reduced phenolphalein to form a red complex, which is monitored at 550 nm. The method gives a linear calibration curve in the range of 0–115 ppb and the sensitivity is 4 μg of silver L^{-1}.

Dai and Jiang[777] describe a similar catalytic method for determining silver in natural waters based on the catalytic oxidation of reduced phenolphthalein by ammonium persulfate in the presence of ethylenediamine. The method has a linear calibration curve in the range of 10–120 μg L^{-1} and a sensitivity of 0.8 μg silver L^{-1}.

3.49 SODIUM

Various techniques have been used for the determination of low levels of sodium in natural waters including atomic adsorption spectrometry,[778] inductively coupled plasma atomic emission spectrometry,[779] inductively coupled plasma mass spectrometry,[780] emission spectrometry,[781] neutron activation analysis,[782] prompt-γ neutron activation analysis,[783] ion-exchange chromatography,[657] and alpha-particle-induced X-ray emission spectrometry.[784] Sodium ion selective electrodes are a simple method for the determination of sodium in natural waters.

Van den Winkel et al.[778] described an Auto Analyzer system for the direct determination of 1–100 mg L^{-1} sodium in river water. Detection is achieved by a sodium-selective electrode. The addition of EDTA to the test solution prevents precipitation of magnesium or sulfate, but this limits the sensitivity of the method to 1 mg L^{-1}. A fivefold excess of potassium can be tolerated, and interference from hydrogen ions is eliminated by working at high pH. The reproducibility and accuracy of the results are good and samples can be analyzed at the rate of 20 per hour.

3.50 STRONTIUM

3.50.1 Neutron Activation Analysis

Neutron activation analysis has been used[785] to determine down to 0.1 mg L^{-1} strontium in natural waters. Strontium is first extracted with chloroform as the 8-hydroxyquinolate from the sample initially adjusted to pH 11.5 with sodium hydroxide.

3.50.2 X-ray Fluorescence Spectroscopy

Nishioka et al.[786] used X-ray fluorescence for the determination of strontium in natural waters after precipitation as the carbonate. A linear calibration curve for strontium in the range of 2–150 μg was obtained with little interference from 2 mg of calcium or 10 mg of magnesium.

3.50.3 Miscellaneous Techniques

Other methods for the determination of strontium in natural waters include atomic absorption spectrometry,[787] inductivity coupled plasma atomic-emission spectrometry,[788] inductivity coupled plasma mass spectrometry,[789] polarography,[790] ion chromatography,[791] and alpha-particle-induced X-ray emission spectrometry.[792]

3.51 TECHNETIUM

3.51.1 Inductively Coupled Plasma Mass Spectrometry

Beals[793] has described an inductively coupled plasma mass spectrometric technique for the determination of down to 0.6 ppt of 99 technetium in natural waters. It is claimed that this technique is faster and less prone to interference than radiometric techniques.

3.51.2 Anodic Stripping Voltammetry

Stripping voltammetry[794] has also been used to determine $\mu g L^{-1}$ levels of technetium in natural waters.

3.52 TELLERIUM

3.52.1 Absorption Spectrometry

Andreae[795] has described a technique for the determination of tellurium(IV) and tellurium(VI) species in a natural water matrix by magnesium hydroxide coprecipitation followed by hydride generation atomic absorption spectrometry. The limit of detection is $0.5 pmol L^{-1}$ and the precision is 10–20%.

3.52.2 Miscellaneous Techniques

Other methods that have been used to determine tellurium include inductivity coupled plasma atomic-emission spectrometry,[796] stripping voltammetry,[797] emission spectrometry,[798] flow-injection analysis,[799] hydride-generation atomic-absorption spectrometry,[135] and kinetic spectrophotometry.[389]

3.53 TERBIUM

See under Lanthanides, Section 3.24.

3.54 THALLIUM

3.54.1 Spectrophotometry

Chandrawanski et al.[800] determined thallium spectrophotometrically by extraction of thallium from natural water as a chlorothallate ion-pair complex of a cationic surfactant and determination at concentrations of $20 \mu g L^{-1}$ by complexation with Brilliant Green.

3.54.2 Atomic Absorption Spectrometry

De Ruck et al.[801] determined thallium in natural waters by electrothermal atomic absorption spectrophotometry. The thallium is oxidized and retained as the tetrachlorothallate(III) ion on an anion-exchange column. The detection limit of the method was 3.3 ng of thallium L^{-1}.

Electrothermal atomic absorption spectrophotometry was used by Welz et al.[802] to determine thallium in natural waters. A palladium modifier and hydrogen purge gas were used to prevent the volatilization of thallium during the pyrolysis stage. Chloride interference on thallium is in part caused by volatilization of thallium chloride in the pyrolysis stage and in part by formation of thallous chloride in the gas phase during the atomization stage. The palladium modifier is not as effective for thallium as it is for other elements. Its stabilizing power can be improved substantially when the modifier is pyrolized at 1000 °C before the sample is pipetted. Use of hydrogen purge gas has a similar effect on the stabilizing power of palladium. It was found necessary to apply both these measures, pyrolysis of the modifier prior to the addition of the sample and use of hydrogen, for interference-free determination of thallium in samples with high chloride content such as sea water. A detection limit (3σ) of 1 μg L^{-1} was obtained for thallium in both sample types with 10 μL sample volumes, and the characteristic mass was 19 ng.

Li and Ma[803] found that pyrolytic-coated graphite platform and palladium as a matrix modifier improved sensitivity. The recoveries ranged from 95% to 103%, the sensitivity was 0.68 μg of Tl L^{-1}, and the relative standard deviation was 1.5–2.7%.

Mohammad et al.[804] found that a mixed buffer of 0.1 M acetate and 0.1 M NH_4Cl at a pH of 10 was most successful for the extraction of thallium quinolin-8-ol from water. The detection limit for flame atomic absorption spectrophotometry was 3 μg L^{-1}.

3.54.3 Laser Exit of Spectrofluorometry

Two studies have been conducted on the application of this technique to the determination of thallium in natural water.[805,806] The method described by Axner et al.[806] employed laser-induced fluorescence in a graphite furnace. The method allowed the direct measurement of thallium without any preconcentration steps. A detection limit was not given, though a range of sample concentrations found in a study of freshwaters was 8–56 ng of Tl L^{-1}.

3.54.4 Scanning Voltammetry

Sun et al.[807] applied differential-pulsed anodic-stripping voltammetry to the determination of thallium. Britton–Robinson buffer solution was used as the thallium in natural waters was stripped at −0.46 V. Preconcentration for 30 min gave a detection limit of 2.3×10^{-2} M and a linear calibration curve was obtained over a range of 5×10^{-12} to 5×10^{-7} M. The relative standard deviation was 2.6%.

Potentionmetric stripping analysis for the determination of thallium in natural waters is described by Zhou et al.[808] Sodium sulfite was added to the sample for removal of oxygen and to serve as a masking agent. The thallium is preconcentrated for 10 min, giving a detection limit of 8 $\mu g L^{-1}$. Thallium was determined with recoveries of 95–105%.

3.54.5 Miscellaneous Techniques

Riley and Siddiqui[809] have described a procedure for the determination of thallium in natural waters that involves preliminary concentration by adsorption onto a strongly basic anion-exchange resin as the tetrachlorothallate ion, followed by elution with sulfur dioxide. After evaporation of the sulfur dioxide, thallium is determined by graphite-furnace atomic-absorption spectroscopy or differential-pulse anodic-stripping voltammetry.

Miloshova et al.[810] used a thallium-selective chalogenide glass sensor to determine thallium in natural water samples.

Lin and Nriagu[811] have shown that thallium in its trivalent form was the most dominant form of thallium in Great Lakes (United States) water samples.

3.55 THORIUM

3.55.1 Spectrophotometry

Arsenazo(III) (chlorophosphonazo(III)) is the only chromogenic reagent reported for thorium. Yamamoto[812] used extraction with 3-methyl-butanol-1 as a means of improving the sensitivity of the method. Cospito and Rigali[813] preconcentrated by co-precipitation of calcium-thorium oxalate and then used extraction with Aliquat 336 dissolved by xylene as a means of removing interferences.

Yang et al.[814] determined thorium down to 1 $mg L^{-1}$ by flow-injection analysis using DBF arseno as a chromogenic reagent.

Guo[815] determined thorium in natural waters by using cetylpyridinium bromide, amino C acid chlorophosphonoazo, which forms a complex with thorium.

3.55.2 Miscellaneous Techniques

Necemers et al.[816] described a method for determination of thorium in water in which the thorium was co-precipitated with iron, the precipitate was dissolved, and the thorium was purified by anion exchange and then electroplated on a stainless steel planchette, and measured by a spectrometry.[817] Ion-exchange phase spectrometry has been used to measure thorium with a linear range of 0–9 μg/50 mL. Yang et al.[818] presented a procedure that used ionizable crown ethers for separation of thorium from the lanthanides by solvent extraction.[817] Gong and Wu[819] described a salting-out extraction of the complex of thorium with *m*-sulfochlorophosphonazo from water using poly(ethylene glycol) 2000 and ammonium sulfate.

3.56 THULIUM

See under Lanthanides, Section 3.24.

3.57 TIN

3.57.1 Spectrophotometry

Valencia et al.[820] used solid-phase spectrophotometry to determine ppb concentrations of tin in natural waters.

3.57.2 Atomic Absorption Spectrometry

Donard et al.[821] have described a method for the speciation of inorganic tin and alkyltin compounds in natural waters by atomic absorption spectrometry using an electrothermal quartz furnace after hydride generation. These workers speciated inorganic tin and methyl- and *n*-butyltin compounds by volatilization from water samples by hydride generation, separation by a chromatographic packing material, and detection by atomic absorption spectrophotometry in an electrothermal quartz furnace at the 224.61 nm wavelength. Absolute detection limits of 20–50 pg (3σ) were reached.

Sturgeon et al.[822] combined hydride generation with electrothermal atomic absorption spectrophotometry for the determination of tin in natural waters. Stannane is generated with sodium borohydride and is trapped on a preheated graphite tube. A detection limit of $2\,ng\,L^{-1}$ based on a 30 mL water sample, was achieved for inorganic tin.

Pinel et al.[823] extracted tin into toluene, and picric acid was added as a matrix modifier, prior to determination by electrothermal atomic-absorption spectrophotometry. This method can be combined with a tin–tropolone extraction to determine both inorganic and butyl tin down to $10\,ng\,L^{-1}$ levels in natural waters.

Jin et al.[824] combined a trioctylphosphine-coated tungsten probe with atomic absorption spectrophotometry for the determination of tin. The detection limit was $0.4\,\mu g$ of $Sn\,L^{-1}$ with a relative standard deviation of 5.7% for samples with a concentration of $5\,\mu g\,L^{-1}$. Gallium phosphate has been used to co-precipitate tin from water samples in the 100–500-mL range.[825] The tin was determined by electrothermal atomic-absorption spectrophotometry with a detection limit of $0.23\,\mu g\,L^{-1}$.

3.57.3 Chemiluminescence Method

A chemiluminescence method using the Sn-catalyzed oxidation of water in *o*-phenanthroline has been described. The detection limit was $0.16\,\mu g$ of $Sn\,L^{-1}$, with a linear working range of $2–700\,\mu g\,L^{-1}$.

3.57.4 Performance Liquid Chromatography

Ebdon et al.[826,827] applied performance liquic chromatography to the determination of tin in nonsaline[826] and estuarine waters.[827]

3.57.5 Gas Chromatography

Abalos and Bayone[828] used a preconcentration technique followed by derivatization and then gas chromatography with conductivity coupled plasma mass spectrometric detection to determine 1–100 $pg\,L^{-1}$ of tin in natural waters.

3.57.6 Anodic Stripping Voltammetry

Macchi and Pettine[829] constructed differential anodic stripping voltammograms of tin(IV) at different pH values in sodium nitrate. A well-defined peak of anodic red dissoluation was apparent at pH below 7, with the height of the peak increasing when the pH was lowered to pH 3.5. A hydroxy complex of tin thus forms during anodic stripping.

Analysis of tin(IV) was carried out at pH 2, where the tin peak is stable enough to give a linear relationship between peak current and tin concentration. However, under such conditions the two peaks of lead and tin practically overlap.

Weber[830] determined tin in river water by voltammetry or atomic preconcentration methods such as ion-exchange and extraction with tropolone/toluene prior to determination by atomic absorption spectrometry. Results obtained by anodic stripping voltammetry were consistent with those from atomic absorption spectroscopy.

Weber[831] also applied differential pulse polarography to the determination of tin in river water. The addition of tropolene to acetate supporting electrolyte at about pH 4.7 enhanced the signal 30-fold, giving a sensitivity comparable to that obtained by anodic stripping voltammetry but without the need for preconcentration. The response was linear over more than two orders of magnitude.

Wang and Zadell[832] used adsorptive stripping voltammetry of the tin-tropolene complex with preconcentration onto a hanging mercury drop electrode. The tin is preconcentrated for 8 min at −0.4 V. The detection limit was 28 $ng\,L^{-1}$ and the standard deviation was 2.6% at 6 $\mu g\,L^{-1}$.

3.58 TITANIUM

3.58.1 Spectrophotometry

An extraction method has been described[833] using a chloroform solution of N-*n*-methoxyphenyl-2-furohydroxamic acid as chromogenic reagent. Interference by iron, molybdenum, chromium, zirconium and tantalum is eliminated by the presence of stannous chloride. The golden-yellow TiIV *N*-*p*-methoxyphenyl-2-furohydroxamine acid extract has a maximum absorbance of 385 nm and obeys Beer's law in the range 0.5 to 10 $mg\,L^{-1}$ titanium(IV), which renders the method of limited value for the examination of uncontaminated natural waters.

Abbasi[834] described spectrophotometric and atomic absorption spectrophotometric methods for titanium. Titanium chelated with *N-p*-methoxyphenol-2-furylacrylohydroxamic acid (MFHA) was extracted (after suitable processing of acid-digested samples) with chloroform or isoamyl prior to spectrophotometric or atomic absorption spectrometric determination. Detection limits were 1 and 200 $\mu g L^{-1}$ respectively, for the spectrophotometric and atomic absorption spectrophotometric methods.

Chen[835] describes a spectrophotometric method using diantipyrine waters. The method has a detection limit of 5 μg of titanium, with recoveries of 86–114% for titanium at 0.01–0.12 $mg L^{-1}$, and a relative standard deviation of 2%.

3.58.2 Atomic Absorption Spectrometry

Abbasi[834] extracted titanium from natural waters with *N-p*-methoxyphenyl-2-furylacrylohydroxamic acid prior to spectrophotometric or atomic absorption spectrophotometric measurement. The sensitivity of the spectrophotometric method with chloroform as the solvent is 1 $\mu g L^{-1}$. Isoamyl alcohol was used as the extracting solvent for the atomic absorption measurement and the sensitivity is reported at 10.2 $mg L^{-1}$.

3.58.3 Miscellaneous Techniques

Other methods that have been used for the determination of tin in natural waters include inductively coupled plasma atomic-emission spectrometry,[788] inductively coupled plasma mass spectrometry,[836] voltammetry,[837] and prompt ɣ-neutron activation analysis.[783]

3.59 TUNGSTEN

3.59.1 Atomic Absorption Spectrometry

Korrey and Goulden[838] have described a method for the determination of down to 100 $\mu g L^{-1}$ of tungsten in natural waters that involves formation of the benzoin anti-oxine derivative and extraction into methyl isobutyl ketone followed by atomic absorption spectroscopy of tungsten at a wavelength of 400.8 μm.

Several metals are known to form chelates with benzoin anti-oxine. Using the above method, no significant interference was observed in this determination of 1.0 mg of tungsten in 1.0 L water in the presence of 1.0 mg of the following ions: copper(II), manganese(III), chromium(VI), 10 mg of potassium, magnesium, 20 mg of sodium, 50 mg of calcium, molybdenum(VI), vanadium(V), and iron(III) did not interfere at the 0.01 $mg L^{-1}$ tungsten level. At the 1 $mg L^{-1}$ tungsten level vanadium(V) and iron(III) suppressed absorption and molybdenum(VI) enhanced it.

Hall et al.[839] examined the relative advantages of inductively coupled plasma atomic-emission spectrometry and inductively coupled plasma mass spectrometry. Adsorption of these analytes onto activated charcoal was used as a preconcentration step for both procedures. The detection limits for ICP-MS were 0.06 $\mu g L^{-1}$ for both elements, and for ICP-AES were 1.2 and 0.4 $\mu g L^{-1}$, respectively, for tungsten and molybdenum.

3.59.2 Catalytic Polarography

Catalytic polarography has been applied to the determination of tungsten as a β-mercapto propionic acid-2,2′-dipyridyl complex. This method was applied in the 0.02–50 $\mu g L^{-1}$ range.[840]

3.59.3 Fluorescent Quenching

Li et al.[841] described a highly sensitive and selective fluorescence-quenching method of trace tungsten in environmental samples using dibromohydroxyphenylfluorone as an emission reagent. In the presence of 0.04 $mol L^{-1}$ sulfuric acid and acetyltrimethylammonium bromide, tungsten(VI) reacts with dibromohydroxyphenylfluorone to form a 1:3 complex within 5.0 min. In order for the dibromohydroxyphenylfluorone–tungsten(VI) complex to form, the fluorescence intensity of the reagent solution was quenched by adding 0.1–1.0 μg of tungsten(VI) in 25 mL of solution. This was measured at 528 nm with excitation of 495 nm. In this work, a standard addition method was used for sample analysis. The decrease in fluorescence intensity of the reagent solution (ΔF) was linear for 0–0.9 μg of tungsten(VI) in 25 mL of solution, and the detection limit (3 s) of the standard addition method was found to be 0.012 $ng mL^{-1}$ of tungsten(VI). The effects of various metal and nonmetal ions were studied in detail. The experiments clearly showed that most foreign ions can be tolerated in considerable amounts, in particular, 50-fold Mo(VI), V(V), Zr(VI), and Ti(IV) do not interfere, and the selectivity of this is better than other previously described methods. Moreover, the method proposed is very stable and simple, and the fluorescence intensity of the solution can remain almost unchanged for 2.0 h at room temperature. The method has been used successfully to determine tungsten in environmental samples.

3.60 URANIUM

3.60.1 Spectrophotometry

Uranium(VI) forms a colored complex with chromatopic acid[842] at pH 7.25 in a 0.1 M triethanolamine-perchloric acid buffer and has strong absorptions at 410, 460, and 500 nm. Many interfering cations can be masked with a mixture of 2,3-diaminocyclohexanetetra-acetic acid (calcium salt), sodium sulfosalicylate, and sodium potassium tartrate. A 100-fold excess of sodium, lithium, ammonium, beryllium, calcium, barium, zinc, lead, nickel, indium, and zirconium and some of the lanthanide elements do not interfere; the interference caused by iron(III) and titanium(IV) is at a minimum at pH 7.25 and at 500 nm.

Korkisch and Koch[843,844] determined low concentrations of uranium by extraction and ion-exchange in a solvent system containing trictyl phosphine oxide. Uranium is extracted from the sample solution (adjusted to be 1 M in hydrochloric acid and to contain 0.5% of ascorbic acid) with 0.1 M trioctylphosphine oxide in ethyl ether. The extract was applied to a column of Dowex 1-X8 resin (Cl^- form). The column is washed with 6 M hydrochloric and uranium is eluted with molar hydrochloric acid and determined fluorometrically or spectrophotometrically with ammonium thiocyanate.

In another method, the sample was passed through a TBP-plasticized dibenzoylmethane-loaded polyurethane foam bed. Uranium was eluted and determined spectrophotometrically using Arsenazo III.[845] Thiocyanate and Victoria blue 4R were used to determine U spectrophotometrically.[846] A liquid–liquid extraction method has been described in which the chelate of uranium with *N*-phenyl-3-styrylacrylohdroxamic acid is extracted into chloroform and determined spectrophotometrically at 410 nm. The system obeyed Beer's law in the range of 1.2–22 ppm.[847] Another spectrophotometric method for determining uranium used 2,2′-bis[(2-hydroxynaphthyl)azo]-4,4′-bithiazole as the color reagent.[848] Uranium and thorium were separated by HPLC and detected spectrophotometrically following postcolumn reaction with Arsenazo III. Detection limits were in the range of 1 ppb for both species.[849]

3.60.2 Spectrofluorometry

Leung et al.[850] and Kim and Zeitlin[851] describe a method for the separation and determination of uranium in water. Thoric hydroxide ($Th(OH)^4$) was used as a collector. The final uranium concentration was measured via fluorescence (at 575 nm) of its rhodamine B complex. The detection limit was about 200 $\mu g\,L^{-1}$.

3.60.3 Inductively Coupled Plasma Mass Spectrometry

Inductively coupled plasma atomic emission spectrometry has been used for the analysis of uranium. However, the technique suffers from spectral interferences and it has relatively poor detection limits.

Inductively coupled plasma mass spectrometry has superior limits of detection over optical methods. Also, this technique has an order of magnitude better detection limit than that obtained for the conventional fluorometric method.

Boomer and Powell[852] have described an inductively coupled plasma method that has a detection limit of 0.1 $\mu g\,L^{-1}$. Calibration is linear from the low limit to 1000 $\mu g\,L^{-1}$. Precision, accuracy, and a quality-control protocol have been established. A comparison with the conventional fluorometric method was performed by these workers.

Water samples are preserved with 1% nitric acid and stored in plastic or glass bottles with plastic-lined caps.

Two other methods have been tried for the determination of uranium by inductively coupled plasma mass spectrometry, one of which was successful at levels to 0.5 fg mL^{-1}[853] with the other giving a detection limit of 0.5 ng L^{-1}.[854]

3.60.4 Polarography

Deutscher and Mann[855] used differential pulse polarography of a triphenyloxine extract to measure uranium in natural water.

3.60.5 Scanning Voltammetry

Zhang et al.[856] used differential pulse stripping voltammetry for the direct determination of uranium in natural waters. The supporting electrolyte was acetic acid-sodium acetate in the presence of cupferron. The uranium was concentrated on the hanging drop mercury electrode at −1.1 V, relative to the sliver/silver chloride electrode. The calibration curve was linear in the range of 0.16–6.6 μg L^{-1} and the detection limit was 0.05 μg L^{-1}.

Van den Berg and Nimmo[857] complexed uranium with 8-hydroxyquinoline prior to adsorptive collection on the hanging mercury drop electrode. The uranium was determined by cathodic stripping voltammetry at an adsorption potential of −0.4 V. The limits of detection in natural waters were 0.2 nM uranium(VI) after 1 min of adsorption and 0.02 M uranium(VI) after 10 min.

Jin et al.[858] also used cathodic stripping voltammetry to determine uranium(VI) in natural waters. The uranyl ion was chelated with thenyltrifluoroacetate prior to adsorption and preconcentration on the mercury electrode. The detection limit was 0.4 μg L^{-1} for 2 min adsorptive preconcentration and the relative standard deviation was 3.3%.

Possie et al.[859] used absorptive cathodic stripping voltammetry with hanging drop mercury electrode and a chloranilic acid ligand in the trace analysis of uranyl ions at pH 2 in low-ionic-strength groundwaters around mining areas. Upon optimization, a limit of detection around 0.10 μg L^{-1} was found with linearity up to 10 μg L^{-1}. In the abandoned mining area of Val Vedello (Orbic Alps, Italy), measured uranium concentrations in water ranged from 0.3 μg L^{-1} above uranium mineralization levels to 145 μg L^{-1} in groundwater percolating from mine galleries. Such uranium concentrations are related to natural weathering effects of carbon dioxide and/or hydrogen carbonate ion on uranium minerlizations under oxic conditions. A marked seasonal dependence was then found, in agreement with literature data on a pre-operational survey back to 1980–1981. No significant chemical impact of the abandoned mining activity on groundwater quality could be found.

3.60.6 Neutron Activation Analysis

This technique has been used fairly extensively for the measurement of uranium in natural waters.[860–864] Anion-exchange resins have been employed for preconcentration of uranium.[860,863,864] Fleischer[862] analyzed individual drops of water for uranium concentrations down to less than 0.01 μg L^{-1}.

Zielinski and McKown[864] concentrated microgram quantities of uranium in natural waters into 10 mL of purified kerosene containing a liquid anion-exchange resin (Amberlite LA-1). The organic phase was then analyzed by a standard delayed neutron counting technique. The technique showed similar precision and sensitivity to standard fluorometric methods and was less sensitive to elemental interferences.

3.60.7 High-Performance Liquid Chromatography

Cassidy and Elchuk[865,866] applied high-performance liquid chromatography to the determination of uranium(IV) in groundwater samples. The best chromatographic separation was obtained on bonded-phase cation exchangers with an α-hydroxylisobutyrate eluent. The metal ions were detected either by visible spectrophotometry of the Arsenazo(III), U(VI) complex at 650 nm, after a post-column reaction with a complexing reagent, or with a polarographic detector. Detection after postcolumn reaction gave the best sensitivity; the detection limit (2 × baseline noise) was 6 ng for 60 μg L^{-1} samples.

Kerr et al.[867] determined uranium in natural waters using high-performance liquid chromatography. Uranium was preconcentrated and separated from the bulk of other constituents by passing through a small reversed-phase enrichment cartridge. The uranium was back-flushed from the enrichment column for separation. Separated species were monitored spectrophotometrically after postcolumn reaction with the chromogenic reagent Arsenazo(III). The system was automated and capable of analyzing 40 samples per day. Detection limits were in the range of 1–2 μg L^{-1}.

3.60.8 Fission Track Analysis

Various workers have applied fission track analysis to the determination of uranium in natural waters.[836,868–870] Zhai and Kang et al.[871] applied the technique to the determination of uranium in natural waters. The detector was a polycarboxylic acid ester film on a plastic strip. The detection limit was in the ppb range for 1–2 drops of water.

Guo et al.[872] compared fission track, laser fluorometry, and fluorocolorimetric methods for the determination of uranium in natural waters in the range of 0.01–10 μg L^{-1}. The laser fluorometry and fluorocolorimetric methods had a detection limit of 0.1 μg L^{-1} and the fission track method had a detection limit of 0.01 μg L^{-1}.

Deng et al.[873] automated the determination of uranium in natural waters by the fission track method. The automated track counting was comparable to the microscopic technique for a track density of 200–50,000 tracks cm^{-2}. The sensitivity is approximately 0.001 μg L^{-1} for a single drop of water.

3.60.9 Alpha-Spectrometry

It is well known than ammunition containing depleted uranium was used by NATO during the Balkan conflict. To evaluate the depleted uranium origin (natural uranium enrichment or spent nuclear fuel reprocessing) it is necessary to check the presence of activation products (^{236}U, $^{239+240}Pu$, ^{241}Am, ^{237}Np, etc.) in the ammunition.

Desideri et al.[874] separated transuranium elements from the uranium matrix by extraction chromatography with microporous polyethylene (Icorene) supporting suitable stationary phases. Plutonium was separated by tri-*n*-octylamine.^{241}Am was separated by tri-*n*-octylamine and di-(2-ethylhexyl) phosphoric acid. Neptunium also was separated by tri-*n*-octylamine using different conditions. After elution, the transuranic elements were electroplated and counted by alpha spectrometry. The transuranic element decontamination factors from uranium were higher than 10^6. The final chemical yields ranged from 50% to 70%. The detection limit was 1 Bq kg^{-1}, for one penetrationammunition $^{239+294}Pu$. The ^{237}Np concentration in one penetrator was 30.1 Bq kg^{-1}.

The presence of these anthropogenic radionuclides in the penetrators indicates that at least part of the uranium originated from the reprocessing of nuclear fuel, although because of their very low concentrations, the radiotoxicological effect is negligible.

3.60.10 Miscellaneous Techniques

Billon et al.[875] carried out a study of some mechanistic aspects associated with uranium release/immobilization and sedimentation in Authie Bay uranium in estuarine oxic waters, porewaters, and sediment solids. These analytical data made it possible to appraise the partitioning of this metal between the liquid phase and the particulate matter/sedimentary material by calculating its distribution coefficient. The findings reveal that the distribution coefficient varies significantly with the depth probably in response to the microbial activities in these sediments. This was confirmed by the geochemical behavior of iron and manganese and Authie Bay sediments. Finally, studies on the thermodynamic characteristics of sedimentary uranium in Authie Bay were undertaken in order to select possible uranium water–mineral equilibria that could be involved in this environment, and to help define conditions of sedimentary uranium bioreduction.

3.61 VANADIUM

3.61.1 Spectrophotometry

Chromogenic reagents that have been used for the determination of vanadium include methylthymol blue[876] (absorption maximum 590 nm), xylenol orange[877](510–520 nm), 4-(2-pyridylazo)resorcinol (545 nm), gallic acid (420 nm),[878] and *N*-(*p*-*N*,*N*-dimethylaniline-3-methoxy-2-naphtho)hydroxamic acid[879] (570 nm).

Agrawal and Mehd[880] evaluated eight different chloro-substituted hydroxamic acids for spectrophotometric determination of vanadium. Catalytic spectrophotometric methods have been described for the determination of nanogram amounts ($\mu g\,L^{-1}$) of vanadium in water.[878] These methods rely on the catalytic effect of vanadium on the oxidation of iodide by bromate or the catalytic effect of vanadium on the ammonium persulfate oxidation of gallic acid.[878,881] In the automated method, slight interference was observed for samples containing chromium, molybdenum, copper, aluminum, and iron. Chloride caused slight interference at concentrations above 200 $mg\,L^{-1}$.

Abbasi[882] extracted vanadium from natural waters with *N*-*p*-methyoxyphenyl-2-furylacrylohydroxamic acid in chloroform. The complex is equilibrated with 3-(*o*-carboxyphenyl)-1-phenyltriazine *N*-oxide at a pH of 1.5. The ternary complex is measured spectrophotometrically at 450 nm.

Sun et al.[883] used 2-(3,5-dibromo-2-pyridylazo)-5-(diethylamino)phenol and hydrogen peroxide as the chromogenic agent for the spectrophotometric determination of vanadium in natural waters. The detection limit is 2 $\mu g\,L^{-1}$ and the method is applicable for vanadium determination in the range of 0.4–560 $\mu g\,L^{-1}$.

3.61.2 Stripping Voltammetry

Li and Smart[884] determined down to 15 M of vanadium in natural waters using square-wave stripping voltammetry.

3.61.3 High-Performance Liquid Chromatography

Mierura[885] has described a method for the determination of traces of vanadium in natural waters with 2-(8-quinolyl azo)-s(dimethylamino)phenol by reversed-phase liquid chromatography-spectrometry.

In this reversed-phase high-performance liquid chromatographic method for neutral and cationic metal chelates with azo dyes, tetraalkylammonium salts are added to an aqueous organic mobile phase. The tetraalkylammonium salts are dynamically coated on the reversed stationary support. As a result of the addition of tetraalkylammonium salts, the retention of the chelates is remarkably reduced. When a 100 mm^3 aqueous sample was injected, sensitivity and precision were as follows: peak height calibration curves of vanadium(V) were linear up to 800 pg at 0.005 absorbance unit full scale (AUFS) and up to 160 pg at 0.001 AUFS; the relative standard deviation for 10 determination at 0.005 AUFS was 2.3% at a level of 320 pg of vanadium(V); the detection limit was 2.6 pg at 0.001 AUFS. Many cations including iron(III) and aluminum(III) do not interfere with the determination.

Nagoasa and Kimata[886] determined down to 1 ppb of vanadium in natural waters by high-performance liquid chromatography with electrochemical detection.

3.61.4 Ion Exchange Chromatography

Orvini et al.[887] described an exchange chromatographic method for the determination of the various forms of vanadium in fresh water. These include tetravalent cationic, pentavalent anionic, and neutral complexed forms of vanadium. Separation is achieved on two columns in series involving the absorption of the sample in Chelex 100 and Dowex 1X8 columns followed by the selective elution of the different vanadium species and the assay by neutron activation analysis. Experiments were carried out using vanadium-48 radiotracer. Recoveries of total, complexed tetravalent, and pentavalent vanadium were 31%, 36%, and 34%, respectively, and total vanadium was 100%.

3.62 YTTRIUM

3.62.1 Spectrophotometry

Elements of the yttrium subgroup in natural water in microgram quantities have been determined as their complexes with boron and catechol violet.[888] Firstly, interfering metals are extracted as their thiocyanate complexes with a chloroform solution of diantipyrylmethane. The pH of the aqueous solution is adjusted to 3–4, catechol violet is added, and the solution is adjusted to pH 8.7 and buffered with borate solution. Spectrophotometry is carried out at 610 nm.

Chea and Zong[889] determined yttrium by electrothermal atomic absorption spectrophotometry. The yttrium was extracted by using Levextrel with recoveries in the 92–101% range. The method was applied to waters in the 5–20 ng of yttrium L^{-1} range in natural waters.

3.63 ZINC

3.63.1 Spectrophotometry

Miller[890] carried out detailed studies of a method for determining zinc using zircon at 620 nm, after selective release from its cyanide complex using cyclohexanone. This method is capable of determining zinc down to $20\,\mu g\,L^{-1}$. In this method, zinc forms a blue complex with 2-carboxy-2′-hydroxy-5′-sulfoformazylbenzene (zircon) in a solution buffered at pH 9.0. Other heavy metals likewise form colored complexes with zircon. Cyanide is added to complex the zinc and the heavy metals present.

Copper, iron, iron III, chromium III, chromium VI, lead, and nickel all interfere at ion extraction above $0.05\,\mu g\,L^{-1}$ while silver and cobalt II interference occurs at concentrations greater than $0.150\,\mu g\,L^{-1}$.

Yoshimura et al.[891] described an ion exchanger colorimetric method for the determination of zinc. Zinc in a water sample can be determined by sorption onto an anion-exchange resin from 2M chloride solution followed by transformation into a colored complex with zircon. The sensitivity of the method is claimed to be 10 times greater than that for conventional colorimetry.

Fan et al.[892] used a zinc complex with 2-(5-bromo-2-pyridylazo)-5-diethylamino) phenol in the presence of a cationic surfactant, cetyltrimethylammonium bromide, to determine zinc by a spectrophotometric method. Recoveries were 98.3–103.6% for zinc in natural waters with a linear calibration range of 0–500 $\mu g L^{-1}$. A detection limit is not given but the relative standard deviation is 0.045%.

Herrador et al.[893] describe a spectrophotometric method for zinc in natural waters using methylglyoxal bis(4-phenyl-3-thio-semicarbazone) in aqueous dimethyl formide. The complex has an absorption maximum of 455 nm and obeys Beer's law in the concentration range of 0.2–4 $mg L^{-1}$.

Zinc is complexed with 5-(6-bromo-2-benzothiazolylazo)-8-hydroxyquinoline in a method described by Sang and Yang[894] for determining zinc in natural waters. Beer's law is obeyed in the range of 0–600 μg of zinc L^{-1}.

Chen and Su[895] describe a spectrophotometric method for the determination of zinc in natural waters using 2-(5-bromo-2-pyridylazo)-5-(diethylamino)phenol in the presence of the nonionic surfactant poly(ethylene glycol) octylphenyl ether. The maximum absorption is at 552 nm and a linear calibration curve is obtained in the range of 0–100 $\mu g L^{-1}$. The use of masking agents to remove interferences from other ions was described.

Camran et al.[896] reviewed the literature for the determination of zinc in natural waters. The review covers UV-visible spectrophotometry, atomic absorption spectrophotometry, and electrochemical methods.

3.63.2 Flow-Injection Analysis

An atomic flow-injection method for the determination of zinc in natural waters is described by Koupparis and Anagnostopoulou.[897,898] The Zircon method with differential demasking of the cyanide metal complex with cyclohexanone was used to determine zinc in the range of 1–10 $mg L^{-1}$. A detection limit of 0.05 $mg L^{-1}$, precision better than 1%, and a sampling rate of 80 per hour were obtained.

Flow injection with fluorometric detection has been used for the determination of zinc in water.[899] Cation exchange was used to separate interfering ions prior to determination in a fluorimeter. The detection limit was 0.06 $\mu g L^{-1}$, and the precision was ±6% at a total zinc concentration of 0.28 $\mu g L^{-1}$. Flow-injection spectrophotometry using 5,10,15,20-tetrakis(3-chloro-4-sulfophenyl) porphine has been used for the determination of zinc and mercury.[900] A detection limit was not given, but the relative standard deviations were <1% for zinc at 1.0 $mg L^{-1}$.

3.63.3 Spectrofluorometry

Zinc is reacted with *meso*-tetrakis(4-hydroxyphenyl)porphyrin, excited at 448 nm, and the fluorescence is measured at 635 nm in a method described by Tong et al.[901] for the analysis of natural waters. The calibration curve is linear over the range of 0–50 μg of zinc L^{-1} and the detection limit is 0.6 ppb.

Tong and Sun[902] described a fluorometric method for the determination of zinc in natural waters using *meso*-tetrakis(3-*N*-methyl-pyridyl)porphyrin. The complex is excited at 430 nm and the fluorescence is measured at 606 nm. The calibration curve is linear in the range of 0–40 μg and the detection limit in natural waters is 1 ppb.

Igarashi and Yotsuvanagi[903] determined down to 5 ppb of zinc in natural waters by a fluorescence spectroscopic method.

Wang et al. have described the development of a ratiometric fluorescent probe based on the quinoline fluorophore for the determination of zinc ions in water.[904]

3.63.4 Molecular Imprinting

Kheteh et al.[905] described a method for the separation of zinc from aqueous samples using a molecular imprinting technique.

Xue and Syg[906] also used this technique to determine zinc in water. Zinc imprinting polymer was prepared by free radical solution polymerization in a glass tube containing zinc sulfate, morin, 4-vinylpyridine as a functional monomer, ethyleneglycoldimethacrylate as a cross-linking monomer, and 2,2′-azobisisobutyronitrile as an initiator. The obtained polymer block was ground and sieved (55–75 μm) and the zinc-morin complex was separated from polymer particles by leaching with 2 M hydrochloric acid. The synthesized polymer particles were characterized by IR and different scanning calorimetric studies either before or after leaching. The effects of different parameters, such as pH, adsorption, and desorption time, type and minimum amount of the eluent for elution of the complex from polymer were evaluated. Extraction efficiency more than 99% was obtained by elution of the polymers with 10 mL of dichloro methylene-dimethyl sulfoxide (1:1, v/v). The detection limit of the method was 2.9 μg L^{-1} was obtained. The relative standard deviation was found to be below 9.2%. In addition, the influence of various cationic interferences on the complex recovery was studied. Various aspects of the method described were studied in detail. The percentage of zinc adsorbed increased by increasing the pH to 4, and then decreased.

Kheteh et al.[905] investigated the optimum amount of morin by the molecular imprinting in which the liquid:zinc ratio was varied between 1 and 6. As seen in results were quantitative at ratios above 2, but to ensure that all the zinc ions are able to produce complex, a liquid:zinc ratio of 4 was adopted.

3.63.5 Atomic Absorption Spectrometry

Onishchenko et al.[907] extracted zinc from water samples using a 1 M solution of capric acid and pyridine or 1,10-phenanthroline in heptane and applied atomic absorption spectrometry to the extract.

3.63.6 Anodic Stripping Voltammetry

Wang and Greene[908] determine zinc in river waters by anodic stripping voltammetry with medium exchange. This involves deposition of the metal from the sample, followed by stripping it into a more suitable electrolyte, which minimizes the interference due to the hydrogen evolution current, which masks the zinc peak in acidified waters. A flow cell with a stationary mercury film disk electrode is employed. Measurement can be performed in the presence of oxygen in the sample solution, utilizing an oxygen-free exchange solution.

Chen and Zhang[909] reported on the use of an anodic stripping voltammetry method for the determination of zinc in natural waters. The zinc is concentrated on a hanging mercury drop electrode at −1.20 to −1.25 V. A linear calibration curve is obtained in the range of 1–100 µg of zinc L^{-1}.

3.63.7 Polarography Differential Pulse

Pardo-Botello et al.[240] investigated the adsorption process of Zn(II) and Cd(II) from aqueous solution from both kinetic and equilibrium standpoints using differential-pulse polarography on a mercury dropping electrode as the analytical technique. With such an aim, adsorption experiments were performed using not only a single metal ion-Zn(II) or Cd(II) solution but also a multicomponent ion metal-Zn(II), Cd(II), and Hg(II) solution. The influence of the pH change in the multicomponent ion metal solution on the adsorption of Zn(II) and Cd(II) was also studied.

The adsorption processes is relatively fast for Zn(II) and Cd(II). The presence of two foreign ions in the solution slightly speeds up the adsorption process for Zn(II) and significantly slows it down for Cd(II). The adsorption isotherms are similarly shaped for Zn(II) and Cd(II). The addition of the foreign ions has a more unfavorable effect on the adsorption for Cd(II) than for Zn(II). At pH 2, neither Zn(II) nor Cd(II) is absorbed practically on the carbon. The voltammetric approach has proved to be a fast and efficient method that, at the same time, enables one to monitor the adsorption of Zn(II) and Cd(II) with potential online application.

Kang et al.[910] investigated a polarographic method for determining zinc in natural waters.

3.63.8 Biosensors

Saal et al.[911] described the development of a heavy-metal biosensor based on either recombinant 6His-Tag glutathione *S*-transfrase (GST-$(His)_6$) or glutathione *s*-transferase theta-2-(GST-theta 2-2), and a capacitive transducer. The dynamic range of the pure bovine liver GST-theta 2-2 biosensor was 1 fM to 1 mM for Zn^{2+}, and 10 pM to 1 mM for Cd^{2+}. The GST-$(His)_6$ biosensor was able to detect Zn^{2+} and Cd^{2+} in the range of 1 fM to 10 μM and Hg^{2+} in the range of 1 fM to 10 mM. The bovine liver GST theta 2-2 biosensor displays an increased selectivity and a wider dynamic range for Zn^{2+} compared with the GST-$(His)_6$ biosensor. Therefore, by using different GST isozymes, it is possible to modulate important characteristics of capacitive biosensors for the detection of heavy metals.

3.63.9 Miscellaneous Techniques

Caneau et al.[912] have reviewed the literature for the determination of zinc in natural waters covering UV-visible spectrophotometric, atomic absorption spectrophotometry, and electrochemical methods.

Lu et al.[52] have discussed the speciation of zinc in river and lake waters.

3.64 ZIRCONIUM

3.64.1 Isotope-Dilution Mass Spectrometry

Boswell and Elderfield[402] determined zirconium and hafnium in natural waters by isotope-dilution mass spectrometry. The elements were extracted by co-precipitation with ferric hydroxide and separated by a single cation-exchange column using hydrochloric acid and nitric acid as eluents. The mass spectrometry technique involved a single rhenium filament with the sample loaded in a mixture of nitric acid and colloidal carbon. Concentrations of 80–2400 pmol zirconium per kg and 3–45 pmol hafnium per kg were measured in samples of estuarine, coastal, and oceanic waters.

3.64.2 Polarography

Wang et al.[913] described an electrochemical stripping procedure for ultatrace measurement of zirconium in natural waters, in which preconcentration is achieved by the adsorption of a zirconium-Solochrome Violet RS complex onto a hanging mercury drop electrode. The detection limit was 2.3×10^{-10} M for a 10-min preconcentration time. The relative standard deviation at 5.5×10^{-8} M was 1.7%.

3.65 MULTIPLE METALS IN RIVER WATERS

In many instances methods described in the literature cover not just the determination of a single element but cover the analysis of several metals in a mixture. Various techniques have been employed to this end, including atomic absorption spectrometry, hydride-generated inductively coupled plasma mass spectrometry, inductively coupled atomic-emission spectrometry, anodic scanning and polarographic techniques, and various forms of chromatography. Space does not allow a detailed discussion of these analyses but, to help the reader, they are reviewed in Table 3.8 .

TABLE 3.8 Review of Published Work on Multiple Metal Analysis in River Water

Methods	Elements	References
Atomic Absorption Spectrometry	Fe, Mn, Cr, Al, Ba, Ca, K, Na	390
	As, Sn, Ga, Sb	391,392,914
	Fe, Zn, Cr, Ag, Mn	915
	Cd, Pt, Cu, Pb	916
	Cu, Zn, Cd, Pb, Cd, Cu, Mn, As, Sb, Se, Tl	917
	Cd, Au	918,919
	Ag, Au	918,919
	16 metals	919
	Cd, Pb, Zu, Cu	920
	Pb, Cd	921
	Ba, V, Mo	922
	Co, Ni, Cu, Cr	
	Sb, As	923
	Ag, Ni, Co, Cr	924
	Cu, Zn	925
	Hg, As, Bi, Pu, Sn	926
	Pb, Hg, Cd	927
	Ca, Fe, Co, Ni, Cu, Zn	928
	Cd, Hg, Pb, Ag	
	Ag, Hg	929
	Cr, Mn, Fe, Co, Ni, Ca	930
Hydride-Generation Atomic-Absorption Spectrometry	As, Sb, Bi, Sc, Te	931
	As, Bi, Sb, Se	932
	As, Se	933,934
	Hg, Sb, As, Se, Te, Bi, Sn, Pb	935
	As, Sb, Bi, Se, Te, As	936

Continued

TABLE 3.8 Review of Published Work on Multiple Metal Analysis in River Water—cont'd

Methods	Elements	References
Hydride-Generation Gas Chromatography	As, Sb, Sn	937
	As, Se, Sn, Sb	938
	Sb, As, Se	939
	As, Sb	940
	As, Sb, Se, Te	941
	Pt, Pd, Rb	942
	As, Sb, Bi, Se	943–945
Inductively Coupled Plasma Mass Spectrometry	Re, Ir, Pt	868
	Co, Fe, Mn, Ni,	869
	Na, Mg, K, Ca, Al, V	
	Cr, Mn, Cu, Zn, Sr, Mo, Sb, Ba, U	870
	Sr, Ti, V	836
	14 rare earths	946–951
Inductively Coupled Plasma Atomic-Emission Spectrometry	Cd, Cu, Pd	952
	26 elements	953,954
	Ca, Ag	788
	Al, Pb, Mn	955
	Al, Be, Cd, Co, Cr	
	Cu, Fe, Mn, Ni, Zn	956
	Hg, Se, As, Sb, Bi	897
	Pb, Zn, Fe, Cu, Ni, Mo, V	957
	Cd, Pb	958
	Cu, Zn, Ni	959
Polarography	Cu, Pd, Cd, Zn	960
	Cu, Mg	961
	Ni, Co	962
	Zn, Fe	963
	Ca, Sr, Ba	964

TABLE 3.8 Review of Published Work on Multiple Metal Analysis in River Water—cont'd

Methods	Elements	References
High-Performance Liquid Chromatography	Cu, Ni, Co, Cr	965
	Cu, Co, Ni, Pb, Fe	966
	Cd, Co, Cu, Pb, Zn	967–969
	Hg, Cu, Ni, Co, Pb	970
	Ni, Fe, Cu, Hg	971
	Al, Fe, Mn	972
	Ca, Mg	973
	Pb,Sn	
	Co, Ni, Co, Cr	974
	Co, Ni, Pb, Fe	975
	Cu, Cd, Pb, Zn, Pb	976
	Ni, Fe, Cd, Hg	977,978
	Al, Fe, Mn	979
	Ca, Mg	980
	Pb, Sn	468
Emission Spectrometry	Alkaline earths	981
	Ag, Bi, Cd, Cu, Mg, Pb, Tl, Bi	982
	As, Bi, Se, Si	983
	Sb, As	984
	Se, Te	985
	Al, Ba, Be, Ca, Cd	
	Cr, Cu, In, K, Mg	
	Pb, Si, Te	
	Mn, Mo, Na, Ni	
	Pb, Si, Te	986
Anodic Stripping Voltammetry	Cd, Pb, Cu, Hg	987
	Cd, Pb, Cu	988
	Cd, Zn, Pb	989
	Fe, Ti	990

Continued

TABLE 3.8 Review of Published Work on Multiple Metal Analysis in River Water—cont'd

Methods	Elements	References
	Cu, Pb, Cd, Zn, Fe, Mn	991
	Cu, Ni, Bi	992
	Se, Hg, Cu, Pb	993
	Platinum group	994
	Pb, Cd	995
	Cu, Sb, Bi, Pb	996
	Cd, Pb, Cu, Zn	
	Ni, Fe, Te, Bi, Sb, Ta	997–1041
	V, Cr	1042
Ion-Exchange Chromatography	Lanthanides	1043
	U, Co, Cd	1044
	NH_4, Ca, Mg	1045
Ion Chromatography	Cu, Ni, Zn, Mn	974
	Na, K, Hi, NH_4, Mg, Ca, Sr	975–979
	Na, NH_4, K, Mg, Ca, Pb, Cu, Cd, Co, Ni, Zn, Fe	468,784,980,1046–1052
	14 lanthanides	468,784,980,1046–1052
Neutron Activation Analysis	Au, Ba, Ce, Cr	
	Eu, Fe, K	1053
	Na, Sb, Si, Se, U	
	Si, Mn, Ce, Mg, Fe, Ni	
	Zn, Sr, Na, K, Al, Sb	1054–1056
	Sm, Eu, Yb	1057
	20 elements	782
	Ca, Fe, Ni, Zn, Sr, Mg	899
	Si, Cd, Gd	1058,1059
Inductivly Coupled Plasma Atomic-Emission Spectrometry	Al, Be, Pb, Bi, Cr, Cu, Fe	1060

REFERENCES

1. Herrichsen A, Poulson MM. *Vallen* 1995;**31**:339.
2. Roseberg JEJ, Herrichsen A. *Vatten* 1985;**41**:48.
3. Noller BN, Cusbert PK, Surry NA, Bradley PH, Tuor M. *Environ Technol Lett* 1985;**6**:381.
4. Dougan WK, Wilson AL. *Analyst* 1974;**99**:413.
5. Seip HM, Muller L, Nass A. *Water Air Soil Pollut* 1984;**23**:81.
6. Rainwater FH, Thatcher LK. *US Geol Surv* 1960;**1963**(88):109. Paper No 1454.
7. Wilson AD, Sargeant GA. *Analyst* 1963;**88**:109.
8. Driscoll CT. *Chemical characterisation of some diluted acidified lakes in the Aaronrdeck region of New York state* [Ph.D. thesis]. Itchica, New York: Connell University; 1980.
9. Wiganowski C, Motomizu S, Toei K. *Anal Chim Acta* 1982;**140**:313.
10. Brown E, Skougstad MW, Fishman MJ. *Methods for collection and analysis of water samples for dissolved metals and gases*. Washington: US Government Printing Office; 1970.
11. Halfon Y, Cognet L, Le Prince A, Courtois G. *Eau Ind Nuis* 1986;**101**:41.
12. Pakalus P, Farrar Y. *Water Res* 1977;**11**:387.
13. *APHA standard methods for examination of water and waste water*. 13th ed. Washington: American Publich Health Authority; 1971.
14. Carter JM. *Proc Anal Div Chem Soc* 1975;**12**:246.
15. Sampson B, Fleck A. *Analyst* 1984;**109**:369.
16. Hydes DJ, Liss PS. *Analyst* 1976;**101**:922.
17. De Pablos F, Ariza JLG, Pino F. *Analyst* 1986;**111**:1159.
18. Rojas Sanchez F, Garcia de Torres AGA, Bosch Ojeda C, Cano Pavon JM. *Analyst* 1988;**113**:1287.
19. Yuan Y. *Huascue Shizi* 1988;**8**:85.
20. Vilchez JL, Navalon A, Avidad R, Garcia-Lopez T, Capitan-Vallvey LF. *Analyst* 1993;**118**:303.
21. Sutheimer SH, Cabaniss SE. *Anal Chim Acta* 1995;**303**:211.
22. Chen L, Shi G. *Fenxi Huaxue* 1992;**20**(8):942. *Chem Abstr* 1993;118:32230w.
23. Lin Q, Guo J, Xu J. *Fenxi Huaxue* 1992;**20**(7):813. *Chem Abstr* 1993;118:11403q.
24. Fairman B, Sanz-Medel A. *Int J Environ Anal Chem* 1993;**50**(3):161. *Chem Abstr* 1993;119:256190s.
25. Bosque-Sendra JM, Valencia MC, Boudra S. *Anal Lett* 1994;**27**(8):1579. *Chem Abstr* 1994;**121**:65100.
26. Ohmori T. *Kogyo Yosui* 1992;**409**:58. *Chem Abstr* 1993;**118**:182200a.
27. Howard AG, Coxhead AJ, Potter IA, Watt HP. *Analyst* 1988;**111**:1379.
28. Zhu X. *Lihua Jianyan Huaxue Fence* 1992;**28**(5):314. *Chem Abstr* 1993;**119**:79348b.
29. Uchino T, Morita M, Fuma FK. *Anal Chem* 1984;**56**:2020.
30. Roeyset O. *Anal Chem* 1987;**59**:899.
31. Roeyset O. *Anal Chem* 1985;**178**:223.
32. Roeyset O. *Anal Chim Acta* 1986;**185**:75.
33. Zolzer D, Schwedt G. *Fresenius Z Anal Chem* 1984;**317**:422.
34. Chung HK, Ingle JD. *Anal Chem* 1990;**62**:2547.
35. Brueggemeyer TW, Fricke FL. *Anal Chem* 1986;**58**:1143.
36. Takahashi Y, Yokoyama T, Tarutani T. *Chietsu* 1987;**24**:261.
37. Huang S, Shong X, Ji X. *Xiamen Daxue Xuebao Ziran Kexveban* 1987;**26**:216.

38. Yan DR, Schwedt G. *Fresenius Z Für Anal Chem* 1985;**320**:252.
39. Satiroglu N, Tokgaz I. *Int J Environ Anal Chem* 2010;**90**:560.
40. Redic N. *Analyst* 1976;**101**:657.
41. Richie GSP, Posner AM, Richie IM. *Anal Chem Acta* 1980;**177**:99.
42. Kitsuki H, Yokayama T, Nakamura Y, Shimiza S, Iamaizumi Y, Taruntani T. *Chiemetsu* 1985;**22**:14.
43. Nordstrum DK, Ball JW. *Source* 1986;**232**:54.
44. Benson RL, Worsfold PJ. *Sci Total Environ* 1999;**135**:17.
45. Bekov GI, Yegorov AS, Letoshov US, Rodayer NV. *Nature* 1983;**301**:410.
46. Capmbell PGC, Bisson M, Bangie R, Tessler A, Vittenveue JP. *Anal Chem* 1983;**55**:2246.
47. Vanden Berg CMG, Murphy K, Riley JP. *Anal Chim Acta* 1986;**188**:177.
48. Cabaniss SE. *Environ Sci Technol* 1987;**21**:209.
49. Miller JR, Andelman JB. In: Lester HN, Perry R, Sterritt RM, editors. *Chemical environmental proceedings, international conference*. London (UK): Selper; 1986. p. 26–35.
50. Alvarez E, Perez A, Calvo R. *Sci Total Environ* 1993;**133**:17.
51. Fairman B, Sanz-Medel A, Gallego M, Quintela MJ, Jones P, Benson R. *Anal Chim Acta* 1994;**286**:401.
52. Lu Y, Chakrabarti CL, Back MH, Gregoire DC, Schroeder WH. *Anal Chim Acta* 1994;**293**:95.
53. Goenaga X, Byrant R, Williams DJA. *Anal Chem* 1987;**59**:2673.
54. Driscoll RC. *Int J Environ Anal Chem* 1984;**16**:267.
55. Solorzano L. *Limnol Oceanogr* 1969;**14**:799.
56. Ivancis I, Degobbis D. *Water Res* 1984;**18**:1143.
57. Huyser DJ. *Water Res* 1970;**4**:501.
58. Stewart RM. *Water Res* 1985;**19**:1443.
59. Verdouw H, VanEchfield CJA, Dekkers EMJ. *Water Res* 1978;**12**:399.
60. Kron MD. *Analyst* 1980;**105**:305.
61. Crowther J, Evans J. *Analyst* 1980;**105**:841.
62. Crowther J, Evans J. *Analyst* 1980;**105**:849.
63. Water Research Centre, Medmenham, Bucks, UK. *Analyst* 1982;**107**:680.
64. Nickels WC. *Bull Environ Contam Toxicol* 1980;**25**:39.
65. Carson FW, Gross RL. *US technical information service, Springfield, Virginia* 1977. Report No. 288301, Colorimetric analysis of ammonia in water.
66. Aoki T, Uemura S, Munemori M. *Anal Chem* 1983;**55**:209.
67. Aoki T, Munemuri M. *Anal Chem* 1983;**55**:209.
68. Krug FJ, Kizicka EH. *Analyst* 1979;**104**:47.
69. Bergamin H, Reis BF, Jacintho AO, Zagatto EAG. *Anal Chim Acta* 1980;**117**:81.
70. Krug FJ, Reis BF, Gine MF, Zagatto EAG, Ferreira JR, Jacintho AO. *Anal Chim Acta* 1983;**151**:39.
71. Van Son M, Scothorst RC, Den Boef G. *Anal Chim Acta* 1983;**153**:271.
72. Application Note ASN 50–02/84. *Determination of ammonia nitrogen in water by flow injection analysis and gas diffusion, 1–10 mg L^{-1} N*. Tecator Ltd; 1984.
73. Application Note ASN 50–04/84. *Determination of ammonia nitrogen in water by flow injection analysis and gas diffuser 100–1000 mg L^{-1} N*. Tecator Ltd; 1984.
74. Application Note ASN 50 03/84. *Determination of ammonia nitrogen in water by flow injection analysis and gas diffuser 100–1000 mg L^{-1} N*. Tecator Ltd; 1984.
75. Application Note ASN 50–01/84. *Determination of ammonia nitrogen in water by flow injection analysis and gas diffuser 100–1000 mg L^{-1} N*. Tecator Ltd; 1984.

76. Application Note ASN 50/83. *Determination of ammonia nitrogen in water by flow injection analysis and gas diffuser*. Tecator Ltd; 1984.
77. Gailani BRM, Greenway GM, McCreedy T. *Int J Environ Anal Chem* 2000;**87**:425.
78. Hara H, Motoke A, Okazaki S. *Analyst* 1988;**113**:113.
79. Hara H, Motoke A, Okazaki S. *Anal Chem* 1987;**59**:1995.
80. Belcher R, Bogdanski SL, Calokerinos AC, Townshend A. *Analyst* 1981;**106**:625.
81. Mizobuchi M, Tamase K, Kitada Y, Sasaki M, Tanigawa K. *Anal Chem* 1984;**56**:603.
82. Conboy JJ, Henion JA, Martin MW, Zweigenbaum JA. *Anal Chem* 1990;**62**:800.
83. Department of the Environment National Water Council Standing Committee of Analysts. *Methods for the examination of waters and associated materials*. London: H.M. Stationery Office; 1982. 47 pp. (RD 22 B: C ENV) Ammonia in waters.
84. Khadro Basma, Namour P, Bessneille F, Leonard D, Jaffrezic-Renault N. *Int J Environ Anal Chem* 2009;**80**:11.
85. Marquez KL, Pyres CK, Sentos JML, Zagatto EAG, Lima JLFC. *Int J Environ Anal Chem* 2007;**87**:77.
86. Sandhu SS, Nelson P. *Anal Chem* 1978;**50**:322.
87. Stauffer RE. *Anal Chem* 1983;**55**:1205.
88. Stauffer RE. *Environ Sci Technol* 1980;**14**:1475.
89. Nyamah D, Torgbor J. *Water Res* 1986;**20**:1341.
90. Li W, Wang B. *Fenxi Huaxue* 1987;**15**:485.
91. Chakraborthi D, Irgolic KJ. *Int J Environ Anal Chem* 1984;**17**:241.
92. Pacey GE, Ford JA. *Talanta* 1981;**28**:935.
93. Duttemans F, Massart DL. *Microchim Acta* 1984;**1**:261.
94. Subramanian KS, Leung PC, Meranger JC. *Int J Environ Anal Chem* 1982;**11**:121.
95. Yasui A, Tsutsumi C, Toda S. *Agric Biol Chem* 1978;**2**:2139.
96. Ficklin WH. *Talanta* 1983;**30**:371.
97. Faust SD, Lemmou NW, Belton T, Tucker R. *J Environ Sci Health* 1983;**A18**:335.
98. Faust SD, Winka A, Belton T, Tucker F. *J Environ Sci Health* 1983;**A18**:389.
99. Tefalidct S, Irgum K. *Anal Chem* 1988;**60**:2031.
100. Bundaleska AJM, Stafilov T, Arpadjan S. *Int J Environ Anal Chem* 2005;**85**:199.
101. Amberger MA, Bings RH, Pohl P, Broekaert JAC. *Int J Environ Anal Chem* 2008;**88**:625.
102. Kumar AR, Ryuzuddin P. *Int J Environ Anal Chem* 2008;**88**:255.
103. Pallier V, Serpaud B, Feuillander G, Bollinger JL. *J Environ Anal Chem* 2011;**91**:1.
104. Chen Y, Qi W, Cao J, Chang M. *J Anal At Spectrosc* 1993;**8**:379.
105. Brovko TA. *Anal Khim* 1987;**42**:1637.
106. Narasaki H. *Anal At Spectrosc* 1988;**3**:517.
107. Andreae MO. *Anal Chem* 1977;**49**:820.
108. Anderson RK, Thompson M, Culbard E. *Anal Lond* 1986;**111**:1153.
109. Shaikh AN, Tallman DE. *Anal Chim Acta* 1978;**98**:251.
110. Arbab Zavar MH, Howard AG. *Analyst* 1980;**105**:744.
111. Aggett J, Aspell AC. *Analyst* 1976;**101**:341.
112. Crecelius EA, Bloom NS, Cowan CE, Jenne EA. *Speciation of selenium and arsenic in natural waters and sediments*. Report EPRI. Palo Alto (CA): Arsenic Speciation Electric Power Storage Research Institute; 1986. EA–4641, 2.
113. Welz B, Melcher M. *Analyst* 1984;**109**:420.
114. Hinners JA. *Analyst* 1980;**105**:751.
115. Anderson RK, Thompson M, Culbard E. *Analyst* 1986;**111**:1143.
116. Matsumoto K, Fuwa K. *Anal Chem* 1982;**54**:2012.

117. Narasaki H, Fuwa K. *Anal Chem* 1984;**56**:2059.
118. Aggett J, Hayashi Y. *Analyst* 1987;**112**:277.
119. Hagen JA, Lovett RJ. *At Spectrosc* 1986;**7**:269.
120. Abe K, Tereshima S. *Christu Chosasho Geppo* 1986;**37**:335.
121. Sun S, Sun S, Xue J. *Yankangye* 1986;**5**:31.
122. Narasaki HJ. *Anal At Spectrosc* 1988;**3**:517.
123. Pierce FD, Brown HR. *Anal Chem* 1976;**48**:693.
124. Pierce FD, Lamoroeaux TC, Brown HR, Fraser RS. *Appl Spectrosc* 1976;**30**:38.
125. Pierce FD, Brown HR. *Anal Chem* 1977;**49**:1417.
126. Davies E, Kempster PL. *Spectrochim Acta Part B* 1986;**41B**:1203.
127. Huang MF, Jiang SJ, Hwang CJ. *J Anal At Spectrosc* 1995;**10**:31.
128. Feng YI, Cao J. *Anal Chim Acta* 1994;**293**:211.
129. Davies FGF, Kempster PL. *Spectrochim Acta Part B* 1986;**41B**:1203.
130. Rubio R, Padro A, Alberti J, Rauret G. *Anal Chim Acta* 1993;**283**:160.
131. Cullen WR, Eigendorf GK, Pergantis SA. *Rapid Commun Mass Spectrom* 1993;**7**:33.
132. Goosens FRJ, Moens I, Dam R. *J Anal At Spectrosc* 1993;**8**:921.
133. Jiang SJ, Lu PL, Huang MF. *J Chin Chem Soc Toipei* 1994;**41**:139.
134. Krystek P, Ritsema R. *Int J Environ Anal Chem* 2009;**89**:331.
135. Vazquez EA, Deano RB, Rodriques MT, Fermandez Espinosa AJ. *Int J Environ Anal Chem* 2011;**91**:462.
136. Yaqoub M, Wagrem A, Nabu A. *Int J Environ Anal Chem* 2008;**88**:603.
137. Elci L, Divrikik Y, Saylak M. *Int J Environ Anal Chem* 2008;**88**:711.
138. Ali I, Jain CK. *Int J Environ Anal Chem* 2004;**84**:947.
139. Le XC, Yalcin S, Ma M. *Environ Sci Technol* 2000;**34**:2342.
140. Gian HF, Tong SL. *Anal Chim Acta* 1977;**89**:1712.
141. Solanki PR, Prabhakar N, Panday MK, Malhotra BD. *Int J Environ Anal Chem* 2009;**89**:49.
142. Amberger M, Bings G, Pohl P, Broekaert J. *Int J Environ Anal Chem* 2008;**88**:625.
143. Liu JM, Gao F, Yong TL, Lai JA, Li M. *Int J Environ Anal Chem* 2008;**88**:613.
144. Pallier V, Serpauod B, Feuillade-Cathal Faud G, Bollinger JC. *Int J Environ Anal Chem* 2011;**91**:1.
145. Hoste J. *Anal Chim Acta* 1948;**2**:402.
146. Hoste J, Gillis J. *Anal Chim Acta* 1955;**12**:158.
147. Lott PF, Cukor P, Moriber G, Solga J. *Anal Chem* 1963;**35**:1159.
148. Ariyoshi H, Kiniwa M, Toei K. *Talanta* 1960;**5**:112.
149. Tereda K, Ooba T, Kiba T. *Talanta* 1975;**22**:41.
150. Bem EM. *Chem Anal Wars* 1979;**24**:155.
151. Ramachandran KN, Kaveeshwar R, Gupta VK. *Talanta* 1993;**40**:781.
152. Hu I. *Fenxi Shiyanshi* 1994;**13**:36. *Chem Abstr* 1994;**121**:91134.
153. Kasterka B. *Chem Anal Wars* 1992;**37**:361. *Chem Absr* 1993;**118**(22):219286m.
154. Lee SH, Choi JM, Choi HS, Kim YS. *Korean Chem Soc* 1994;**38**:351. *Chem Abstr* 1994;**121**:98684.
155. Li D, Zhang Q, Qiao C. *Lihua Jianyan Huaxue Fence* 1992;**28**:212. *Chem Abstr* 1993;**118**(16):160068m.
156. Jin S, Cheng Y, Che X, Wu Z. *Yejin Fenxi* 1993;**13**:24. *Chem Abstr* 1993;**119**(22):240580h.
157. Ali I, Kimar Jain C. *Int J Environ Anal Chem* 2004;**84**:947.
158. Vian SH, Fry RC. *Anal Chem* 1988;**60**:485.
159. Tye CT, Haswell SJ, O'Neill P, Bancroft KCC. *Anal Chim Acta* 1985;**169**:195.

160. Stosanovic RS, Bond AM, Butler ECV. *Anal Chem* 1990;**62**:2692.
161. Liu FAYM, Fernandez-Sanchez ML, Gonzales EB, Sonz Medel A. *J Anal At Spectrosc* 1993;**8**:815.
162. Butler ECV. *J Chromatogr* 1988;**450**:353.
163. Stary J, Zeman A, Kratzell K, Prasilova J. *Int J Environ Anal Chem* 1980;**8**:49.
164. Bidewig FG, Valenta P, Nurnberg HW. *Fresenius Z Für Anal Chem* 1982;**311**:187.
165. Hemens CM, Elson CM. *Anal Chim Acta* 1986;**188**:311.
166. Wanatabe L, Toshiyuku T. *Kankyo Kagaku* 1993;**3**:404.
167. Tanaka S, Makamura M, Hasimoto Y. *Bunsaki Kagaka* 1987;**36**:114.
168. Mok WM, Shah NW, Fhiu Y, Wai CM. *Anal Chem* 1986;**58**:110.
169. Orvini E, Delfonti R, Gallarini M, Speziali M. *Anal Proc Lond* 1981;**18**:237.
170. Abu Hilal AH, Riley JP. *Anal Chim Acta* 1981;**131**:175.
171. Yonehara N, Nischimoto L, Kamada M. *Anal Chim Acta* 1985;**172**:183.
172. Stauffer RE. *J Res US Geol Surv* 1977;**5**:807.
173. Bertine KK, Lee DS. Trace metals in seawater. In: Wond CS, editor. *Proceedings of a NATO advanced research institute on trace metals in seawater 30/–33/4/81/Sicily*. Italy: Plenum Press; 1981.
174. Xu Z, Fang Z. *Chin Chem Lett* 1992;**3**:915.
175. Xu Z, Wan-Dxa Y. *Fenxi Huaxue* 1992;**20**:1321.
176. De la Calle-Guntinas MB, Madrid Y, Camara C. *Mikrochim Acta* 1992;**109**:149. *Chem Abstr* 1993;118:72733(a).
177. De la Calle-Guntinas MB, Madrid Y, Camara C. *J Anal At Spectrom* 1993;**8**:745.
178. Capodaglio G, Van der Berg CGM, Scarponi G. *J Electro Anal Chem Interfacial Electro Chem* 1987;**235**:275.
179. Sun YC, Yang JY, Lin YF, Yang MH, Alfassi ZB. *Anal Chim Acta* 1993;**276**:33.
180. Sharma M, Patel KS. *Int J Environ Anal Chem* 1993;**50**:63.
181. Calle Guntinas MB, De la Madrid Y, Camara C. *Tech Instrum Anal Chem* 1995;**17**:263.
182. Rollenberg MCE, Curtius AJ. *Mikrochim Acta* 1982;**2**:441.
183. Sun S. *Fenxi Huaxue* 1986;**14**:949.
184. Ferrus R, Torrades F. *Analyst* 1985;**110**:403.
185. Measures CI, Edmond JM. *Anal Chem* 1986;**58**:2065.
186. Tao H, Mlyazaki A, Bansho K. *Anal Sci* 1988;**4**:299.
187. Ueda J, Kitadani T. *Analyst* 1988;**113**:581.
188. Pal BK, Baksi K. *Mikrochim Acta* 1992;**108**:275.
189. Tao D, Xue Y. *Shanhai Huanjing Kexue* 1987;**6**:24.
190. Lai EPC, Statham BD, Ansell K. *Anal Chim Acta* 1993;**276**:393.
191. Arai N, Minamisawa H, Hirota O, Okutani T. *Bunseki Kagaku* 1992;**69**:317.
192. Shijo Y, Mitshuhashi M, Shimizu T, Sakurai S. *Analyst* 1992;**117**:1929.
193. Lee DS. *Anal Chem* 1982;**54**:1682.
194. Nakahara T, Nakanashi K, Utasa T. *Spectrochim Acta Part B* 1987;**42B**:119.
195. Froediro R, Dupree B, Polve M. *Eur Mass Spectrom* 1995;**1**:283.
196. Mal'kov EM, Fedoseeva AG. *Zavod Lab* 1970;**36**:912.
197. Mal'kov EM. *Zavod Lab* 1968;**34**:504.
198. Uchikawa S, Sato S. *Kuwamoto Daiaku Kyokugakuba Kiyo Shizan Kogatum* 1986:29.
199. Oshima M, Shibata K, Gyouten T, Motomizu S, Toei K. *Talanta* 1988;**35**:351.
200. Ma Z. *Yankuang Ceshi* 1987;**6**:126.
201. Zou J, Motomizu S, Oshima M, Fukutami H. *Anal Sci* 1992;**8**:719.
202. Balogl J, Andruck V, Kador M, Posta J, Szebova E. *Int J Environ Anal Chem* 2009;**89**:449.

203. Usenko JI, Prorok MH. *Zavod Lab* 1992;**58**:6.
204. Miyakaki A, Bansho K. *Anal Sci* 1986;**2**:451.
205. Takahashi Y. *Bunseki Kagaku* 1987;**36**:693.
206. Inoue Y, Date Y. *Bunseki Kagaku* 1994;**43**:365.
207. Yakimova VP, Markova O. *Zh Anal Khim* 1992;**47**:2033.
208. Kang XIM, Peng Y, Li P. *Lihua Jianyan Huaxue Fence* 1992;**28**:231.
209. Duchateau NL, Verbruggen A, Hendrickx F, De Bievre P. *Anal Chim Act* 1987;**196**:41.
210. Heumann KG, Sewaki H. *Fresenius' Z Anal Chem* 1987;**329**:485.
211. Mo M. *Fenxi Huaxue* 1987;**15**:414.
212. Zheng K. *Yankuang Ceshi* 1987;**2**:100.
213. Lukionets LG, Kulish NG. *Soviet J Water Chem Technol* 1985;**7**:40.
214. Chen G, Guo L, Wong L. *Fenxi Huaxue* 1994;**22**:583.
215. Lorserna JJ, Navas A, Sanches Gracia. *Anal Lett Lond* 1981;**14**:833.
216. Kabasakalis V, Tsitouridou R. *Fresenius Environ Bull* 1992;**1**:494.
217. Shevchuk IA, Makhno AY. *Khim Tekhnol Vody* 1992;**14**:740. *Chem Abstr* 1992;**117**:257825y.
218. Ohta K, Nakajima N, Iniu S, Wineforner JD, Mizuno T. *Talanta* 1992;**39**:1643.
219. Zhang G, Li J, Fu D, Hao D, Xiang P. *Talanta* 1993;**40**:409.
220. Sawada K, Ohgake S, Kobayashi M, Suzuki T. *Bunseki Kagaku* 1993;**42**:741.
221. Chaung H, Huang S. *Spectrochim Acta Part B* 1994;**49B**:283.
222. Ebdon L, Goodall P, Hill SJ, Stockwell PB, Thompson KCJ. *Anal At Spectrom* 1993;**8**:723.
223. Zhang Z, Liu J, Lin R, Yang X, He H. *Zhongshan Daxue Xuebao Ziran Kexueban* 1986;**301**:109.
224. Han H, Le X, Ni Z. *Huanjing Huaxue* 1986;**5**:34.
225. Ybanez N, Montoro R, Catala R, Cervera ML. *Rev Agroquim Technol Aliment* 1987;**27**:270.
226. Isozaki A, Ueki K, Sasaki H, Utsumi S. *Bunseki Kagaku* 1987;**36**:672.
227. Hasan MZ, Kumar A. *Ind J Environ Health* 1983;**25**:161.
228. Analytical Quality Control (Harmonized Monitoring) Committee. *UK analyst*, vol. 110. Medmenham: Water Research Centre; 1985. p. 247.
229. Committee for Analytical Medmenham, UK Report No. TR 220. *Accuracy of determination of trace concentrations of cadmium in river waters* 1985.
230. Okutani T, Arai N. *Bunseki Kagaku* 1988;**37**:426.
231. Committee for Analytical Quality Control (Harmonized Monitoring). *Report no. TR 220. Accuracy of determination of trace concentrations of cadmium in river waters*. Medmenham (UK): Water Research Centre; 1985.
232. Filik H, Dondurmacioglu F, Pak RA. *Int J Environ Anal Chem* 2008;**88**:637.
233. Sun H, Suo R. *Int J Environ Anal Chem* 2009;**89**:347.
234. Hu X. *Int J Environ Anal Chem* 2011;**91**:263.
235. Lum K, Callaghan M. *Anal Chim Acta* 1986;**187**:157.
236. Baysal A, Tokman N, Akman S. *Int J Environ Anal Chem* 2008;**88**:141.
237. Stewart EE, Smart RB. *Anal Chem* 1984;**56**:1131.
238. Kemula W, Zawadowska J. *Fresenius Z Anal Chem* 1980;**300**:39.
239. Muhlbaier J, Stevens C, Graczyk D, Tisue T. *Anal Chem* 1982;**54**:496.
240. Pardo-Botello R, Pinella-Gil E, Fernandez-Gonzles C, Gomez-Serrano V. *Int J Environ Anal Chem* 2005;**85**:1051.
241. Javar AJ, Hoek M. *Environ Sci Technol* 2010;**44**:2570.

242. Frigieri P, Trucco R, Ciaccolii I, Pampurini G. *Analyst* 1980;**105**:651.
243. Molero J, Moran A, Sanchez-Cabeza JA, Blanco M, Mitchell PI, Vidal-Quadra A. *Radiochim Acta* 1993;**62**:159.
244. Buguslawska K, Cyganski AZ. *Anal Chem* 1972;**261**:392.
245. Liu X, Wu X. *Xiangtan Daxue Ziran Kexuc Xuebao* 1987;**2**:75.
246. Miller YM, Chupakhin MS, Zotov NP. *Zhur Anal Khim* 1978;**23**:1765.
247. Liebenberg CJ, Vonstoerien F. *Anal Chim Acta* 1968;**43**:465.
248. Jackson LL, Osteryoung J, O'Dea J, Osteryoung RA. *Anal Chem* 1980;**52**:71.
249. Jackson LL, Osteryoung J, Osteryoung JA. *Anal Chem* 1980;**52**:76.
250. Qui XC. *Anal Chim Acta* 1983;**149**:375.
251. Song X, Wu B. *Huanujing Hauxue* 1986;**5**:77.
252. Kempster PL, Van Staden JF, Van Vliet HR. *Frezenius Z Anal Chem* 1988;**332**:153.
253. Application Note ASN 48–03.84. *Determination of calcium in water by flow injection analysis. 500–100 mg L*$^{-1}$. Tecator Ltd; 1984.
254. Application Note ASN 48–01/84. *Determination of calcium in water by flow injection analysis. 0.2–5 mg L*$^{-1}$. Tecator Ltd; 1984.
255. Application Note ASN 48–02/84. *Determination of calcium in water by flow injection analysis. 1–20 mg L*$^{-1}$. Tecator Ltd; 1984.
256. Application Note ASN 48/84. *Determination of calcium based on orthocresolphthalein complexone reaction using flow injection analysis*. Tecator Ltd; 1984.
257. British Standard Institution BSI, Milton Keyens, BS 6068, Section 2, 30, 1987.
258. Hulanicki M, Trojanowicz M, Pbozy E. *Analyst* 1982;**107**:1356.
259. Li X, Zeng B, Xin Z, Ning A. *Zhognan Kuangye Xueyuan Xueynan* 1987;**18**:93.
260. Gardner WS, Lundrum PF, Yates DA. *Anal Chem* 1982;**54**:1198.
261. Rho YS, Choi GG. *Arch Pharmacol Res* 1986;**9**:211.
262. Jones O. *Anal Chem* 1994;**66**:6765.
263. Wang OP, Chen ZL, Chen GN, Lin JM. *Int J Environ Anal Chem* 2011;**91**:255.
264. Bilikova A, Bilik V. *Chem Zvesti* 1968;**22**:873.
265. Pettine M, La Noce T, Liberatori A, Loreti L. *Anal Chim Acta* 1988;**209**:315.
266. Goa R, Yuan D. *Talanta* 1993;**40**:637.
267. Piying G, Xuexin G, Tianze Z. *Anal Lett Lond* 1996;**29**:651.
268. Rubio Luor S, Perez-Bendito D. *Int J Environ Anal Chem* 1994;**56**:213.
269. Zaitoun MA. *Int J Environ Anal Chem* 2005;**85**:399.
270. Application Short Note ASTN 26/85. *Determination of the sum of chromium(III) and chromium(VI) by flow injection analysis. 1–10 mg L*$^{-1}$. Tecator Ltd; 1985.
271. Al Y, Xing D. *Fenxl Huaxue* 1988;**16**:478.
272. Abdallah AM, El-Defray MM, Mostefa MA. *Anal Chim Acta* 1984;**165**:105.
273. Muzzucotelli A, Minoia C, Pozzoli L, Ariati L. *At Spectrosc* 1983;**4**:182.
274. He M. *Yankuang Ceshi* 1987;**6**:250.
275. Kubrakova I, Kudinova T, Formanousky A, Kuz'min N, Tsysin G, Zolotov Y. *Analyst* 1994;**119**:2477.
276. Wu H, Zheng Z, Hu G, Wang F. *Ferxi Huaxue* 1992;**20**:1445.
277. Alc R, Sarab R. *Int J Environ Anal Chem* 2007;**87**:375.
278. He Q, Chang X, Zhang H, Jiang N, Wong X. *Int J Environ Anal Chem* 2008;**88**:373.
279. Brydy FA, Olsen LK, Vela NP, Caruso JA. *J Chromatogr* 1995;**713**:311.
280. Ruan Y, Wang Z. *Huanujing Hauxue* 1986;**5**:74.
281. Wang L, Wen S, Chen R. *Huanujing Hauxue* 1987;**6**:76.
282. Yang S, Tang S. *Xiaglan Daxue Ziran Kexue Xucbao* 1986;**4**:64.

283. Pratt KW, Koch WF. *Anal Chem* 1986;**58**:124.
284. Su X, Hong L, Yang S. *Halyyang Yu Hozhav* 1987;**18**:280.
285. Zelensky I, Zelensha V, Kaniansky D. *J Chromatogr* 1987;**390**:111.
286. Baumann RA, Schreurs M, Cooljer C, Velhorst MH, Frei RW. *Can J Chem* 1987;**65**:965.
287. Posta J, Berndt H, Luo SK, Schaldach G. *Anal Chem* 1993;**65**:2590.
288. Powell MJ, Boomer DW, Wiederin DR. *Anal Chem* 1995;**67**:2474.
289. Tanaka H, Kuono M, Morita H, Okamoto K. *Anal Sci* 1992;**8**:857.
290. Gong G, Lu Y, Wang H. *Lanzhou Daxue Xuebao Ziran Kexueban* 1992;**27**:84. *Chem Abstr* 1993;118:4529u.
291. Kabasakalis V. *Anal Lett* 1993;**26**:2269. *Chem Abstr* 1993;119:278227m.
292. Escobor R, Lin Q, Guirani A, De las Roas FF. *Analyst* 1993;**118**:643.
293. Mugo RK, Orians K. *Anal Chim Acta* 1992;**271**:1.
294. Nusko R, Heumann KG. *Anal Chim Acta* 1994;**286**:283.
295. Bailey JR, Julian DH, Armstrong AT, Richardson JN. *Int J Environ Anal Chem* 2008;**88**:119.
296. Izumi N, Yanida A, Mendor Y, Azumi T. *Kenkhu Hokoku – Kogvo Diaobu* 1987;**40A**:100.
297. Zeng GM, Tang L, Glen GL, Huang GH, Nin CG. *Int J Environ Anal Chem* 2004;**84**:761.
298. Linar ML, Rubio S, Perez-Bendito D. *Int J Environ Anal Chem* 1994;**56**:219.
299. Sule PA, Ingle JS. *Anal Chim Acta* 1996;**386**:85.
300. Bercerio-Gonzalez E, Bermejo-Barrea P, Barciela-Garcia J, Barceila-Alonso C. *J Anal At Spectrosc* 1993;**8**:649.
301. Vidal JC, Sanz JM, Castillo JR. *Fresenius J Anal Chem* 1992;**344**:234.
302. Nusko R, Heumann KG. *Anal Chim Acta* 1994;**286**:283.
303. Messman JD, Churchwell ME, Wong ME, Lofthouse J. Report EPA/600/4–86/039 Order NO. PB87-1409 27/GAR. 126 pp. Available NTIS.
304. Hasegawa H, Sohrin Y, Matsui M, Hojo M, Kawashima M. *Anal Chem* 1994;**66**:3247.
305. Abbasi SA, Ahmed J. *Int J Environ Health* 1980;**22**:296.
306. Gharehboghi I, Loghi H, Shemirani F, Bagldadi M. *Int J Environ Anal Chem* 2008;**88**:513.
307. Boyle EA, Handy B, Van Green A. *Anal Chem* 1987;**59**:1499.
308. Rodionova TV, Isanov VM. *Zhur Anal Khim* 1986;**41**:2181.
309. Koizumi H, Yasuda O, Katayama M. *Anal Chem* 1977;**49**:1106.
310. Ophel IL, Judd JM. *60 cobalt and 90 strontium in Perch Lake, atomic energy commission Canada*. Health Science Division; 1974. Progress Report P102 AECL-4911.
311. Ueda J, Yamasaki H. *Analyst* 1987;**112**:283.
312. Bradshaw S, Gascoigne AJ, Headbridge JB, Moffett JH. *Anal Chim Acta* 1987;**197**:323.
313. Savitskii VN, Peleskensko VI, Osadchii VI. *Zhur Anal Khim* 1987;**62**:677.
314. Souza JMO, Tarley CRT. *Int J Environ Anal Chem* 2009;**89**:489.
315. Hao Z, Virc JC, Patirache GJ, Wollast R. *Anal Lett Lond* 1988;**21**:1409.
316. Okashita H, Tanaka T. *Shimadzu Hyron* 1987;**44**:165.
317. Schaller H, Neeb R. *Fresenius Z Für Anal Chem* 1987;**327**:170.
318. Arvard M, Gholizadeh TM, Mohmoodi N. *Int J Environ Anal Chem* 2009;**89**:153.
319. Moffett JW, Zika PJ, Petasue RG. *Anal Chim Acta* 1985;**175**:171.
320. Themelis DG, Vasilikiotis GS. *Analyst* 1987;**112**:797.
321. Itoh J, Komata M, Oka H. *Bunseki Kajaku* 1988;**37**:T1.
322. Yoshimura K, Matsuoka S, Inokura Y, Hase U. *Anal Chim Acta* 1992;**268**:325.
323. Sun G. *Fenxi Shiyanshi* 1992;**11**:44.
324. Wang Q, Wang J, Zhang Z. *Fenxi Huaxue* 1993;**21**:215.

325. Dong C, Dong Y, Zi Y. *Fenxi Shiyanshi* 1992;**11**:46.
326. Wang X. *Lihua Tianyan Huaxue Fence* 1992;**28**:343.
327. Zi Y, Luo C, Lu L. *Yankuang Ceshi* 1992;**11**:284.
328. Brazil JL, Martins LC, Rev R, Dupont J, Dias SLP, Sales JAA, et al. *Int J Environ Anal Chem* 2005;**85**:475.
329. Yamada M, Suzuki S. *Anal Chim Acta* 1987;**193**:337.
330. Li G, Yu Z. *Huaxue Tongbao* 1992;**42**:40.
331. Cao Q, Xu Q, Zhoa J. *Fenxi Huaxue* 1993;**21**:682.
332. Ejaz M, Zuha S, Dit W, Akhtar A, Chaudhri SA. *Talanta* 1981;**28**:441.
333. Silva M, Valcarcel M. *Analyst* 1982;**107**:511.
334. Matsunaga K, Fukase S, Hasebe K. *Geochim Cosmochim Acta* 1980;**44**:1615.
335. Sweileh JA, Lucky D, Kratochvil B, Cantwell FF. *Anal Chem* 1987;**59**:586.
336. Nishoika H, Maeda Y, Asumi T. *Nippon Kalsui Gahkaishi* 1986;**40**:100.
337. Ueda J, Yamazaki H. *Analyst* 1987;**112**:283.
338. Zhang G, Fu D, Hao D, Xiang P, Tian G. *Guangpuxue Yu Guangpu Fenxi* 1992;**12**:79.
339. Madrid Y, Wu M, Jin Q, Hiefje G. *Anal Chim Acta* 1993;**277**:1.
340. Santelli RE, Gallego M, Valcarcel M. *Talanta* 1994;**41**:817.
341. Hulanicki A, Trojanowicz M, Krawczyk TK. *Water Res* 1977;**11**:627.
342. Gulens J, Leeson PK, Seguin L. *Anal Chim Acta* 1982;**156**:99.
343. Stella R, Gazerle-Valenti MT. *Anal Chem* 1979;**51**:2148.
344. Tong H, Yang S. *Goodeng Huaxue Xuebao* 1986;**7**:503.
345. Peng T, Tang Z, Wang G. *Fenxi Huaxue* 1993;**21**:221.
346. Sanchez-Pedreno C, Ortuno JA, Martinez-Rodenas J. *Fresenius J Anal Chem* 1992; **344**:100.
347. Singh AK, Jain AK, Singh J, Mehtab S. *Int J Environ Anal Chem* 2009;**89**:1081.
348. Brasil JL, Martins LC, Pev R, Dupont J, Dias SLP, Sales JAA, et al. *Int J Environ Anal Chem* 2005;**85**:7.
349. Odashima T. *Mizu-Shorigisutsu* 1986;**27**:637.
350. Farias PA, Ferreira SL, O'Hara AK, Bastos MB, Goulart MS. *Talanta* 1992;**39**:1245.
351. Bonelli JE, Skogerboe RK, Taylor HE. *Anal Chim Acta* 1978;**101**:437.
352. Van Denaberg CMG. *Anal Lett Lond* 1984;**17**:2142.
353. Batley GE. *Anal Chim Acta* 1986;**189**:371.
354. Del Cleven RFMJ, Castilho P, Wolfs PM. *Environ Technol Lett* 1988;**9**:869.
355. Haapakka K, Kankare J, Kulmala S. *Anal Chim Acta* 1988;**211**:105.
356. Becker G, Oestvold G, Paul P, Seip PM. *Chemosphere* 1983;**12**:1209.
357. Du Bois HR, Sharma GM. *Anal Chem* 1979;**51**:1702.
358. Van den Berg CMG, Kramer JR. *Anal Chim Acta* 1979;**106**:113.
359. Mackay DG. *Mar Chem* 1982;**11**:169.
360. Parthasarthy N, Buffle J. *Anal Chim Acta* 1993;**284**:649.
361. Itabashi H, Kawamoto H, Akaiwa H. *Anal Sci* 1994;**10**:341.
362. Hill C, Street KW, Philipp WH, Tanner SP. *Anal Lett* 1994;**27**:2589.
363. Bigalke M, Wilcke S, Wicke W. *Environ Sci Technol* 2010;**44**:5496.
364. Chouyyck W, Shin Y, Davidson J, Samuels WD, Lafemina NH, Rutledge RD, et al. *Environ Sci Technol* 2010;**44**:4390.
365. Babecsanyi I, Imfield G, Granet M, Chabaux F. *Envrion Sci Technol* 2014;**48**:5520.
366. Tenteno GM, Callejon MM, Ternero RM, Guiram PA. *Mikrochim Acta* 1992;**109**:301.
367. Honda T, Nozaki T, Ossaka T, Oi T, Kokhara H. *J Radioanal Nucl Chem* 1988;**122**:143.

368. Sharipov EB, Khudaibergov U. *Akad Nauk Uzb SSR Ser Fiz Watem Nauk* 1970;**6**:55. Ref. Zhur Khim 19GD (10) 1971, 10, Abstract No. 10G186, 1971.
369. Wang J, Zadell JM. *Anal Chim Acta* 1986;**185**:229.
370. Burton JD, Culkin F, Riley JP. *Geochim Cosmochim Acta* 1959;**16**:151.
371. Wardani GA. *Geochim Cosmochim Acta* 1958;**15**:237.
372. Johnson DJ, West TS, Dagnall RM. *Anal Chim Acta* 1973;**67**:79.
373. Pollock EN, West TS. *At Absorpt Newsl (Perkin-Elmer Ltd)* 1973;**12**:6.
374. Thomerson DR, Tompson DC. *Am Lab* 1974;**6**:53.
375. Braman RS, Tompkins MA. *Anal Chem* 1978;**50**:1088.
376. Zheng Y, Zhang D. *Anal Chem* 1992;**64**:1656.
377. Guo J, Fang T, Xu Z, Chen Y. *Guangpuxue Yu Guangpu Fenxi* 1993;**13**:57.
378. Tao G, Fang Z. *J Anal At Spectrosc* 1993;**8**:577.
379. Andrae MO, Frohlich PN. *Anal Chem* 1981;**53**:287.
380. Jin K, Shibito Y, Morita M. *Anal Chem* 1991;**63**:981.
381. Choi WH, Lee JS, Kim JS, Kim DH. *Anal Sci Technol* 1992;**5**:7.
382. Choi WH, Lee JM. *Anal Sci Technol* 1992;**5**:17.
383. Neissner R. *Trends Anal Chem* 1991;**10**:310.
384. Vydra F, Stulikova M, Petak P. *J Electro Anal Chem* 1972;**40**:99.
385. Sobrin Y. *Anal Chem* 1991;**63**:811.
386. Tobias RS. *Organomet Chem Rev* 1966;**1**:93.
387. Sobrin Y. *Bull Chem Soc Jpn* 1991;**64**:3633.
388. Rochow RG, Allred AL. *J Am Chem Soc* 1955;**77**:4489.
389. Ensaft AA, Keyvarsfard M. *Int J Environ Anal Chem* 2003;**83**:397.
390. Smith R, Bexunderhout EM, Van Heerden AM. *Water Res* 1993;**17**:1483.
391. Cai Y, Catanas M, Fernandez S, Turiel JL, Abalos M, Bayone JM. *Anal Chim Acta* 1995;**314**:183.
392. Howard AG, Grey MR, Waters AJ. *Anal Chim Acta* 1980;**111**:87.
393. Sazaki T, Sobrin Y, Hasegawa H, Kokine H, Kilhara S, Matsui M. *Anal Chem* 1994;**66**:271.
394. Feingerg JS, Bowyer WJ. *Microchem J* 1993;**47**:72.
395. Schvova OP, Kuchava GP, Kurbrakova IV, Myasoedova GV, Savvin SB, Bannykh LN. *Zh Anal Khim* 1986;**41**:2186.
396. Hall GEM, Vaive JE, Ballantyne SB. *J Geochem Explor* 1986;**26**:191.
397. Faulkner, Edmond. Private Communication.
398. Gomez MM, McLeod CW. *J Anal At Spectrosc* 1993;**8**:461.
399. Ol'khovich PF. *Sov J Water Chem Technol* 1983;**5**:59.
400. Asamov KA, Abdullah AA, Zakhidov AS, Korshunov YF, Sultanov A. *Dolk Acad Nauk Uzb SSR* 1969;**3**:26. Reference: *Zhur Kim* 19GD. Abstract No 23G, 163, (23):1969.
401. Turaev KK, Khudacbargenov U, Gainer AG. *Dokl Acad Nauk Resp Ugb* 1992;**12**:36.
402. Boswell SM, Elderfield H. *Mar Chem* 1988;**25**:197.
403. Kagaya S, Ueda J. *Bull Chem Soc Jpn* 1994;**67**:1965. *Chem Abstr* 1994;121:117127.
404. Wei J, Liu Q, Akutani T. *Anal Sci* 1994;**10**(3). *Chem Abstr* 1994;121:72673.
405. Ueda J, Matsui C. *Anal Sci* 1988;**4**:417.
406. Pastor E, De Pablos F, Gomez-Ariza JL. *Analyst* 1987;**112**:1041.
407. Kroik AA. *Khim Tekhnol Vody* 1992;**14**:903. *Chem Abstr* 1993;118:131620e.
408. Staeescu SP, Spiridon S. *J Radioanal Nucl Chem* 1992;**163**:301.
409. Du Y, Feng R, Huang Z. *Fenxi Huaxue* 1987;**15**:240.
410. Gibbs MM. *Water Res* 1979;**13**:295.

411. Macaldy DL, Granland CP, Granland JG, Vervacke L. *Water Res* 1982;**16**:1277.
412. Yeda K, Kaboyashi N, Yaramoto Y. *Analysis* 1986;**111**:731.
413. Nigo S, Yashimura Y, Tarutani T. *Talanta* 1981;**26**:669.
414. Pakalns P, Farrer YJ. *Water Res* 1987;**13**:987.
415. *APHA, AWWA and WPCT. Standard methods for the examination of water and wastewaters.* 13th ed. American Public Health Association; 1971.
416. Box JD. *Water Res* 1984;**18**:397.
417. Obraztov AA, Bocharova VG. *Tr Voronezh Gas Univ* 1971;**82**:182. Ref. *Zhur Khim* 19GD (13) Abstract No. 13G242:1971.
418. Kuselman I, Low O. *Talanta* 1993;**40**:749.
419. Kanti Bhadra A, Bauer B. *Ind J Chem* 1969;**7**:936.
420. Korenaga T, Motomizu S, Toei K. *Anal Chim Acta* 1979;**104**:369.
421. Wright RC. *Freshwater Biol* 1983;**13**:293.
422. Department of the Environment and National Water Council. *Methods for the examination of waters and associated materials, iron in raw and potable waters by spectrophotometry (using 2,4,6-tripyridyl-3,5-trianzine).* London: HM Stationary Office; 1977. 1978.
423. Abe S, Saito T, Sude M. *Anal Chim Acta* 1986;**181**:203.
424. Salinas F, Berzas-Nevado JJ, Vallente P. *Talanta* 1987;**34**:321.
425. Inoue H, Iti K. *Microchem J* 1994;**49**:249.
426. Xu J, Yinfa Ma. *J Microcolumn Sep* 1996;**8**:137.
427. Tanaka T, Higaishi K, Kawahara A, Wikide S, Yamari M, Thro K. *Talanta* 1995;**40**:605.
428. Kanada T, Takanu S. *Bunseki Kagaka* 1987;**36**:109.
429. Themalis DG, Vasilikiotis GS. *Analyst* 1987;**112**:791.
430. Nishioka H, Assadamongkol S, Maeda Y, Asumi T. *Nippon Kaisui Gakkaishi* 1987;**40**:286.
431. Nakamura T, Sato J. *Onsen Kogakkaishi* 1986;**20**:37.
432. Yu Y. *Fusheu Fanghu* 1988;**8**:146.
433. Wang S, Zhang S. *Huanjing Huaxue* 1986;**5**:48.
434. Sawamoto H. *Bunsaki Kagaka* 1988;**37**:212.
435. Nishoika H, Maeda Y, Azumi T. *Nippon Kaisui Gakkaishi* 1986;**40**:100.
436. Odashimia T. *Mizu Shori Gijutsu* 1986;**27**:637.
437. Burguerra JL, Burguerra M. *Anal Chim Acta* 1984;**161**:375.
438. Mortatti J, Krug FJ, Pessenda LCR, Zagatto EAG, Jorgensen S. *Analyst* 1982;**107**:659.
439. Ohno N, Sakai T. *Analyst* 1987;**112**:1127.
440. Rehman Atlig-ur, Yagoob M, Wassem A, Nabi A. *Int J Environ Anal Chem* 2009;**89**:1071.
441. Zhou M, Xie Y, Zhao L, Zhu Y. *Huaxue Shizi* 1987;**9**:352.
442. Urasa IT, O'Reilly AM. *Anal Chim Acta* 1986;**33**:593.
443. Yamane T, Watanabe K, Mottola HA. *Anal Chim Acta* 1988;**207**:331.
444. Hao Z, Vire JC, Patriarche GJ, Wollast R. *Anal Lett* 1988;**21**:1409.
445. Xia L, Wu Y, Jiang Z, Li S, Hu B. *Int J Environ Anal Chem* 2003;**83**:953.
446. Fassett JD, Powell LJ, Moore LJ. *Anal Chem* 1984;**56**:2228.
447. Thorburn-Burns D, Dalgarno BG, Flockhart BD. *Anal Proc Lond* 1985;**22**:379.
448. Burns DT, Dalgarno BG, Flockhart BD. *Anal Proc Lond* 1985;**22**:24.
449. Askeland RA, Skogerboe RK. *Anal Chim Acta* 1987;**192**:133.
450. Yang S, Tong H. *Huaxue Xuebao* 1987;**45**:711.
451. Borrero JM, Moreno-Bondi MC, Perez-Conde MC, Camera C. *Talanta* 1993;**40**:1619.
452. Ito I, Ueki O, Nakamura S. *Anal Chim Acta* 1996;**299**:401.

453. Zhou M, Xia Y, Zhao L, Zu Y. *Huaxue Shizi* 1987;**9**:352.
454. Huynk NL, Whitehead NH. *Oceonosr Acta* 1986;**9**:433.
455. Schaller H, Neeb R. *Fresenius Z Für Analyistch Chem* 1987;**327**:170.
456. Stroh A. *At Spectrosc* 1992;**13**:89.
457. Moeller P, Dulski P, Luciz J. *Spectrochim Acta Part B* 1992;**47B**:1379.
458. Hall GEM, Vaive JE, McConnell JW. *Chem Geol* 1995;**120**:91.
459. Aggarwal JK, Shabani MB, Palmer MR, Ragnar KV, Dottir S. *Anal Chem* 1996;**68**:4418.
460. Klinkenhammer G, German ER, Elderfield H, Greaves MJ, Mitra A. *Mar Chem* 1994;**45**:179.
461. Elderfield H, Greaves MJ, Trace IN. *Metals in seawater*. New York: Plenum Press; 1983. pp. 427–445.
462. Muller P, Dulski P, Hucic J. *Spectrochim Acta* 1992;**47B**:1379.
463. Weiss D, Panelkert T, Rubeska I. *J Anal At Spectrosc* 1990;**5**:171.
464. Esser BK, Volpe A, Kenneally JM, Smith DK. *Anal Chem* 1994;**66**:1736.
465. Kubitz J, Uebel U, Anders A. *Proc SPIE Int Soc Opt Eng* 1995;**2503**:14.
466. Moulin N, Tils J, Moulin C, Decambox P, Mauchien B, De Ruty O. *Radiochim Acta Part 1* 1992;**58**:121.
467. Brina R, Miller AG. *Spectrosc (Eugene Org)* 1993;**8**:25.
468. Rubin RB, Heberling SS. *Int Lab* September 1987:54.
469. Abassi SA, Ahmed J. *Ind J Environ Health* 1980;**22**:296.
470. Xiao B. *Fenxi Hauxue* 1987;**15**:45.
471. Lieser KH, Calmano W, Huess E, Neitzert V. *J Radioanal Anal Chem* 1977;**37**:717.
472. Nevoral VZ. *Anal Chem* 1974;**268**:189.
473. Anubaker MA, Harrington K, Von Wondelruszke R. *Anal Lett* 1992;**26**:1681.
474. Fan S, Fang Z. *Guangpuxue Yu Guangpu Fenxi* 1992;**12**:63.
475. Panegrahi BS, Peter S, Viswanathan KS, Matthews OK. *Anal Chim Acta* 1993;**282**:117.
476. Kisfaludi G, Henry C, Jourdain JL. *Chim Anal* 1971;**53**:388.
477. Breueggemeyer TW, Caruso JA. *Anal Chem* 1982;**34**:872.
478. Sinemus HW. *Fresenius Z Anal Chem* 1984;**317**:259.
479. Bertenshaw MP, Gelsthrope D, Wheatsone KC. *Analyst* 1982;**107**:163.
480. Cheam V, Lechner J, Desrosieres R, Sekerka I, Nriagu J, Lawson G. *Int J Environ Anal Chem* 1993;**53**:13.
481. Cheam V, Lochner J, Sekerka I, Desrosiers R, Nriagu J, Lawson G. *Anal Chim Acta* 1992;**269**:129.
482. Cheam V, Lechner J, Sekerka I, Desroiers R. *J Anal At Spectrom* 1994;**9**:315.
483. Kumar A, Aggarwal AL, Hosan MZ, Zeshmukh BT. *Indian J Pure Appl Phys* 1987;**25**:193.
484. Chen H, Zhu R, Wu J. *J Environ Sci Health Part A* 1994;**A29**:867.
485. Zhang Y, Riby P, Cox AG, McLeod CW, Date AR, Cheung YY. *Analyst* 1988;**113**:125.
486. Martinez-Jiminez P, Gallego M, Valcarcel M. *Analyst* 1987;**112**:1233.
487. Hasseini MS, Hassan-Abodi R. *Int J Environ Anal Chem* 2008;**88**:199.
488. Carbrara C, Lopez ME, Gallego C, Lorenzo ML, Lilto E. *Sci Total Environ* 1995;**159**:17.
489. Granadillo VA, Navarro JA, Romero RA. *J Anal At Spectrom* 1993;**8**:615.
490. Ohta K, Suzuki M. *Fresenius Z Für Anal Chem* 1979;**298**:140.
491. Vandegans J, Roseels P, Verplanken W. *Z Fur Anal Chem* 1987;**193**:169.
492. Sgiyal AN, Elei L, Memon SQ, Akdogan A, Hall A, Kartal AA, et al. *Int J Environ Anal Chem* 2014;**94**:743.
493. Wang X, Viczian M, Lasztity A, Barnes RM. *J Anal At Spectrosc* 1988;**3**:821.

494. Javanbaktit M, Bakht H, Schrabe MR, Attaran AM, Badei A. *Int J Environ Anal Chem* 2010;**90**:1014.
495. Apte VR, Badke SN. *Ind J Environ Health* 1979;**21**:67.
496. Benes P, Koc J, Stulik K. *Water Res* 1979;**13**:967.
497. Ferrier NJ, Buftle J. *Anal Chem* 1996;**68**:3670.
498. Pandya GH. *Ind J Environ Health* 1982;**24**:237.
499. Ansell RO, McAleer H, McNaughton A, Pugh JR. *Sci Total Environ* 1993;**135**:95.
500. Ramos JA, Bermej E, Zapardiel A, Perez JA, Hernandez J. *Anal ChimActa* 1993;**273**:219.
501. Wang J, Tiao B. *Anal Chem* 1993;**65**:1529.
502. Goa Y, Wei W, Goa X, Yin Y. *Int J Environ Anal Chem* 2007;**87**:521.
503. Shpigun LK, Eremine ID, Zolotov YA. *Zh Anal Khim* 1987;**41**:1557.
504. Sanchez-Pedreno O, Ortuno JA, Martinez J. *Fresenius J Anal Chem* 1992;**344**:100.
505. Phillips R, Greenlaw P, Bath RJ. *Environ Test Anal* 1993;**2**:34.
506. Rapsomanikis S, Donard OF, Weber JH. *Anal Chem* 1986;**58**:35.
507. Vasil F, Hamplova V. *Sb Vys Sk Chem-Technol Praze Anal Chem* 1985;**67**:115.
508. Melser JE, Jordan JL, Sutton DG. *Anal Chem* 1980;**52**:348.
509. Catanzaro EJ. *J Water Pollut Control Fed* 1975;**47**:203.
510. Morgen EA, Vlazov NA. *Zhur Prikl Khim Leningr* 1971;**44**:2752.
511. Chen Y, Li S, Li X. *Zhangguo Dizhi Kexueyuan Kuangchan Dizhi Yanjiuso Sokan* 1987;**20**:163.
512. Khakhanina TO, Kaplin AA, Kashkan GV. *J Anal Chem* 1987;**42**:5. part 1.
513. Chan LH. *Anal Chem* 1987;**59**:2662.
514. Yang JY, Tseng CL, Lo JM, Yang MH. *Fresenius Z Anal Chem* 1985;**321**:141.
515. Itoh M, Yamada Y, Kiriyama N, Komura K, Sakanoue M. *J Radioanal Nucl Chem* 1993;**172**:289.
516. Chao JH, Tseng CL. *Appl Radiat Isotopes* 1995;**46**:211.
517. Hoshika Y, Murayama N, Muto G. *Bunseki Kagaku* 1987;**36**:174.
518. Qui X, Zhu Y, Yan J. *Chem Anal Wars* 1987;**32**:285.
519. Forteza R, Cerda V, Maspoch S, Blanco M. *Analusis* 1987;**15**:136.
520. Downrad AJ, Hart JB, Kipton H, Powell J, Xu S. *Anal Chim Acta* 1992;**269**:41.
521. Dolmanova IF, Yatsimirskaya NT, Poddubienko VP, Peshokova VM. *Zhur Anal Khim* 1971;**26**:1540.
522. Maly J, Fadrus H. *Anal Lond* 1974;**99**:128.
523. Janic TJ, Milovanovic GA, Celap MB. *Anal Chem* 1970;**42**:27.
524. Kessick HA, Vuceta J, Morgan J. *J Environ Sci Technol* 1972;**6**:642.
525. Li C, Li Z. *Fenxi Huaxue* 1986;**14**:682.
526. Wang Z, Zheng Z, Hu X. *Fenxi Huaxue* 1987;**15**:145.
527. Morgan EA, Vlasov NH, Kozhemyakina LA. *Zhur Anal Khim* 1972;**27**:2064.
528. Nikolelis DP, Hadjiiannou TP. *Anal Chim Acta* 1978;**97**:111.
529. Nikolelis DP, Hadjiiannou TP. *Analyst* 1977;**102**:591.
530. Zhang G, Cheng D, Feng S. *Talanta* 1993;**40**:1041.
531. Tanaka T, Higashi K, Kawahara A, Wakida S, Yamane M, Hiro K. *Osaka Koava Gilutsu Shikensho Kibo* 1987;**38**:212.
532. Hydes DJ. *Anal Chim Acta* 1987;**199**:221.
533. Kumar A, Hasa MZ, Deshmukh BT. *Indian J Pure Appl Phys* 1986;**24**:4651.
534. Shijo Y, Watenabe J, Akiyama S, Shimizo T, Sakai K. *Buneski Kagaku* 1987;**36**:59.
535. Abbasi SA. *Int J Environ Anal Chem* 1988;**33**:113–21.
536. Corsini A, Wade G, Wn CC, Prasad S. *Can J Chem* 1987;**65**:18.

537. Themelis DG, Veseiliiotis GS. *Anal Lond* 1987;**112**:791.
538. Yang S, Tong H. *Huaxue Xuabao* 1987;**45**(7):711–4.
539. Choi N, Sakai T. *Analyst* 1987;**112**:1127.
540. Technical Report TR 141 Water Research Centre Medmenham UK. *Multi element analysis using an inductively coupled plasma spectrometer* June 1980. Part 1, commission and evaluation of performance.
541. Thompson M, Ramsers MH, Pahlavanpour B. *Analyst* 1982;**107**:1330.
542. Kumar A, Hasan MZ, Deshmukh BT. *Indian J Pure Appl Phys* 1986;**24**:485.
543. Gine MF, Zagatto EAG, Filho B. *Anal Lond* 1980;**104**:371.
544. Maggi L, Valentini MTG, Stella R. *Anal Lond* 1987;**112**:1617.
545. Chiswell B, Makhtar MP. *Talanta* 1987;**34**:307.
546. Cheng G. *Fenxi Huaxue* 1987;**15**:920.
547. Beinrohr E, Csemi P, Rojas FJ, Hofbayerova H. *Analyst* 1994;**119**:1355.
548. Committee for Analytical Quality Control (Harmonised Monitoring). *The accuracy of determination of total mercury in river waters: results of water authority tests made for the harmonised monitoring scheme of the department of the environment.* Medmenha: Water Research Centre; 1984. Technical Report TR 219.
549. Analytical Quality Control (Harmonised Monitoring) Committee. Accuracy of determination of total mercury in river water. Analytical quality control in the harmonised monitoring scheme. *Analyst* 1985;**110**:103.
550. Iskander K, Syers JK, Jacobs LW, Keeney DR, Gilmore JT. *Analyst* 1972;**97**:388.
551. Thompson KC, Godden RG. *Analyst* 1975;**100**:544.
552. Mahan KI, Mahan SE. *Anal Chem* 1977;**49**:662.
553. Lutze RL. *Analyst* 1979;**104**:979.
554. Mercury Analysis Working Party of the Bureau International Technique Du Chlore. *Anal Chim Acta* 1979;**109**:209.
555. Pinstock H, Umland F. *Fresenius Z Anal Chem* 1985;**320**:237.
556. Anderson PJ. *At Spectrosc* 1984;**5**:101.
557. Bricker JL. *Anal Chem* 1980;**52**:492.
558. Pratt LK, Elrick KA. *At Spectrosc* 1987;**8**:170.
559. Harsanyi E, Polos L, Pungor E. *Anal Chim Acta* 1973;**67**:229.
560. Joensun OI. *Appl Spectrosc* 1971;**25**:526.
561. Temmerman E, Dumary R, Dams R. *Stud Environ Sci* 1986;**29** (*Chem Prot Environ* 1985), 745:1986.
562. Boehnke MCLB. *Chem Labor Betr* 1986;**37**:619.
563. Aoki T, Kajikawa M, Munemori M. *Chem Express* 1987;**2**:463.
564. Churchwell ME, Livingston RL, Sgontz DL, Messman JD, Beckert WF. *Environ Int* 1987;**13**:475.
565. Korenaga T, Yameda E, Hara Y, Sakamoto H, Chohji J, Nakagawa C, et al. *Buneski Kagaku* 1987;**36**:194.
566. Gill GA, Fitzgerald WF. *Mar Chem* 1987;**20**:227.
567. Welz B, Schuber-Jacobs M. *Fresenius Z Für Anal Chem* 1988;**331**:324.
568. Birnie GF. *J Autom Chem* 1988;**10**:140.
569. Yan X, Ni Z, Guo Q. *Anal Chim Acta* 1993;**272**:105.
570. McIntosh. *At Spectrosc* 1993;**14**:47.
571. Hanna CP, McIntosch SA. *At Spectrosc* 1995;**16**:106.
572. Streufert D. *Z Chem* 1997;**27**:200.
573. Tanaka H, Morita H, Shimomora S, Okamoto K. *Anal Sci* 1993;**9**:859.

574. Kusakal Z, May K, Branica M. *Sci Total Environ* 1994;**15**:463.
575. Wang I. *Lihua Jianyan Huaxue Fence* 1993;**29**:273.
576. Ma Y. *Fenxi Huaxue* 1993;**21**:303.
577. Nojiri Y, Otsuki A, Fuwa K. *Anal Chem* 1986;**58**:544.
578. Anderson KA, Isaacs B, Tracy M, Moeller G. *J AOAC Int* 1994;**77**:473.
579. Tong HI, Giblin DF, Lapp RL. *Anal Chem* 1991;**63**:1772.
580. Borgnon J, Cadet JL. *Analusis* 1988;**16**:LXXVII.
581. Camunan-Aguilar JF, Pereiri-Garcie R, Sanchez-Uria JE, Sanz Medel A. *Spectrochim Acta Part B* 1994;**49B**:475.
582. Krull IS, Bushee DS, Schleicher RG, Smith SB. *Analyst* 1986;**111**:345.
583. Bushee D, Krull SIS, Smith SB. *Liq Chromatogr* 1984;**7**:861.
584. Smith RG. *Anal Chem* 1993;**65**:2485.
585. Jyh-Myng Z, Mu-Jye Chung. *Anal Chem* 1995;**67**:3571.
586. Huang H, Jagner D, Renman L. *Anal Chim Acta* 1987;**201**:1.
587. Zhang S, Wang L. *Fenxi Huaxue* 1993;**21**:76.
588. Hosseini M, Rahimi M, Sodeghi BH, Taghvaei-Ganjali S, Abkanar SD, Gangali MR. *Int J Environ Anal Chem* 2009;**89**:407.
589. Gharchbaghi M, Shemirani F, Baghdadi M. *Int J Environ Anal Chem* 2009;**89**:21.
590. Theranlaz F, Thomas OP. *Mikrochim Acta* 1994;**113**:53.
591. Wang YC, Whang CW. *J Chromatogr* 1993;**628**:133.
592. Yu A, Fan F, Jin Q. *Fenxi Shiyanshi* 1992;**11**:36.
593. Chan CY, Sadana RS. *Anal Chim Acta* 1993;**282**:109.
594. Cossa P, Sanjuan J, Cloud J, Stockwell PB, Corns WT. *Water Air Soil Pollut* 1995;**80**:1279.
595. Jian W, McLeod CW. *Talanta* 1992;**39**:1537.
596. Seifres J, Wasko M, McDaniel W. *Am Environ Lab* 1995;**7**:34.
597. Emteborg H, Sinemus HW, Radzuik B, Baxter DC, Frech W. *Spectrochim Acta Part B* 1996;**51B**:829.
598. Sarzanini C, Srackero C, Aceto M, Abollino O, Mentasti E. *J Chromatrogr* 1992;**626**:151.
599. Fang G, Tang Z. *Fenxi Huaxue* 1987;**15**:46.
600. Ho MH, Guilbault GG, Scheide EP. *Anal Chim Acta* 1981;**130**:141.
601. Sauerbrey GZ. *Physics* 1966;**155**:193.
602. Stockbrdige D. *Vac Microboil Technol* 1966;**5**:193.
603. Chen N, Guo R, Lao EPC. *Anal Chem* 1988;**60**:2345.
604. Lai ERC, Wong B, Van der Noot VA. *Talanta* 1993;**40**:1097.
605. Van der Noot VA, Lac EPC. *Anal Chem* 1992;**64**:3187.
606. Wu Q, Chang X, He Q, Zhai Y, Lui Y, Huang X. *Int J Environ Anal Chem* 2008;**88**:245.
607. Somerset V, Leance J, Mason R, Iwuoha E, Morrin A. *Int J Environ Anal Chem* 2010;**90**:671.
608. Liu JM, Huang Y, Cai WL, Phi P, Li XH, Lin SQ. *Int J Environ Anal Chem* 2005;**85**:387.
609. Li H. *Fenxi Huaxue* 1986;**14**:846.
610. Capitan F, Capitan-Vallvey L, Gomez MC. *Quim Anal Barc* 1987;**6**:343.
611. Zheng Z, Zheng Y, Sun Y. *Fenxi Huaxue* 1988;**16**:260.
612. Lu J, Zhang X, Feng M, Zhang Z. *Fenxi Huaxue* 1993;**21**:1000.
613. Wafling RJ, Watling HC. *Spectrochim Acta* 1980;**35B**:451.
614. Vilchez JL, Sanchez-Palencia G, Blanc R, Navalon R. *Anal Lett* 1994;**27**:2355.
615. Zhang W. *Fenxi Huaxue* 1994;**22**:373.
616. Qi W, Chen X. *Yejin Fenxi* 1992;**12**:17.

617. Yoshimura K, Hiraoka S, Tarutani T. *Anal Chim Acta* 1982;**142**:101.
618. Emerick RJ. *At Spectrosc* 1987;**8**:69.
619. Kuroda A, Matsumato N, Ogume K. *Fresenius Z Für Anal Chim* 1981;**330**:111.
620. Beceiro GE, Bermejo BF, Bermejo PA. *Quin Anal Barc* 1992;**11**:17.
621. Guo L, Zhang G, Fang Z. *Guangpuxue Yu Guangpu Fenxi* 1993;**13**:73.
622. Xu L. *Fenxi Huaxue* 1994;**22**:556.
623. Willie SN, Berman SS, Page JA, Vanhoon GW. *Can J Chem* 1987;**65**:957.
624. Wei X. *Yankkuang Ceshi* 1994;**13**:74.
625. Yu Z, Li Y. *Feuxi Huaxue* 1987;**15**:841.
626. Sun H, Suo R. *Int J Environ Anal Chem* 2008;**88**:791.
627. Saraji M, Yousefi H, Meghdadi S. *Int J Environ Anal Chem* 2009;**89**:305.
628. Lu X, Lu M, Xhao G. *Toxicol Enrion Chem* 1993;**38**:73. *Chem Abstr* 1993;**119**:13059 4h.
629. Lu X, Lu M, Zhu I. *Guangpuxue Yu Guangpu Fenxi* 1993;**13**:113. *Chem Abstr* 1994;**120**:279611q.
630. Li H, Yu Z. *Fenxi Huaxue* 1993;**21**:1052. *Chem Abstr* 1993;**119**:278213d.
631. McCurdy EJ, Lange JD, Haygarth PM. *Sci Total Environ* 1993;**135**:131. *Chem Abstr* 1993;**19**(22):233548(a).
632. Wang S, Zhang Z. *Huanjing Huaxue* 1986;**5**:48.
633. Sawamoto H. *Bunscki Kagaku* 1988;**37**:312.
634. Farisa PAM, Ohara AK, Takase I, Ferreira SL, Gold JS. *Talanta* 1993;**40**:1167.
635. Marques ALB, Chierice GO. *J Braz Chem Soc* 1994;**4**:7.
636. Yoshimura K, Toshimitsue Y, Ohashi S. *Talanta* 1980;**27**:693.
637. Wilson RL, DiNunzio JE. *Anal Chem* 1981;**53**:692.
638. Aliakbar A, Jalali M. *Int J Environ Anal Chem* 2014;**94**:562.
639. Abbasi SA. *Int J Environ Anal Chem* 1988;**33**:43.
640. Ensafi AA, Rezaei B. *Anal Lett Lond* 1993;**26**:1771.
641. Abbasi SA. *Anal Lett Lond* 1987;**20**:1013.
642. Ma Y. *Yu Guanpuxue Guangpu Fenxi* 1988;**8**:53.
643. Aoki H, Munakami H, Chikuma M. *Bunseki Kagaku* 1993;**42**:T147.
644. Wang J, Varmghese K. *Anal Chim Acta* 1987;**199**:185.
645. Abbasi SA, Wang J. *Anal Lett* 1987;**20**:1016.
646. Guo R, Chem N, Silundka C, Lai EPC. *Analyst* 1988;**113**:1105.
647. Orlandini KA, Penrose WR, Nelsson DM. *Mar Chem* 1986;**18**:49.
648. Suutarinen R, Jaakkola T, Paatero J. *Sci Total Environ* 1993;**130**:65.
649. Begg JD, Zavarin M, Zhao P, Turnay SJ, Powell B, Kersting AR. *Environ Sci Technol* 2013;**47**:5146.
650. Motomizu S, Onoda M, Oshima M, Iwachido T. *Analyst* 1988;**113**:743.
651. Iwachido T, Ishimuur K, Toei K. *Bunseki Kagaku* 1986;**35**:892.
652. Ward GK. *Test and evaluation of potassium sensors in fresh and saltwater*. Natural Oceanic and Atmospheric Administration; March 1997. Interagency energy (environment RRD programme report) EPA 600/779057.
653. Hulanicki A, Trojanowiz M, Pbozy E. *Analyst* 1982;**107**:1356.
654. Sarkouhi M, Yamini Y, Zanjani MRK, Alsharanderi A. *Int J Environ Anal Chem* 2007;**87**:603.
655. Lazarev AL, Gerko VV. *Zavod Lab* 1993;**59**:7.
656. Farnworth TG. Private Communication.
657. Small H, Stevens TS, Bauman WC. *Anal Chem* 1975;**47**:1801.

658. Areceneau A, Mehra MC, Campanella L. *Rass Chim* 1986;**38**:269.
659. Vashal GM, Koshceva I, Ya Morozova RP, Konopleva OVT. *Anal Chem (USSR)* 1971;**26**:829.
660. Yu Z, Li Y. *Fuste Fanghu* 1988;**8**:41.
661. Agrawal YK, Nagar HK. *Indian J Chem Sect A* 1986;**25A**:1065.
662. Kim A. *Geokhimiya* 1972;**1**:124.
663. Dessi GR, Paul J. *Microchem J* 1977;**22**:76.
664. Chen Y, Liu Y. *Fenxi Huaxue* 1993;**2**:102.
665. Yu S, Wang G, Goa. *Fenxi Huaxue* 1993;**21**:331.
666. Santosa SJ, Sato N, Tanaka S. *Anal Sci* 1993;**9**:657.
667. Lund W, Bye R. *Anal Chim Acta* 1979;**114**:279.
668. Zang M, Yuan L. *Guang Puxue Yu Guangpu Fenxi* 1993;**13**:65.
669. Cobo Fernandez MG, Palacios MA, Camara C. *Anal Chim Acta* 1993;**283**:386.
670. Oernemark U, Olin A. *Talanta* 1994;**41**:67. *Chem Abstr* 1994;**120**:94160y.
671. Tao G, Hansen EH. *Analyst* 1994;**119**:333. *Chem Abstr* 1994;**120**:172993 pp.
672. Guo Y, Wang B, Shi W. *Guangpuxue Yu Guangpu Fenxi* 1993;**13**(5):46. *Chem Abstr* 1994;**120**(22):279639e.
673. Ediger RD. *At Absorpt Newsl* 1975;**14**:127.
674. Stein BV, Canelli E, Richards AH. *At Spectrosc* 1980;**1**:61.
675. Vickery TM, Buren MS. *Anal Lett Lond* 1980;**13**:1465.
676. Neve J, Hanocq M, Molle L. *Int J Environ Anal Chem* 1980;**8**:177.
677. Fisher RP. *Tappi* 1978;**61**:63.
678. Brodie KG. *Am Lab* 1977;**9**:73.
679. Seaed K, Thomassen Y. *Anal Chim Acta* 1981;**130**:281.
680. Ishizaki M. *Talanta* 1978;**25**:167.
681. Ihnat M. *Anal Chim Acta* 1976;**82**:292.
682. Kamada T, Kamamura T, Yamamoto Y. *Bunseki Kagaku* 1987;**24**:89.
683. Henn EL. *Anal Chem* 1975;**47**:428.
684. Kamada T, Shiraishi T, Yamamoto Y. *Talanta* 1978;**25**:15.
685. Kamada T, Yamamoto Y. *Talanta* 1980;**27**:473.
686. Ahta K, Suzuki MZ. *Anal Chem* 1980;**302**:177.
687. Kirkbright GF, Hsiad-Chaun S, Snook RD. *At Spectrosc* 1980;**1**:85.
688. Baird RB, Abrielian SM. *Appl Spectrosc* 1974;**28**:273.
689. Martin TD, Kopp IF, Edger RD. *At Absorpt Newsl* 1975;**14**:109.
690. Subrimanian KS, Meranger JD. *Anal Chim Acta* 1981;**124**:131.
691. Montaser A, Mehrabzaheh AA. *Anal Chem* 1978;**50**:1697.
692. Goulden PD, Brooksbank P. *Anal Chem* 1974;**46**:1431.
693. Nakahara T, Kabayashi S, Wakisaka T, Musha S. *Appl Spectrosc* 1980;**34**:194.
694. Schmidt FJ, Royer JL, Muir SM. *Anal Lett* 1975;**12**:3.
695. Hahn MH, Mulligan KJ, Jackson ME, Caruso JA. *Anal Chim Acta* 1980;**118**:115.
696. Caldwell JS, Lishka RJ, McFarren EM. *J Am Water Works Assoc* 1973;**5**:731.
697. Lansford M, McPherson EM, Fishman MJ. *At Absorpt Newsl* 1974;**13**:103.
698. Pollock FN, West SL. *At Absorpt Newsl* 1973;**12**:6.
699. Schmidt ES, Roycr JL. *Anal Lett* 1973;**6**:17.
700. Corbin DR, Barnard WM. *At Absorpt Newsl* 1976;**15**:116.
701. Fernandez FJ. *At Absorpt Newsl* 1973;**12**:93.
702. Sinemus HW, Melcher M, Weiz B. *At Spectrosc* 1981;**2**:81.
703. Nakashima S. *Anal Chem* 1979;**51**:654.

704. Rodan DR, Tallman DE. *Anal Chem* 1982;**54**:307.
705. Krivan V, Petrick K, Welz B, Melcher M. *Anal Chem* 1985;**57**:1203.
706. Cox DH. *J Assoc Off Anal Chem* 1981;**64**:265.
707. Jackwerth E, Willmer PG, Han R, Berndt H. *At Absorpt Newsl* 1979;**18**:66.
708. Reichert JK, Gruber H. *Vom Wasser* 1978;**51**:191.
709. Cheam V, Ageman H. *Anal Chim Acta* 1980;**113**:237.
710. Pierce FD, Lamoreaux TC, Brown HR, Fraser RS. *Appl Spectrosc* 1976;**30**:38.
711. Kock K, Lautenshlager W, Maassen J. *Vom Wasser* 1976;**47**:233.
712. Maier D, Sinemus HW, Wiedeking EZZ. *Z Anal Chem* 1979;**296**:114.
713. Verlinden M, Deelstra H. *Z Für Anal Chem* 1976;**296**:253.
714. Meyer AD, Hofer Ch, Tolg G, Raptis S, Knapp G. *Z Anal Chem* 1979;**296**:337.
715. Vijan PN, Leung D. *Anal Chim Acta* 1980;**120**:141.
716. Pyen G, Fishman M. *At Absorpt Newsl* 1978;**17**:47.
717. Cutter GA. *Anal Chim Acta* 1978;**98**:59.
718. Koelbl G. *Mar Chem* 1995;**48**:185.
719. Winge RF, Fassel VA, Kinsely RN, DeKalb E, Haas WJ. *Spectrochim Acta* 1977;**32B**:327.
720. Cai Y, Cabanas M, Fernandez S, Turiel JL, Abalos M, Bayona JM. *Anal Chim Acta* 1995;**314**:183.
721. Howard AG, Gray MR, Water AJ, Oromiehie AR. *Anal Chim Acta* 1980;**118**:87.
722. Nguyen VD, Valenta P, Nurnberg HW. *Sci Total Environ* 1979;**12**:151.
723. Henze G. *Mikrochim Acta* 1981;**11**:343.
724. Deldime P, Hartman JP. *Anal Lett* 1980;**13**:105.
725. Denis BL, Moyers JH, Wilson GS. *Anal Chem* 1976;**48**:1611.
726. Breyer P, Gilbert BP. *Anal Chim Acta* 1987;**201**:23.
727. Mattsson G, Nyholm L, Olin A, Ornemark U. *Talanta* 1995;**42**:817.
728. Potin-Gautier M, Saby F, Astruc M. *Fresenius J Anal Chem* 1995;**351**:443.
729. Campanella J, Ferri T, Petronio BM. *Analysis* 1996;**24**:35.
730. Seby F, Potin-Gantier M, Castethon AJ. *J Hydrol* 1993;**24**:81.
731. Clementi GF, Mastinu GG. *J Radioanal Anal Chem* 1979;**20**:707.
732. Clementi GF, Mastinu GG, Santaroni GP. *Symposium nucl. Techniques comparative studies of food environment contamination IAEA–511–175/31/Otaniemi Finland* 1973.
733. Rossi LC, Clementi GF, Santaroni G. *Arch Environ Health* 1976;**31**:160.
734. Salbu B, Steinnes G, Papas AC. *Anal Chem* 1975;**47**:1011.
735. Kim JI, Stark H, Fiedler I. *Nucl Instrum Methods* 1980;**177**:557.
736. Kubota J, Cary EE, Gissel-Nielsen G. In: Hemphill DD, editor. *Trace substances in environmental health, – IX*. Columbia: University of Missouri; 1975. p. 123.
737. Navarre JL, Ronneau C, Priest P. *Water Air Soil Pollut* 1980;**14**:207.
738. Schutyser P, Maenhaut M, Dams R. *Anal Chim Acta* 1978;**100**:75.
739. Harrison SH, Lafleur PD, Zoller WH. *Anal Chem* 1975;**47**:1685.
740. Lieser KH, Neitzert V. *J Radioanal Anal Chem* 1976;**31**:397.
741. Habib S, Minski MJ. *J Radioanal Anal Chem* 1981;**63**:379.
742. Sakai Y, Tomura K, Ohshita K. *Daido Kogyo Daigaku Ktyo* 1993;**29**:63–8. *Chem Abstr* 1994;**121**:91131.
743. Haygarth PM, Rowland AP, Sturup S, Jones KC. *Analyst* 1993;**118**:1303. *Chem Abstr* 1994;**120**(2):14526j.
744. Tamari Y, Chayama K, Tsuji H. *Biomed Res Trace Elem* 1993;**4**:263. *Chem Abstr*. 1994, 212, 49287.
745. Aomo T. *Hoshase Kagaku (Tokyo)* 1993;**36**:341. *Chem Abstr* 1994;**120**(18):225890m.

746. Itoh K, Chikuma M, Tanaka H. *Kogyo Yosui* 1993;**412**:3. *Chem Abstr* 1993;**119**(2):14789b.
747. Shimoishi Y. *Bull Chem Soc Jpn* 1971;**44**:3370.
748. Young J, Christian GD. *Anal Chim Acta* 1973;**65**:127.
749. Shimioishi Y, Toei K. *Anal Chim Acta* 1978;**100**:65.
750. Monteil A. *Analytsis* 1981;**9**:112.
751. Talmi Y, Andren AW. *Anal Chem* 1974;**46**:2122.
752. Young JW, Christian GD. *Anal Chim Acta* 1973;**65**:127.
753. Measures CI, Burton JD. *Nature (London)* 1978;**273**:293.
754. Uchida H, Shimoishi Y, Toei K. *Environ Sci Technol* 1980;**14**:541.
755. Measures CI, Burton JD. *Anal Chim Acta* 1980;**120**:177.
756. Measures CI, Burton JD. *Earth Plant Sci Lett* 1980;**46**:385.
757. Measures CI, McDuff RE, Edmond JM. *Earth Plant Sci Lett* 1981;**49**:102.
758. Flinn CG, Aue WA. *J Chromatogr* 1978;**153**:49.
759. Johansson K, Oernemark U, Olin A. *Anal Chim Acta* 1993;**274**:129.
760. De La Calle Guntinas MB, Lobinski R, Adams FC. *J Anal At Spectrosc* 1995;**10**:11.
761. Gallus SM, Henmann KG. *J Anal At Spectrosc* 1996;**11**:887.
762. Sarkouhi M, Yamini Y, Zonjani MRK, Afsoumeh A. *Int J Environ Anal Chem* 2007;**87**:603.
763. Nakagawa T, Aoyama E, Hasegawa N, Kobayashi N, Tanaka H. *Anal Chem* 1989;**61**:233.
764. Que M, Kimura K, Shono T. *Analyst* 1988;**113**:551.
765. White MN, Lisk DJ. *J Assoc Off Anal Chem* 1979;**53**:1055.
766. McHugh J. *At Spectrosc* 1984;**5**:123.
767. Chau TT, Fishman MJ, Ball TW. *Anal Chim Acta* 1969;**43**:189.
768. Guo Y, Zhang Y, Zhou, Wei Q. *Jiijn Daxue Ziran Kexue Xuebao* 1988;**1**:98.
769. Wu Q, Yu X, Liu Y, Zhang S, Ying S, Liu K. *Lihua Jianyan Huaxue Fence* 1992;**28**:233.
770. Robinson R, Bell M, Burns C, Knab D. Los Alamos National Lab Report, LA-11095-MS: Order NO. DE88001629, 45 pp. Avail NTIS from *Energy Res Abstr* 1983;**13**(4):Abstrct No. 7853.
771. Jin L, Xu J, Qian J, Tong W. *Fenxi Ceshi Tongbao* 1992;**11**:56. *Chem Abstr* 1993;**118**:160077 pp.
772. Hiraide M, Zhou SH, Kawaguchi H. *Anal Camb UK* 1993;**118**:1441. *Chem Abstr* 1994;**120**:61810g.
773. Oreshkin VN, Malofeeva GI, Vnukovskaya GL. *Okeanol Mosc* 1993;**33**:784. *Chem Abstr* 1994;**120**:199934g.
774. Tao S, Shijo Y, Wu L, Lin L. *Analyst* 1994;**119**:1455. *Chem Abstr* 1994;**121**:141084.
775. Whitlow SI, Rice DL. *Water Res* 1985;**19**:619.
776. Jiang Z, Liang A, Dai G. *Huanjing Kexue* 1987;**8**:72.
777. Dai G, Jiang Z. *Gaodong Xuaxue Xeubae* 1987;**8**:703.
778. Van den Winkel P, Mertens J, De Baenst G, Massart DL. *Anal Lett Lond* 1972;**5**:567.
779. Smith S, Bexnieterboni EM, Van Heerden AM. *Water Res* 1983;**17**:1483.
780. Thompson M, Ramsay MH, Paklavanpour R. *Analyst* 1982;**107**:1330.
781. Beanchemin D, McLaren JW, Mykytink AP, Berman SS. *Anal Chem* 1987;**59**:778.
782. Bart G, Van Gunten HR. *Int J Environ Anal Chem* 1979;**6**:25.
783. Sueki T, Kabyayashi K, Sato W, Nakahara H, Tomizawa T. *Anal Chem* 1996;**68**:2203.
784. Brodsky SM, Varvaritsa VP, Mambauan SV, Filatov VI. *J Radioanal Nucl Chem* 1984;**81**:155.
785. Abdullah AA, Khankinov S, Khasanov AS. *Sov Radio Chem* 1972;**14**:509.
786. Nishioka H, Yoneda A, Maeda Y, Asumi T. *Nippon Katsui Gakaishi* 1986;**34**:393.

787. Jaffar M, Ashraf M, Tarig N. *At Spectrosc* 1986;**7**:961.
788. Thompson M, Ramsey MH, Paklavanpour B. *Analyst* 1982;**107**:1330.
789. Beanchimia P, McLaren JW, Mykytink AP, Merman SS. *Anal Chem* 1987;**59**:778.
790. Zhang Q, Huang Y. *Talanta* 1987;**10**:149.
791. Gyras N, Gorene B. *Chromatographia* 1994;**39**:448.
792. Brodskey SM, Varvarita VP, Mankouyan SV, Filatov VI. *J Radioanal Nucl Chem* 1984;**81**:155.
793. Beals DM. *J Radioanal Nucl Chem* 1996;**204**:253.
794. Weidenauer M, Lieser KH. *Fresenius Z Für Anal Chem* 1985;**330**:550.
795. Andreae MD. *Anal Chem* 1984;**56**:2064.
796. Thompson M, Ramsey MH, Pahlavanpour B. *Analyst* 1982;**107**:1330.
797. Schliekmann F, Umland F. *Fresenius Z Für Anal Chem* 1983;**314**:21.
798. Hayrynen H, Lajunen LHJ, Peramaki P. *At Spectrosc* 1985;**6**:88.
799. Yamamoto M, Yashuda M, Yamamioto Y. *Anal Chem* 1985;**57**:1382.
800. Chandrawanski S, Sharma Y, Patel KS. *Fresenius Z Für Anal Chem* 1995;**351**:305.
801. De Ruck A, Vandescastelle C, Dams R. *Mikrochim Acta* 1987;**2**:187.
802. Welz B, Schlemmer G, Mudakavi IR. *Anal Chem* 1988;**60**:2567.
803. Li H, Ma B. *Guangpuxue Yu Guangpu Fenxi* 1993;**13**:127.
804. Mohammad B, Ure AM, Littlejohn D. *Mikrochim Acta* 1994;**113**:325.
805. Chram V, Lechner J, Desrosieu R, Azgura J, Mundrach A. *Fresenius Z Für Anal Chem* 1996;**355**:336.
806. Axner O, Chekalin N, Ljungberg P, Malmsten Y. *Int J Environ Anal Chem* 1993;**53**:185.
807. Sun C, Liu X, Xu K. *Fenxi Huane* 1986;**14**:684.
808. Zhou Z, Ruan X, Yao W. *Xiangtan Daxue Ziran Kexue Xuebao* 1986;**3**:34.
809. Riley JP, A Siddiqui S. *Anal Chim Acta* 1986;**181**:177.
810. Miloshova MS, Seleznev BL, Bychkov EA. *Sens Acutators B* 1994;**19**:373.
811. Lin LS, Nriagu J. *Environ Sci Technol* 1999;**33**:3394.
812. Yamamoto T. *Anal Chim Acta* 1973;**63**:65.
813. Cospito M, Rigali L. *Anal Chim Acta* 1979;**106**:385.
814. Yang Y, Cai R, Zeng Y. *Chin J Chem* 1993;**11**:66.
815. Guo N. *Fenxi Huaxue* 1987;**15**:105.
816. Necemers M, Byrne AR, Juznik K. *Sci Total Environ* 1993;**130**:261.
817. Du H,S, Wood DJ, Elshani S, Wai EM. *Talanta* 1993;**40**:173.
818. Yang Y, Hsu C, Pan J. *Microchem J* 1993;**48**:178.
819. Gong J, Wu Q. *Lihua Jianyan Huaxue Fence* 1992;**28**:199.
820. Valencia MC, Gimino D, Capitan-Vallvey LF. *Anal Lett* 1993;**26**:1211.
821. Donard OFX, Rapsomanskis S, Weber JH. *Anal Chem* 1986;**58**:772.
822. Sturgeon RE, Willie SN, Berman SS. *Anal Chem* 1987;**59**:2441.
823. Pinel R, Beneabdallah MZ, Astruc A. *J Anal At Spectrom* 1988;**3**:475.
824. Jin C, Xu B, Xu T, Fang Yl. *Fenxi Ceshi Xuebao* 1993;**12**:64.
825. Kagay S, Ueda J. *Bull Chem Soc Jpn* 1993;**66**:1404.
826. Ebdon L, Hill SJ, Jones P. *Analyst* 1985;**110**:515.
827. Ebdon L, Alonso JI. *Analyst* 1987;**112**:1551.
828. Abalos M, Bayone JM. *J Chromatogr* 2000;**891**:287.
829. Macchi G, Pettine M. *Environ Sci Technol* 1980;**14**:814.
830. Weber G. *Anal Chim Acta* 1986;**186**:49.
831. Weber G. *Fresenius Z Für Anal Chem* 1985;**322**:311.
832. Wang L, Zadell J. *Talanta* 1987;**34**:909.

833. Abbasi SA. *Int J Environ Anal Chem* 1982;**11**:1.
834. Abbasi SA. *Anal Lett Lond* 1987;**20**:1697.
835. Chen Y. *Zhonghua Yufangixue Zazhi* 1986;**20**:299.
836. Yang KL, Jiang SJ, Hwong TJ. *J Anal At Spectrosc* 1996;**11**:139.
837. Wong L, Chen R, Wen S, Kong D. *Huaxue Shizi* 1988;**29**:120.
838. Korrey JS, Goulden PD. *At Absorpt Newsl* 1975;**14**:33.
839. Hall GEM, Jefferson CW, Michel FA. *J Geochem Explor* 1988;**30**:63.
840. Wei X. *Yankukuong Cheshi* 1993;**12**:105.
841. Li Z, Yulung Y, Wei Q, Liu Z, Tang J. *Int J Environ Anal Chem* 2004;**84**:789.
842. Skedlik M, Havel J, Sommer L. *Scr Chem* 1977;**1**:99.
843. Korkisch J, Koch W. *Mikrochim Acta* 1973;**1**:157.
844. Korkisch J. *Mikrochim Acta* 1972;**1**:687.
845. Aziz M, Beheir G, Shakir K. *J Radioanal Nucl Chem* 1993;**172**:319. *Chem Abstr* 1993;**119**(24):256176(s).
846. Liu S, Lui Z. *Yankuanz Ceshi* 1992;**11**:311. *Chem Abstr* 1994;**120**(10):123661v.
847. Agrawal YK, Upadhyaya DB, Chudasama SP. *J Radioanal Nucl Chem* 1993;**170**:79. *Chem Abstr* 1993;**119**(2):19454y.
848. Sun W, Zhang S, Qi W, Tang F. *Fenxi Huaxue* 1993;**21**:93. *ChemAbstr* 1993;**118**:175346z.
849. Lee CH, Joe KS, Suh M, Lee W. *J Korean Chem Soc* 1994;**28**:502.
850. Leung G, Kim YS, Zeitlin H. *Anal Chim Acta* 1972;**60**:229.
851. Kim YS, Zeitlin H. *Anal Abstr* 1972;**22**:4571.
852. Boomer DW, Powell MI. *Anal Chem* 1987;**59**:2870.
853. Bailey EH, Kemp AJ, Ragnarsdottir KV. *Anal At Spectrom* 1993;**8**:551.
854. Crain JS, Mikesell BL. *Appl Spectrosc* 1992;**46**:1498.
855. Deutscher RL, Mann AW. *Analyst* 1977;**102**:929.
856. Zhang W, Shao K, Li S, Xia Y. *He Huane Yu Fangshe Huaxue* 1986;**8**:102.
857. Van den Berg CMG, Nimmo M. *Anal Chem* 1987;**59**:924.
858. Jin J, Bao Y, Bai Z. *Fenxi Huaxue* 1987;**15**:620.
859. Possie C, Lannugali G, Abradaro A, Gambillara R, Roberro M, Pozzie A. *Int J Environ Anal Chem* 2007;**87**:361.
860. Brits RJS, Smit MCB. *Anal Chem* 1977;**49**:67.
861. Steinnes E. *Radiochem Radioanal Lett* 1973;**16**:25.
862. Fleischer RL. *Anal Chem* 1976;**48**:642.
863. Gladney ES, Owens JW, Starner JW. *Anal Chem* 1976;**48**:973.
864. Zielinski RA, KcKown J. *J Radioanal Nucl Chem* 1984;**84**:207.
865. Cassidy RM, Elchuk S. *Int J Environ Anal Chem* 1981;**10**:1876.
866. Cassidy RM, Elchuk S. *J Chromatogr Sci* 1980;**18**:217.
867. Kerr A, Kupferschmidgt W, Attas M. *Anal Chem* 1988;**60**:2729.
868. Colodner DC, Boyle E, Edmond JA. *Anal Chem* 1993;**65**:1419.
869. Rosenberg R, Zilliacui R, Manninen PR. *J At Spectrosc* 1994;**9**:713.
870. Beauchemin D, Mcharen JW, Mykytink AP, Berman SS. *Anal Chem* 1987;**59**:778.
871. Zhai P, Kang T. *Fenxi Huaxue* 1986;**14**:821.
872. Guo S, Deng X, Sun X, Ment W, Zhang P, Hao X. *Nucl Tracks Radiat Meas* 1986;**12**:801.
873. Deng X, Guo S, Meng W, Sun S, Zhang P, Hao X. *Hejishu* 1987;**10**:17.
874. Desideri D, Meli MA, Roselli C, Testa C. *J Environ Anal Chem* 2004;**54**:331.
875. Billon G, Ouddane B, Proix N, Abdelhour Y, Boughriet A. *Int J Environ Anal Chem* 2005;**85**:1013.
876. Wakamatsu Y, Otomo M. *J Chem Soc Jpn* 1969:595. Sect 90.

877. Doadrio A, Diaz MG. *Inferion Quim Anal Pura Apl Ind* 1973;**27**:247.
878. Basson AT, Kempster PL. *Water S Afr* 1980;**6**:88.
879. Abbasi SA. *Int J Environ Stud* 1981;**6**:88.
880. Agrawal YK, Mehd GD. *Int J Environ Anal Chem* 1981;**10**:183.
881. Weiguo Q. *Anal Chem* 1983;**55**:2043.
882. Abbasi SA. *Anal Lett Lond* 1987;**20**:1347.
883. Sun H, Zheng L, Xi Z. *Lanzhou Daxue Xuebao Ziran Kexueban* 1987;**23**:167.
884. Li H, Smart RB. *Anal Chim Acta* 1996;**333**:131.
885. Mierura J. *Anal Chem* 1990;**62**:1424.
886. Nagoasa Y, Kimata Y. *Anal Chim Acta* 1996;**327**:203.
887. Orvini E, Ladola L, Sabbioni E, Pietra R, Goetz L. *Sci Total Environ* 1979;**13**:195.
888. Norwiki JL, Johnson KS, Coale KH, Elrod VA, Lieberman SH. *Anal Chem* 1994;**66**:2732.
889. Chea Shang D, Zong Y. *Huaxue Tongbao* 1986;**8**:27.
890. Miller DG. *J Water Pollut Control Fed* 1979;**51**:2402.
891. Yoshimura K, Waki H, Ohashi S. *Talanta* 1978;**25**:579.
892. Fan J, Jiang C, Wang S. *Shandong Daxue Xuebao Ziran Kexueban* 1986;**21**:84.
893. Herrador MA, Jiminez AM, Asuero AG. *Analyst (London)* 1987;**112**:1237.
894. Sang X, Yang G. *Yejin Fenxi* 1987;**7**:18.
895. Chen A, Su J. *Huanjing Huaxue* 1988;**7**:65.
896. Camran AM, Herrador MA, Asuero AG, Marques ML. *Aliment Madr* 1986;**23**:74.
897. Borgnon J, Cadet JL. *Analysis* 1988;**16**:LXXVII.
898. El-Ashgar NM, El-Nahhal IM, Chehimi MM, Babonneau F, Livage J. *Int J Environ Anal Chem* 2009;**89**:1057.
899. Herman AW, McNelles LA, Campbell A. *Int J Appl Radiat Isotope Chem* 1973;**24**:677.
900. Peng X, Mao Q, Cheng J. *Fresenius J Anal Chem* 1994;**348**:644.
901. Tong S, Zhao F, Sun G. *Fenxi Huaxue* 1988;**16**:337.
902. Tong S, Sun G. *Huaxue Xuebao* 1988;**46**:812.
903. Igarashi S, Yotsuvanagi T. *Anal Chim Acta* 1993;**281**:347.
904. Wang Y, Wang SH, Yang HM, Xue Tan JH, Guo G, Xiao XK. *Int J Environ Anal Chem* 2011;**91**:87.
905. Kheteh M, Kaykhaii M, Mirmoghadd M, Hashemi H. *Int J Environ Anal Chem* 2009;**89**:981.
906. Xue HB, Syg L. *Anal Chim Acta* 1993;**284**:505.
907. Onishchenko TA, Onishenko YK, Sukkan VV, Kayazevc K, Ukr V. *Khim Zu* 1987; **53**:855.
908. Wang J, Greene B. *Water Res* 1983;**17**:1635.
909. Chen X, Zhang S. *Huanjing Huaxue* 1987;**6**:52.
910. Kang J, Hi Y, Gao J, Bai G. *Xibei Shifan Doxue Xuebao Zuron Kexueban* 1992;**28**:30.
911. Saal E, Nistor M, Gaspoar G, Csoregi E, Iscan M. *Int J Environ Anal Chem* 2007;**87**:745.
912. Caneau AM, Herrador MA, Asuero AG, Marquez ML. *Allmentaria Madr* 1986;**23**:74.
913. Wang J, Peng T, Varughese A. *Talanta* 1987;**34**:561.
914. Ngugen VD, Valenta P, Nurnberg HW. *Sci Total Environ* 1979;**12**:151.
915. West MH, Melina JF, Yuan CL, Davies DG. *Anal Chem* 1979;**51**:2370.
916. Savitskii VN, Peleshenko VI, Osadchii VI. *Zhur Anal Khim* 1987;**42**:677.
917. Welz B, Schlemmer G, Mudakavi JR. *J Anal At Spectrscopc* 1988;**3**:695.
918. Schoeva OP, Kuchava GP, Kubrakova IV, Myasoedova GV, Savvini SB, Bannkyh LN. *Zhur Anal Khim* 1986;**41**:2186.
919. Oki Y, Tashiro Y, Maeda M, Honda C. *Anal Chem* 1993;**65**:2096.

920. Fordham AW. *J Geochem Explor* 1978;**10**:41.
921. Favretto LG, Marletta GP, Favretto L. *Fresenius Z Anal Chem* 1984;**318**:434.
922. Logos P. *Anal Chim Acta* 1978;**98**:201.
923. Haring BJA, Van Delft W, Born CM. *Fresenius Z Anal Chem* 1982;**310**:217.
924. Bozsai G, Melegh M. *Microchem J* 1995;**51**:39.
925. Murty K, Kamas D. *Indian J Chem* 1994;**334**:180.
926. Saler T, Sperling M, Welz B. *Analyst* 1992;**117**:1735.
927. Zolfongun E, Rouhallahi A, Semnani A. *Int J Environ Anal Chem* 2008;**88**:327.
928. Soliman EM, Ahmed S. *Int J Environ Anal Chem* 2009;**89**:389.
929. Sharma R, Perul P. *Int J Environ Anal Chem* 2009;**89**:389.
930. Sharma R, Pant D. *Int J Environ Anal Chem* 2009;**89**:503.
931. Sinemus HW, Mecher M, Weir B. *At Spectrosc* 1981;**2**:81.
932. Knudson EJ. *Anal Lett* 1973;**6**:1059.
933. Weiz B. *Chem Br* February 1986:132.
934. Brovko. *Zhur Anal Khim* 1987;**42**:1637.
935. Brodie KG. *Am Lab* 1977;**9**:23.
936. Yamamoto M, Yasuda M, Yanamoto Y. *Anal Chem* 1985;**57**:1382.
937. Hard PRG, Bruchenstein S. *Anal Chem* 1980;**52**:1028.
938. Vien SH, Fry RC. *Anal Chem* 1988;**60**:465.
939. Haraldsson C, Pollak M, Dehuan P. *J Anal At Spectrosc* 1992;**7**:1183.
940. Cutter LS, Cutter GA, Maria LC, McGlove SD. *Anal Chem* 1991;**63**:1138.
941. Kumar AM, Riyazuddin P. *Int J Environ Anal Chem* 2007;**87**:469.
942. Sutherland RA. *Int J Environ Anal Chem* 2007;**87**:501.
943. Pyen GS, Beowner PL. *Spectrochim Acta Part B* 1986;**413**:1203.
944. Thompson M, Pahlavanpour B, Walton SJ, Kirkbright GF. *Analyst* 1978;**103**:705.
945. Thompson M, Pahlavapour B, Walton SJ, Kirkbright GF. *Analyst* 1978;**103**:568.
946. Hall GEM, Vaive JE, Mcconnell JW. *Chem Geol* 1995;**120**:91.
947. Aggarwal JK, Shabani MB, Palmer MR, Ragnarsdottir KV. *Anal Chem* 1996;**68**:4418.
948. Klinkenhammer G, German CR, Elderfield H, Greave MJ, Mitra A. *Mar Chem* 1994;**45**:179.
949. Elderfield H, Greaves MJ. In: *Trace metals in seawater*. New York: Plenum Press; 1983. p. 427–45.
950. a. Moller P, Dulski P, Lucic J. *Spectrochim Acta* 1992;**47B**:1379.
b. Weiss D, Paulkert T, Rubeska I. *J Anal At Spectrosc* 1990;**5**:317.
951. Shabani MB, Akagi T, Shimuzu H, Masuda H. *Anal Chem* 1990;**62**:2709.
952. Rubio R, Hugnet J, Rauret G. *Water Res* 1984;**18**:423.
953. Zavaras JT, Shenddrikon AD. *American Lab* 1984:1659.
954. Water Research Centre Medmenham UK. Technical Report No. TR141. Multi element analysis using inductivley coupled plasma.
955. Goulden RD, Anthony DHJ, Austen KD. *Anal Chem* 1981;**53**:2027.
956. Janssens E, Schutxyser P, Dams R. *Environ Technol Lett* 1982;**3**:35.
957. Miyazaki A, Kimura A, Bansho K, Uimezaki Y. *Anal Chim Acta* 1982;**144**:213.
958. Cui Y, Chang ZX, Zhu X, Zou X. *Int J Environ Anal Chem* 2008;**88**:857.
959. Zou X, Cui Y, Chang X, Zhu X, Hu Z, Yang D. *Int J Environ Anal Chem* 2009;**89**:1043.
960. Zhang Z, Huoi G, Cao Z, Zhang B. *Huan Daxue Xuebao* 1993;**20**:47.
961. Wang C, Luo D. *Mikrochim Acta* 1993;**111**:257.
962. Sherma P, Kumat S, Rawat C. *Int J Environ Anal Chem* 1992;**48**:201.
963. du Rosario Crato G. *Revta Port Quim* 1969;**10**:149.

964. Zhang O, Huang Y. *Talanta* 1987;**34**:555.
965. Florence JM, Batley GE. *Talanta* 1976;**23**:179.
966. Riley JP, Taylor D. *Anal Chim Acta* 1968;**41**:175.
967. Abdullah MJ, Royle LG. *Anal Chim Acta* 1972;**58**:283.
968. Corsini A, Crang S, Francia RD. *Anal Chem* 1982;**54**:1433.
969. Wan C, Chon S, Corsini A. *Anal Chem* 1985;**57**:719.
970. Duinker JC, Kramer CMJ. *Mar Chem* 1977;**5**:207.
971. Clanch F, Delangele R, Popoft G. *Water Res* 1981;**15**:591.
972. Everarts FM, Verbeggen M, Reijenga JC, Abel GUA. *J Chromatogr* 1985;**320**:263.
973. Pakalns P. *Anal Chim Acta* 1980;**120**:289.
974. Vasioncebos MJ, Gomez GAR. *J Chromatogr* 1995;**696**:227.
975. Iwachido T, Ishimaru K, Motomizu S. *Anal Sci* 1988;**4**:81.
976. Gros N, Gorene B. *Chromatographia* 1994;**39**:448.
977. Hill R, Leiser KH. *Fresenius Z Anal Chem* 1987;**327**:165.
978. Smith DL. Ames Lab. Report, IS–T–1318, Order No. DE870 13601, 73 pp. Avail. NTIS from *Energy Res Abstr* 1997;**12**(21) Abstract No. 43657 1987.
979. Smith DL, Fritz JS. *Anal Chim Acta* 1988;**204**:87.
980. Jen J, Chen C. *Anal Chim Acta* 1992;**270**:55.
981. Johnson GW, Taylor HE, Skogerbie RK. *Spectrochim Acta* 1979;**34B**:197.
982. Na HC, Niemczyk TM. *Anal Chem* 1982;**54**:1839.
983. Nrasa JT. *Anal Chem* 1984;**56**:904.
984. Zhang Z, Pugh J, Yu Wuton. *Huabau* 1986;**10**:230.
985. Hagrynen H, Lajunea LHJ, Permki P. *At Spectrosc* 1985;**6**:88.
986. Johnson GW, Taylor HE, Skogerboe RK. *Spectrochim Acta* 1979;**340**:197.
987. Bettagi AM, Ghoneim EM, Ghoneim MA. *Int J Environ Anal Chem* 2011;**91**:17.
988. Koynaves SP, Deng W, Hallack PR, Kovacs GTA. *Anal Chem* 1994;**66**:418.
989. Ikeda S, Motonaka J, Kiyoky H, Masuda T. *Bunseki Kagaku* 1993;**42**:183.
990. Jadner D, Kenman L, Stefandottir SH. *Anal Chim Acta* 1993;**281**:305.
991. Panneli MG, Voulgaropoulos AN. *Fresenius J Anal Chem* 1994;**348**:837. *Chem Abstr* 1994;**212**:147903.
992. Ni Y, Kokot S, Selby M, Hodgkinson M. *Fenxi Huaxue* 1994;**22**:431–5. *Chem Abstr* 1994;**121**:163511.
993. Gil EP, Ostapezuk P. *Anal Chim Acta* 1994;**293**:55.
994. Ulakhovich NA, Budnikov GK, Medy Antsevg EP. *Zhur Anal Khim* 1992;**47**:1546.
995. Wang C, Chen J, Jan E. *Zhaowohan Doxue Wuhan Doxue Xuebao Ziran Kexueban* 1993;**1**:61.
996. Wong F, Li S, Liu S, Zhang Y, Liu Z. *Anal Lett* 1994;**27**:1779.
997. Poldoski JE, Glass GE. *Anal Chim Acta* 1978;**101**:79.
998. Figura P, McDuffie B. *Anal Chem* 1980;**52**:1433.
999. Analiitia TU, Pickering WF. *Water Res* 1986;**20**:1397.
1000. Valenta P, Sios L, Kramer I, Krumpen P, Rutzel H. *Fresenius Z Anal Chem* 1982;**312**:101.
1001. Wang J, Dewald HD. *Anal Chem* 1983;**55**:933.
1002. Hu A, Dessy RE, Graneli A. *Anal Chem* 1983;**55**:320.
1003. Hsi T, Johnson DC. *Anal Chim Acta* 1985;**175**:23.
1004. Schliekmann F, Umland F. *Fresenius Z Anal Chem* 1983;**314**:21.
1005. Wang J, Dewald HD, Green N. *Anal Chim Acta* 1983;**146**:45.
1006. Wang J. *Talanta* 1984;**31**:703.
1007. Wang E, Sun W, Yany Y. *Anal Chem* 1984;**56**:1903.

1008. Nurnberg HW. *Sci Total Environ* 1984;**37**:9.
1009. Private communication.
1010. Weidenauer M, Lieser KH. *Fresenius J Anal Chem* 1985;**330**:550.
1011. Chau YK, Cahu KLS. *Water Res* 1974;**8**:383.
1012. Davidson W, De Mora R, Harrison RM, Wilsin S. *Sci Total Environ* 1987;**60**:35.
1013. Mann KJ, Florence TM. *Sci Total Environ* 1987;**60**:67.
1014. Pratt KW, Koch WF. *Anal Chim Acta* 1988;**215**:21.
1015. Almeida Mota AM, Buffle J, Kounaves SP, Simoes Goncalves ML. *Anal Chim Acta* 1985;**172**:19.
1016. Beverage A, Pickering WF. *Water Res* 1984;**18**:1119.
1017. Brainina KZ, Khanina M, Roitman LI. *Anal Lett Lond* 1985;**18**:117.
1018. Florence TM, Mann KJ. *Anal Chim Acta* 1987;**200**:305.
1019. Bernard JP, Buffle J, parthasarthy N. *Anal Chim Acta* 1987;**200**:191.
1020. Apte SC, Gardner MJ, Ravenscroft JE. *Anal Chim Acta* 1988;**212**:1.
1021. Kyle JH. *Environ Technol Lett* 1987;**8**:181.
1022. Yang X, Risinger L, Jokausson G. *Anal Chim Acta* 1987;**192**:1.
1023. Analiitia T, Pickering WF. *Talanta* 1987;**34**:231.
1024. Williams G, D'Silva C. *Analyst* 1994;**119**:2337.
1025. Acherberg EP, Van den Berg CMG. *Anal Chim Acta* 1994;**291**:213.
1026. Valenta P, Mart L, Rutzel H. *J Electro Anal Chem* 1977;**82**:327.
1027. Wang J, Ariel M. *Anal Chim Acta* 1978;**99**:89.
1028. Lewis BL, Luther GW, Lane H, Church TM. *Electrolysis* 1995;**7**:166.
1029. Adeljou SB, Sahara E, Jagner D. *Anal Lett Lond* 1996;**29**:283.
1030. Belmont-Hebert C, Tercier ML, Buffle J, Fiaccabrino GC, et al. *Anal Chem* 1998;**70**:2949.
1031. Tercier Cambell ML, Buffle J. *Electroanalysis* 1993;**5**:187.
1032. Cambell PGC. Metal speciation and bioavailability in aquatic systems. In: Tessier A, Turner DR, editors. *IUPAC series on analytical and physical chemistry of environmental systems*, vol. 3. New York: Wiley; 1995. p. 45. Chapter 2.
1033. Buffle J, Vuilleumier JJ, Tercier M-L, Parthasarathy N. *Sci Total Environ* 1987;**60**:75.
1034. Sagberg P, Lund W. *Talanta* 1982;**29**:457.
1035. Tercier M-L, Buffle J. *Anal Chem* 1996;**68**:3670.
1036. Aldstadt JH, Dewald HD. *Anal Chem* 1993;**65**:922.
1037. Morrison GMP, Florence TM. *Electroanalysis* 1990;**2**:383.
1038. Dam MER, Thomsen KN, Pickup PG, Schroder KH. *Electroanalysis* 1995;**7**:70.
1039. Wang J, Taha Z. *Electroanalysis* 1990;**2**:383.
1040. Wang J, Gala A, Zimmer H, Mark HB. *Electroanalysis* 1992;**4**:77.
1041. Montenegro MI, Queiros MA, Dashbar JL. Microelectrodes, theory and applications. In: *NATO ASI series*, vol. 497. Dordrecht (The Netherlands): Kluwer; 1991.
1042. Pretly JP, Bluhaugh FA, Caruso JA, Davidson TM. *Anal Chem* 1994;**66**:1540.
1043. Nevoral VC. *Anal Chem* 1974;**268**:189.
1044. Kerisch J, Godl L. *Talanta* 1974;**12**:1035.
1045. Small H, Stevens TS, Bauman WC. *Anal Chem* 1975;**47**:1801.
1046. Frenzel W, Schepers D, Schultz G. *Anal Chim Acta* 1993;**277**:103.
1047. Small H. *Ion chromatography*. New York: Penum; 1989. p. 132.
1048. Blasius E, Janzen K-P, Adrian W, Klautke G. *Z Für Chemie* 1977. *Anal Chem* 1977;**284**:337.
1049. Small H, Soderquist ME, Pische JW. US Patent 4,732,686, 1988.
1050. Hu W, Takeuchi T, Haraguchi H. *Anal Chem* 1993;**65**:2204.

1051. Hu W, Tao H, Haraguchi H. *Anal Chem* 1994;**66**:2514.
1052. Hu W, Mujasaki A, Tao H, Itoh A, Unemura T, Haraguchi H. *Anal Chem* 1995;**67**:3713.
1053. Lieser KH, Calmond W, Heuss E, Neitzert V. *J Radioanal Chem* 1997;**37**:717.
1054. Moore PJ. *Trans Inst Miner Metall Sect B* 1970;**79**:107.
1055. Kusaka Y, Tsuji H, Figimoto Y, Ishidi K. *J Radioanal Chem* 1982;**71**:7.
1056. Kusaka JY, Tsuji T, Fugimoto Y, Ishida K. *Bull Inst Chem Res Kyoto Univ* 1980;**58**:111.
1057. Honda T, Nozaki T, Ossaka T, Oi T, Kokihara H. *J Radioanal Nucl Chem* 1988;**122**:143.
1058. Anderson DL, Cunningham WC, Mackey WC. *Fresenius J Anal Chem* 1990;**338**:554.
1059. Anderson DL, Foiley MD, Zola WH, Walters WB, Fordon GE, Lindstrom RM. *J Radioanal Chem* 1981;**63**:97.
1060. Zhang L, Liu X, Xia W, Zhang Y, Zhang W. *Int J Environ Anal Chem* 2014;**94**:728.

Chapter 4

Metals in Surface, Ground, and Mineral Waters

Chapter Outline

4.1 SURFACE WATER

4.1.1 Miscellaneous Metals

The levels of several metals present in groundwater have been determined, including antimony,[1–4] arsenic,[4,5] barium,[6,7] cadmium,[5,8] chromium,[9,10] copper,[11] mercury,[10,12,13] platinum,[14] rhenium,[14] selenium,[2] thorium,[15] uranium,[15] uranylion,[7] and vanadium.[16,17]

Various workers[2,15–21] have discussed a novel ion-exchange chromatographic method for the determination of a wide range of elements in surface water.

Thus Small[18] showed that ion exchange resins have a well-known ability to provide excellent separation of ions, but the automated analysis of the eluted species is often frustrated by the presence of the background electrolyte used for elution. By using a novel combination of resins, these workers have succeeded in neutralizing or suppressing this background without significantly affecting species being analyzed, which in turn permits the use of a conductivity cell as a universal and very sensitive monitor of all ionic species, either cationic or anionic. Using this technique, automated analytical schemes have been devised for lithium, sodium, rubidium, caesium, ammonium, and magnesium.

Henshaw et al.[22] analyzed numerous samples of surface water from lakes in the eastern United States by inductively coupled plasma mass spectrometry for 49 elements. Standard calibrations were used for 21 elements, and surrogate standards were used for 28 elements. The system detection limits, evaluated by using field blanks carried through the entire sampling and pretreatment process, were less than $0.1\,\mu g\,L^{-1}$ for most elements. Contamination during sampling and pretreatment was often the limiting factor. The accuracy of the determinations, as determined from the analysis of NBS SRM 1643b samples and by recoveries for spiked water samples, was typically better than ±10% for the elements determined by using standard calibration and better than ±25% for the elements determined by using surrogate standards.

O'Day et al.[23] have applied X-ray absorption spectrometry to determine levels of zinc, cadmium, and lead in surface water. They used the technique to identify the local molecular coordination of metals in contaminated untreated stream sediments. Methods have been developed for the determination of a wide range of the elements including arsenic,[4] barium,[7] cadmium,[5] chromium, mercury,[10] platinum, antimony,[4] selenium, and uranyl ion.[7]

4.2 GROUNDWATER

4.2.1 Miscellaneous Metals

Harper[5] give details of a surface-water sampling device designed to eliminate accidental pick-up of trace metals originating from, for example, anti-fouling paints used in marine structures or on the hulls of ships from which the samples might be taken. The device comprised a buoy, a vertically positioned retaining frame below a float, and a length of Teflon tubing attached to the frame extending at last 1 m below the frame to avoid possible contamination. The buoy was deployed at a distance of not less than 2.5 m from the ship. Samples could be pumped at a rate up to 1 L per minute. The system was practicable under all but the worst weather conditions.

Beard[10] have discussed the design and construction of PVB 6.4-cm-diameter well casings.

Parker[24] have reviewed the literature and issued guidance for the use of PTFE, PVC, and stainless steel in groundwater samplers.

Puls et al.[25] discussed sampling procedures for the determination of inorganic groundwater contaminants and assessment of their transport by colloidal mobility.

Reynolds and Zemo[26] have discussed methods of in-field design of monitoring groundwater wells in heterogeneous fine-grained formations.

Van de Kamp and Keller[27] have discussed the detection, confirmation, and prevention of leaks in the case of groundwater-monitoring wells.

Reilly and Gibbs[28] have described hypothetical numerical experiments and chemical analyses and used these to illustrate the impact of physical and chemical heterogeneity in an aquifer on groundwater samples drawn from wells.

Pohlmann et al.[29] evaluated selected groundwater sampling and filtering methods to determine their effects on trace metal concentrations. They showed that filtration may be important in the transport of trace metals.

Johnson et al.[30] and Szucs and Jordan[31] have used empirical analysis of chemical data to determine sampling schedules for monitoring groundwater pollutants.

Kearl et al.[32] and Shanklin[33] have used micropurge, low-flow sampling techniques for collecting representative groundwater samples.

Weisbrod et al.[34] used dialysis cells with 10-μm pore membrane for passive sampling of groundwater colloids.

Puls[35] have reviewed sampling methods used in groundwater analysis.

Several articles have been written about groundwater sampling. Monitoring of groundwater contaminants has been reviewed.[36] Empirical analysis of chemical data has been used to determine sampling schedules for monitoring groundwater pollutants.[30,31] Dialysis cells with 10-μm pore membranes have been used for passive sampling of groundwater colloids.[34] Selected groundwater sampling and filtering methods have been evaluated to determine their effects on trace metal concentrations; filtration may remove colloids that may be important in the transport of trace metals.[37] Sampling for trace metals has also been the subject of a review with 59 references.[38] Micropurge, low-flow sampling has been found to collect representative groundwater samples.[32,33] Cone penetrometer testing combined with discrete-depth groundwater sampling were found to reduce the time and money required to characterize a large site with multiple aquifers.[39]

In a seven-year study, tritium has been used to monitor groundwater movement through a nuclear test site.[40] Sulfur lexa fluoride has been used as a tracer to detect the presence of a dense, nonaqueous liquid in an aquifer.[41] ^{234}U, ^{238}U, and ^{36}Cl have been used to determine the mixing of water from different aquifers.[42] Boron isotope ratios have been used as tracers to determine groundwater sources.[43–45]

Stetzenbach et al.[46] used inductively coupled plasma mass spectrometry to determine 54 metals at precipitate concentrations in groundwater.

Meyer et al.[47] Barcelona and Helfrich[48] and others[49] have reviewed the determination of metals in groundwater, wells, reservoirs, and springs.

Melloul and Goldenberg[50] have reviewed the monitoring of metallic contaminants in groundwater.

Gibbons[51] give an overview of statistical methods for groundwater monitoring.

Krepster and Ritsema[52] have presented a validation assessment of the determination of arsenic, cadmium, chromium, and iron in groundwater by high-resolution inductively coupled plasma mass spectrometry (HR-ICPMS). The groundwater samples were used for the measurements of this validation assessment had the following concentration ranges: As, 0.2–7.8 $\mu g L^{-1}$; Cd, 0.05–0.6 $\mu g L^{-1}$; Cr, 0.2–1 $\mu g L^{-1}$; Cu, 1.1–7.5 $\mu g L^{-1}$; Fe, 0.08–2.6 $mg L^{-1}$).

They originated from different locations in the Netherlands. Besides the methodological aspects and the obtained analytical results, relevant performance characteristics (limit of detection, recovery, repeatability, reproducibility, measuring range, trueness, lack of fit, expanded uncertainty of measurement, robustness, and selectivity) are defined, calculated, and discussed. The validated method was routinely applied in monitoring programs. This work serves as a guideline for a complete validation assessment in environmental matrices.

4.2.2 Arsenic

Yokoyama et al.[53,54] used ion-exclusion chromatography and continuous hydride atomic absorption spectrometry to study arsenic speciation (ASIII, ASV) in geothermal waters. Arsenic was determined in the range 0.01–10 $mg\,L^{-1}$.

Freeney and Konnaves[55] carried out on-site measurement of arsenic in groundwater using square-wave anodic stripping voltammetry with a microfabricated gold ultra-microelectrode.

4.2.3 Barium

Minola et al.[56] detected barium in amounts between 10 and 925 $\mu g\,L^{-1}$ in Italian well water using Zeeman graphite-furnace atomic-absorption spectrometry.

4.2.4 Chromium

Wang and Lu[57] used absorptive catalytic stripping voltammetry to determine chromium in groundwater samples in amounts down to 1 ptt.

Roddatz et al.[58] have pointed out that at Idaho National Laboratory, Cr(VI) concentrations in a groundwater plume once exceeded regulatory limits in some monitoring wells but have generally decreased over time. This study used chromium stable isotope measurements to determine if part of this decrease resulted from removal of Cr(VI) via reduction to insoluble Cr(III). Although water in the study area contains dissolved oxygen, the basalt host rock contains abundant Fe(II) and may contain reducing microenvironments or aerobic microbes that reduce Cr(VI). In some contaminated locations, $^{53}Cr/^{52}Cr$ ratios are close to those of the contamination source, indicating a lack of Cr(VI) reduction. In other locations, ratios are elevated. Part of this shift may be caused by mixing with natural background Cr(VI), which is present at low concentrations but in some locations has elevated $^{53}Cr/^{52}Cr$. Some contaminated wells have $^{53}Cr/^{52}Cr$ ratios greater than the maximum attainable by mixing between the inferred contaminant and the range of natural background observed in several uncontaminated wells, suggesting that Cr(VI) reduction has occurred. Definitive proof of reduction would require additional evidence. Depth profiles of $^{53}Cr/^{52}Cr$ suggest that reduction occurs immediately below the water table, where basalts are likely least weathered and most reactive, and is weak or nonexistent at greater depth.

4.2.5 Copper

Yoshimura et al.[59] have developed a method for the determination of down to 80 ppt of copper in groundwater based on complexation with 4.7-diphenyl-2,9, dimethyl-1,10 phenanthrolinedisulfonate, followed by concentration on an ion exchanger packed in a flow-through cell and detection by spectrophotometry.

4.2.6 Iron

Baedecker and Cozzarelli[60] have reviewed the determination and the fate of unstable iron(II) in groundwater.

Resinsch et al.[61] studied chemical transformation occurring during aging of zero-valent iron nanoparticles in the presence of common groundwater-dissolved constituents.

Nano-zerovalent iron (NZVI) that had been aged in simulated groundwater was evaluated for alterations in composition and speciation over 6 months. This was to understand the possible transformations that NZVI could undergo in natural waters. NZVI was exposed to 10 mN of various common groundwater anions (Cl^-, NO_3^-, SO_4^{2-}, HPO_4^{2-}, and HCO_3^-) or to dissolved oxygen (saturated, ~9 mg L^{-1}). Fresh and exposed NZVI samples, along with Fe-oxide model compounds, were then analyzed using synchrotron radiation X-ray absorption spectroscopy (XAS) to yield both relative oxidation state, using the X-ray absorption near-edge structure (XANES), and quantitative speciation information regarding the types and proportions of mineral species present, from analysis of the extended X-ray absorption fine structure (EXAFS). Over 1 month of aging the dissolved anions inhibited the oxidation of the NZVI to varying degrees. Aging for 6 months, however, resulted in average oxidation states that were similar to each other regardless of the anion used, except for nitrate. Nitrate passivated the NZVI surface such that even after 6 months of aging, the particles retained nearly the same mineral and Fe^0 content as fresh NZVI. Linear least-squares combination-fitting (LCF) of the EXAFS spectra for 1-month-aged samples indicated that the oxidized particles remain predominantly a binary-phase system containing Fe^0 and Fe_3O_4, while the 6-month-aged samples contained additional mineral phases such as vivianite ($Fe_3(PO_4)_2 \cdot 8H_2O$) and iron surface species, possibly schwertmannite ($Fe^{3+}{}_{16}O_{16}(OH,SO_4)_{12\text{-}13} \cdot 10\text{–}12H_2O$). The presence of these additional mineral species was confirmed using synchrotron-based X-ray diffraction (XRD). NZVI exposed to water saturated with dissolved oxygen showed a rapid (<24 h) loss of Fe^0 and evolved both magnetite and maghemite (δ-Fe_2O_3) within the oxide layer. These findings have implications concerning the eventual fate, transport, and toxicity of NZVI used for groundwater remediation.

4.2.7 Neptunium

Clark et al.[62] used atomic fluorescence spectroscopy to study the structure of the main forms of pentavalent neptunium carbonate complexes.

4.2.8 Lanthanides

Stroh[63,64] determined lanthanides in groundwater in amounts down to 0.05 ppt by inductively coupled plasma. Davis et al.[65] used resin extraction followed by scintillation counting. This was an easy screening technique for lanthanides in groundwater.

Systematically varying properties and reactivities have led to focused research on the environment forensic capabilities of rare earth elements. Increasing anthropogenic inputs to natural systems may permanently alter the natural signatures of rare earth elements. Increasing anthropogenic input to natural systems may permanently alter the natural signatures of rare earths, motivating characterization of natural rare earth. Nonyck et al.[66] compiled and analyzed reported dissolved rare-earth element concentration data over a wide range of natural water types (ground, ocean, river, and lake water) and groundwater chemistries (e.g., fresh, brine, and acidic) with the goal of quantifying the extent of natural rare-earth element variability, especially for groundwater systems. Quantitative challenges presented by censored data were addressed with nonparametric distributions and regressions. Reported measurements of rare earth elements in natural waters range over nearly 10 orders of magnitude, though the majority of measurements are within 2 to 4 orders of magnitude, and are highly correlated with one another. Few global correlations exist among dissolved abundance and bulk properties in groundwater, indicating the complex nature of source-sink terms and the need for care when comparing results between studies. This collection, homogenization, and analysis of a disparate literature facilitates interstudy comparison and provides insight into the wide range of variables that influence rare-earth element geochemistry.

4.2.9 Rhenium

Rhenium is one of the rarest elements in the earth's crust, possessing an abundance of approximately 1 $\mu g\,kg^{-1}$. The abundance of rhenium has been established at about 0.01 $\mu g\,L^{-1}$. It has been established that the perrhenate ion $Re^{Viii}O_4^-$, is the only significant rhenium species present in most aqueous environments.

Ketterer[67] has described a procedure for the determination of rhenium in groundwater utilizing inductively coupled plasma mass spectrometry.

The method is capable of determining rhenium in groundwater samples that contain up to 4000 $mg\,L^{-1}$ dissolved solids. A cation-exchange membrane cartridge is used online to exchange cationic species for equivalent quantities of hydrogen ion; rhenium, which is present as the perrhenate anion, remains on the upstream side of the membrane and is transported directly into the inductively coupled plasma. The arrangement successfully alleviates matrix-related sample introduction difficulties and permits direct determination of rhenium in water with a detection limit of 0.03 $\mu g\,L^{-1}$ using a Meinhard-type nebulizer. Removal efficiencies of up to 100% are achieved for sodium, magnesium, aluminum, potassium, and calcium ions, while perrhenate is transmitted with 100% efficiency. Results are presented for the determination of rhenium in groundwater samples from the vicinity of a metal sulfide tailings impoundment in the western United States.

4.2.10 Selenium

Zhang et al.[68] have studied the speciation of selenium in agricultural drainage waters and aqueous sediment extracts. The method was developed to determine organic selenium(-II) in sediment extracts and agricultural drainage water by using persulfate to oxidize organic selenium(-II) and using manganese oxide as an indicator for oxidation completion.

Results showed that organic selenium(-II) can be quantitatively oxidized to selenite without changing the selenite concentration in the sediment extract and agricultural drainage water and then quantified by hydride generation atomic absorption spectrometry. Recoveries of spiked organic selenium and selenite were also 105% in the sediment extracts and 96–103% in the agricultural drainage water.

Concentrations of soluble selenium in the sediment extracts were 0.0534–2.45 $\mu g\,g^{-1}$, of which organic selenium(–II) accounted for 4.5–59.1%. Selenate is the dominant form of selenium in agricultural drainage water, accounting for about 90% of the total selenium. In contrast, organic selenium(–II) was an important form of selenium in the wetlands. These results showed that wetland sediments are more active in reducing selenite compared to evaporated pond sediments.

4.2.11 Uranium

Kerr et al.[69] employed high-performance liquid chromatography for the determination of uranium in groundwater. The sample was passed through a small reversed-phase enrichment cartridge to separate the uranium from the bulk of the dissolved constituents. The uranium was then back-flushed from the cartridge onto a reversed-phase analytical column. The separated species were monitored spectrophotometrically after reaction with arsenazo(III). The detection limit was in the 1–2 $\mu g\,L^{-1}$ range, with a precision of approximately 4%.

Wu et al.[70] have shown that it is possible to determine as little as 50 ppt of uranium in groundwater, without sample pretreatment, using laser-induced fluorescence spectroscopy.

Cassidy and Elchuck[71,72] applied high-performance liquid chromatography to the determination of uranium(IV) in groundwater samples. They studied conventional cross-linked and bonded-phase ion exchangers, both cation and anion, with aqueous mobile phases containing tartrate, citrate, or α-hydroxylisobutyrate. The best chromatography was obtained on bonded-phase cation exchangers with an α-hydroxylisobutyrate eluate. The metal ions were detected either by visible spectrophotometry of the arsenazo(III);VI complex at 650 nm, after a postcolumn reaction with a complexing reagent; or with a polarographic detector. Detection after postcolumn reaction gave the best sensitivity; the detection limit (2 × baseline noise) was 6 ng. In-line trace enrichment was used to decrease detection limits, and linear calibration curves were observed in the range of 0.5–50 μg L^{-1} for groundwater.

The other metal ions that exhibited an appreciable reaction with arsenazo(III) at 650 nm under the separation and detection conditions used, were iron(III), zirconium(IV), thorium(IV), and the lanthanides. The lanthanides, iron(III), and zirconium(IV) were eluted at or near the solvent front before uranium(VI) and thorium(IV) was eluted after uranium(VI).

Dossi et al.[73] used sorptive cathodic stripping voltammetry to study the release and separation of uranium in low-ionic-strength groundwater from an abandoned uranium mine in Val Vedelioi, Italy.

4.2.12 Heavy Metals

Komy[74] have applied differential pulse-stripping voltammetry to the determination of cadmium, lead, copper, and zinc in groundwater.

Letterer and Miench[75] determined aluminum, arsenic, cadmium, chromium, copper, manganese, nickel, lead, and zinc simultaneously in groundwater by inductively coupled plasma mass spectrometry. These workers studied the effect of matrix interferences. Inductively coupled plasma atomic emission spectrometry has been used to monitor levels of copper, nickel, and lead in municipal landfill leachates.[76]

Kaplan et al.[77] used synchrotron X-ray fluorescence spectroscopy and energy-dispersive X-ray analysis to identify chromium, nickel, copper, and lead.

4.2.13 Actinides and Transuranic Elements

Nitsch et al.[78] studied the dependence of actinide solubility and speciation on carbonate concentration and ionic strength at pH 6, 7, and 8.5 in the case of the actinides americium, neptunium, and plutonium in groundwater.

4.2.14 Multimetal Analysis

4.2.14.1 Chromium, Arsenic, Copper, Lead, and Cadmium

Ray and Bickerton[79] demonstrated a screening approach for the detection of groundwater contaminants along urban streams within unconsolidated beds. It involves the rapid acquisition of groundwater samples along urban stream reaches at a spacing of about 10 m and from depths of about 25–75 cm below the streambed, with analyses for a suite of potential contaminants. This screening approach may serve two functions: (1) providing information for assessing and mitigating the toxicity and eutrophication risks to aquatic ecosystems posed by groundwater contaminants and (2) detecting and identifying groundwater contamination in urban settings more rapidly and inexpensively compared to land-based well installations. The screening approach was tested at three urban streams, each affected by a known chlorinated-solvent plume. All three known groundwater plumes were detected and roughly delineated. Multiple, previously unknown, areas or types of groundwater contamination were also identified at each stream. The newly identified contaminants and plumes included petroleum hydrocarbons, naphthalene, 1,4-dioxane, nitrate and phosphate, road salt, and various metals (including arsenic, cadmium, chromium, copper, and lead) at elevated concentrations compared to background values and relevant Canadian water quality guidelines. These findings suggest that this screening approach may be a useful tool for both ecologists performing ecological assessments and stream restorations and for hydrogeologists undertaking groundwater protection activities. Given the numerous contaminants detected, it may be appropriate to apply this technique proactively to better determine the pervasiveness of urban groundwater contaminants, especially along urban streams.

4.2.14.2 Cadmium, Zinc, Aluminum, Copper, Chromium, and Uranium

Roy and Bickerton[80] screened levels of common groundwater contaminants along eight urban stream reaches (100s–1000s of meters) at approximately 25–75 cm below the streambeds. Four sites had known or suspected chlorinated-solvent plumes; otherwise no groundwater contamination was known previously. At each site, between 5 and 22 contaminants were detected at levels above guideline concentrations for the preservation of aquatic life, while several others were detected at lower levels, but which may still indicate some risk. Contaminants of greatest concern include numerous metals (Cd, Zn, AI, Cu, Cr, U), arsenic, various organics (chlorinated and petroleum), nitrate and ammonium, and chloride (road salt likely), with multiple types occurring at each site and often at the same sampling location. Substantial portions of the stream reaches (from 40–88% of locations sampled) possessed one or more contaminants above guidelines. These findings suggest that this diffuse and variable-composition urban groundwater contamination is a toxicity concern for all sites and over a large portion of each study reach. Synergistic toxicity both for similar and disparate compounds may also be important. It was concluded that groundwater contaminants should be considered a genuine risk to urban stream aquatic ecosystems, specifically benthic organisms, and may contribute to urban stream syndrome.

4.3 WELL, RESERVOIR, SPRING, AND MINERAL WATERS

The analysis of these types of water has been reviewed by several workers.[81–83] Results of monitoring programs have been reported.[84,85] An overview of statistical methods for monitoring these types of water has been presented by Gibbons. [51]

Early-time transient electromagnetic sounding and dc-resistivity sounding have been used to monitor changes in a groundwater solution plume; however, much improvement is needed in the operation and interpretation of transient electromagnetic soundings.[86] Several articles discussed monitoring wells and sampling and are summarized in Table 4.1.

General information about the determination of metals in different types of water are reviewed in Table 4.2.

TABLE 4.1 Groundwater Monitoring Wells and Sampling Information

Sample Design	References
Guidelines for the use of PTFE, PVC, and stainless steel in samplers reviewed with 5 references.	87
Design and construction of PVC, 6.6-cm diameter, well casings.	88
Detection, confirmation, and prevention of leaks in the casings of monitoring wells.	89
Dual-walled screen made of wire-wrapped screens and filter packs was an effective well design for fine-sand aquifers.	90
Method for in-field design of monitoring wells in heterogeneous fine-grained formations.	91
Piezometric cone penetration testing and penetrometer groundwater sampling for volatile organic-contaminant plume location aids in placement of monitoring wells.	92
Combined applications of cone penetrometer, hydropunch sampling, and borehole geophysics were used for site assessment to determined placement of groundwater wells.	93
Sampling	
Monitoring with low-flow, dedicated pumping devices for purging and sampling of wells; dissolved oxygen and specific conductance were used to decide when to sample for volatile organic compounds.	94
Collection of mobile colloids described; slow pumping of groundwater samples yielded most representative in situ colloid populates.	95
Sampling procedures for the determination of aqueous inorganic contaminants and assessment of their transport by colloidal mobility evaluated.	96
Differences in water chemistry between the casing and screened interval volumes of four wells were studied; tracer experiments were used to study the differences in natural flushing between the casing and screened interval volumes.	97
Multiport sampler with seven screened intervals was used to study vertical variations in water chemistry.	98
Hypothetical numerical experiments and chemical analyses illustrates the impact of physical and chemical heterogeneity in an aquifer on samples drawn from wells.	99
Nitrate and atrazine measured in irrigation wells, in transmissive formations; samples may be taken after 15 min of pumping.	100

TABLE 4.2 Determination of Metals in Mineral, Spa, and Spring Waters

Type of Water	Elements	Technique	Detection Limit ($\mu g L^{-1}$)	Interferences	References
Individual Metals					
Mineral	Arsenic	Amperometric titration with standard potassium bromate	50	–	101
Mineral	Calcium	Ring colorimetry	–	Cerium, phosphate	102
Hot spring water	Lithium	Ion-exchange with hydrogen flame ionization detection	–	–	103
	Sodium				
	Potassium				
	Rubidium				
	Caesium				
Mineral	Sodium	Autoanalyzer system with sodium selective electrode	1000	Excess potassium i.e., greater than 5-fold excess	104
Mineral	Nickel	Ion-exchange chromatography on Dowex A-I then Dowex I-X10	0.05 μg	Removed	105
Hot spring water	Copper	Cation-exchange in ammoniacal pyrophosphate medium	–	Iron interference	106

Continued

TABLE 4.2 Determination of Metals in Mineral, Spa, and Spring Waters—cont'd

Type of Water	Elements	Technique	Detection Limit ($\mu g L^{-1}$)	Interferences	References
Mineral	Molybdenum	Graphite furnace atomic absorption spectrometry		Detection limit $0.19 \mu g L^{-1}$	107
Mineral	Molybdenum	Electrothermal atomic absorption spectrometry		Preconcentration on Amberlite IRA-400	108
Mixtures of Metals					
Hot spa water	32 elements	Mass spectrometry	$<2 mg L^{-1}$	–	109
Hot mineral water	Arsenic	Atomic absorption spectrometry	1	–	110
	Antimony				
	Germanium				
	Selenium				
	Cadmium				
Mineral	Silver	Emission spectrography	–	–	111
	Beryllium				
	Copper				
	Germanium				
	Manganese				
	Molybdenum				
	Nickel				

Medical mineral	Potassium	Atomic absorption spectrometry	–	–	112
	Lithium				
	Magnesium				
	Strontium				
	Chromium				
	Manganese				
	Nickel				
	Copper				
	Zinc				
Mineral water	Mercury	Atomic absorption spectrometry	–	–	113
	Lead				
	Cadmium				
	Chromium				
Thermal water	Arsenic	Hydride generation atomic absorption spectrometry	As 0.2 ng	–	114
	Antimony		Bi 0.05 ng		
	Selenium		Se 0.3 ng		
	Tellurium		Te 0.1 ng		
Mineral water	Boron	Ion exchange on Dovex 50WX-8 cation exchange resin mass spectrometry	–		115

Continued

TABLE 4.2 Determination of Metals in Mineral, Spa, and Spring Waters—cont'd

Type of Water	Elements	Technique	Detection Limit ($\mu g\,L^{-1}$)	Interferences	References
Mineral water	Silver	Inductively coupled atomic emission spectrometry	0.3–80 $ng\,L^{-1}$		116
	Cadmium				
	Vanadium				
	Chromium				
	Arsenic				
	Selenium				
Mineral water	Boron	Spectrophotometer	–		117
Well water	Cr		–		118
Well water	Radon		–		119
	Uranium				66

REFERENCES

1. Brondi M, Gragnani R, Presperi M. Appl Zeeman graphite furnace atomic spectroscopy. *Chem Lab Toxicol* 1992:143–54.
2. Haraldson C, Pollak M, Oehman P. *J Anal At Spectrosc* 1992;**7**:1183.
3. Huang C, Jiang S. *Anal Chim Acta* 1994;**289**:205.
4. Brondy M, Gragnani R, Prosperi M. Applied Zeeman graphite furnace, atomic absorption sepctrometry. *Chem Lab Toxicol* 1992:43–6.
5. Harper DJ. *Mar Chem* 1987;**21**:183.
6. Yamakagi K, Yoshii M, Yamada K. *Anal Sci* 1993;**9**:423.
7. Polezal P, Kahle V, Krejei M. *Fresenius J Anal Chem* 1993;**345**:762.
8. Ohta K, Nakasima N, Inui S, Winefordner JD, Mizuno T. *Talanta* 1992;**39**:1643.
9. Fung WG, Sham WC. *Analyst* 1994;**119**:1029.
10. Beard LD. (*Current practice groundwater vadose zone investigation*). ASTM Special Technical, Publications STP1; 1992. p. 256–69.
11. Chow CWK, Davey DE, Mulcahy DE. *Anal Lett Lond* 1994;**27**:113.
12. Janjic J, Kiurski J. *Water Res* 1994;**28**:233.
13. Jian W, McLeod CW. *Talanta* 1992;**39**:1537.
14. Calodner DC, Boyle EA, Edmond JM. *Anal Chem* 1993;**65**:1419.
15. Martinez-Aguirre A, Garcia-Leon M, Ivanovich M. *Nucl Instrum Methods Phys Res Sect A* 1994;**339**:287.
16. Farias PAM, Takase I. *Electroanalysis* 1992;**4**:823.
17. Kawakubo S, Lian B, Iwatsuki M, Fukaswawa T. *Analyst* 1994;**119**:1391.
18. Small A, Stevens TS, Bauman WC. *Anal Chem* 1975;**47**:1801.
19. Dankert T, Sirotek Z. *Chem Geol* 1993;**107**:133.
20. Pettine M, La Noce T, Liberator A. Applied Zeeman graphite furnace atomic absorption spectrometry. *Chem Lab Toxicol* 1992;**165**:77.
21. Inhat M, Gable DS, Gilchirst GFR. *Int J Environ Anal Chem* 1993;**53**:63.
22. Henshaw JM, Heitherman EM, Hinners TA. *Anal Chem* 1989;**61**:335.
23. O'Day PA, Carroll SA, Waychunas GA. *Environ Sci Technol* 1998;**32**:943.
24. Parker LV. (*Current practice groundwater vadose zone investigation*). ASTM Special Technical Publication STP 1118; 1992. p. 217–29.
25. Puls RW, Clark DA, Bledose B, Powell RM, Paul CJ. *Hazards Waste Hazard Matter* 1992;**9**:149.
26. Reynolds SD, Zemo DA. *Current practice groundwater vadose investigation*. ASTM Special Technical Publications STP 1118; 1992. p. 230–40.
27. Van der Kamp G, Keller G. *Groundwater Monit Rem* 1993;**13**:136.
28. Reilly TE, Gibbs J. *Groundwater* 1993;**31**:201.
29. Pohlmann KE, Icopini GA, McArthur RD, Rosal CG. Report EPA/600/R-94/119, EMSL-LV-94–1180.Environmental Protection Agency; 1994 [Order No PB 94-201993 Available NTIS].
30. Johnson VM, Tuckfield RC, Ridley NN, Anderson RA. *Environ Sci Technol* 1995;**30**:355.
31. Szucs A, Jordan G. *Water Sci Technol* 1994;**30**:37.
32. Kearl PM, Korte NE, Stites M, Baker J. *Groundwater Monit Rem* 1994;**14**:183.
33. Shanklin DE, Sidle WC, Ferguson ME. *Groundwater Monit Rem* 1995;**15**:168.
34. Weisbrod N, Ronen D, Nativ R. *Environ Sci Technol* 1996;**30**:3094.
35. Puls RW. *Environ Sampling Trace Anal* 1994:287.
36. Melloul AJ, Goldenburg LC. *Environ Sci Pollut Control Ser* 1994;**11**:529.

37. Pohlmann KF, Icopini GA, McArthur RD, Rosal CG, Report EPA/600/R-94/119, EMSL-LV-94–1180; 1994 [Order No. PB94-201993, Avail. NTIS].
38. Puls RW. *Environment sampling trace anal* 1994. 287.
39. Zemo DA, Pierce YG, Gallinatti JD. *Groundwater Monit Rem* 1994;**14**:176.
40. Li-Xing Z, Ming-shun Z, Gou-Rong T. *Sci Total Environ* 1995;**173/174**:47.
41. Nelson NT, Brusseau ML. *Environ Sci Technol* 1996;**30**:2859.
42. Herczeg AL, Love AJ, Allan G, Fifield LK. Isotopes water research management. In: *Proceedings of the symposium*, vol. 2. Vienna, Austria: International Atomic Energy Agency; 1996. p. 123–33.
43. Bassett RL, Buszka PM, Davidson GR, Chong-Diaz D. *Environ Sci Technol* 1995;**29**:2915.
44. Eisenhut S, Heumann KG, Vengosh A. *Fresenius' J Anal Chem* 1996;**354**:903.
45. Porteous NC, Walsh JN, Jarvis KE. *Analyst* 1995;**120**:1397.
46. Stetzenbach KJ, Amono A, Kreaner DK, Hodge VF. *Groundwater* 1994;**32**:976.
47. Meyer AS, Radibeau AM, Mitchell RJ. *Water Environ Res* 1993;**65**:486.
48. Barelonl MJ, Helfrich JA. (*Curr. prac. ground water vadose zone invest.*). ASTM Specifiction Technical Publications, GPT 1118; 1992. 3–23.
49. Lesage S, Jackson RE, editors. *Ground water contamination and analysis at hazardous waste sites. Environmental science pollution control series*, vol. 4. New York: M. Dekker; 1992.
50. Melloul AJ, Goldenberg IC. *Environ Sci Pollut Control, Ser II* 1994;**11**:529.
51. Gibbons RD. Ground water contamination and analysis at hazardous waste sites. *Environ Sci Pollut Controil Ser* 1992;**4**:199–243.
52. Krepstek P, Ritsema R. *Int J Environ Anal Chem* 2009;**89**:331.
53. Yakoyama T, Takahashi Y, Tarutani T. *Chem Geol Part 1* 1992;**48**:27.
54. Yakoyama Y, Takahashi Y, Tarutani T. *Chem Geol* 1993;**103**:103.
55. Freeney R, Kaunoves SP. *Anal Chem* 2000;**72**:2222.
56. Minola C, Canedeli S, Vaescovi L, Rizzio L, Pietra R, Manzo L. Applied Zeeman, graphite. Furnace absorption atomic absorption. *Spectrom Chem Lab Toxicol* 1992;**00**:179.
57. Wong J, Lu J. *Analyst* 1992;**117**:1913.
58. Raddatz AL, Johnson TM, McLing TL. *Environ Sci Technol* 2011;**45**:502.
59. Yoshimura E, Matsuoka S, Inakhura Y, Hose U. *Anal Chim Acta* 1992;**268**:225.
60. Baedecker MJ, Cozzareli LM. *Environmental science pollution control services*, vol. 4. 1992. [Groundwater contamination and analysis at waste sites]. p. 425–61.
61. Reinsch BC, Forsberg B, Renn PL, Kim CS, Lowry GV. *Environ Sci Technol* 2010;**44**:3455.
62. Clark DL, Conradson SD, Ekberg SA. *J Am Chem Soc* 1996;**118**:2089.
63. Stroh A. *At Spectrosc* 1992;**13**:59.
64. Stroh A. *At Spectrosc* 1992;**50**:191.
65. Davis TM, Nelson DM, Thornton EG. *Radioact Radiochem* 1993;**4**:16.
66. Noack CW, Dzombak DA, Karamelidiz AK. *Environ Sci Technol* 2014;**48**:4317.
67. Ketterer ME. *Anal Chem* 1990;**62**:2522.
68. Zhang Y, Moore JN, William T, Frankenberger JR. *Environ Sci Technol* 1999;**33**:1652.
69. Kerr A, Kupterschmidt W, Atlas M. *Anal Chem* 1988;**60**:2729.
70. Wu J, Yuan Z, Li J, Zhang C, Ren L. Report 1. Stig-t 94064 under No. PB 94–189999; 1994 [available NTIS].
71. Cassidy RM, Elchuck S. *Int J Environ Anal Chem* 1981;**10**:1876.
72. Cassidy RM, Elchuck S. *J Chromatogr Sci* 1980;**18**:217.
73. Dossi C, Cairugati G, Credaro A, Gambilliara R, Martin S, Monticelli D, et al. *Int J Environ Anal Chem* 2007;**87**:361.
74. Komy ZR. *Microchim Acta* 1993;**111**:239.

75. Letterer M, Miench U. *Fresenius Zietschrift Für Anal Chem* 1994;**350**:204.
76. Papini MD, Mozone M, Sonofonte O, Caroli S. *Microchem J* 1994;**50**:191.
77. Kaplan DI, Hunter DB, Bertsil PM, Bajt S, Andrioano DC. *Environ Sci Technol* 1994;**28**:1186.
78. Nitsch H, Muller A, Standyet EM, DienHammer RC. *Radio Chim Acta* 1992;**58**:27.
79. Roy JW, Bickerton G. *Environ Sci Technol* 2010;**44**:6088.
80. Roy JW, Bickerton I. *Environ Sci Technol* 2012;**48**:729.
81. Mayer AS, Rabideau AJ, Mitchell RJ, Imhoff PT, Lowry MI, Miller CT. *Water Environ Res* 1993;**65**:486.
82. Barcelona MJ, Helfrich JA. (*Curr. pract. ground water vadose zone invest.*). ASTM Spec. Tech Publ STP 1118; 1992. p. 3–23.
83. Leage S, Jackson RE, editors. *Groundwater contamination and analysis at hazardous waste sites. Environ Sci Pollut Control Ser*, vol. 4. New York: Dekker; 1992.
84. Westinghouse Savannah River Co. Exploration Resources Inc. Report ESH-EMS-900134, Order No. DE92011908 (Avail. NTIS); 1991.
85. Hall S, Juracich SP. Report PNL-7836, Order No. DE92004452, (Avail. NTIS); 1991.
86. Barber C, Davis GB, Buseli G, Height M. *Int J Environ Pollut* 1991;**1**(1–2):97–112.
87. Parker L. (*Curr. pract. ground water vadose zone invest.*). ASTM Spe. Tech. Publ. STP1118; 1992. p. 217–29.
88. Beard LD. (*Current pract. ground water vadose zone invest.*). ASTM Spec. Tech. Publ. STP 1118; 1992. p. 256–69.
89. Van der Kamp G, Keller CK. *Groundwater Monit Rem* 1993;**13**(4):136–41.
90. Gillespie GA. (*Curr. pract. ground water vadose zone invest.*). ASTM Spec. Tech. Publ. STP1118; 1992. p. 241–55.
91. Reynolds SD, Zemo DA. (*Curr. pract. ground water vadose zone invest.*). ASTM Spec. Tech. Publ. STP1118; 1992. p. 230–40.
92. Strutynsky AI, Sainey TJ. (*Curr. pract. ground water vadose zone invest.*). ASTM Spec. Tech. Publ. STP1118; 1992. p.199–214.
93. Manchon B. *J Soil Contam* 1992;**1**:321.
94. Barcelona MJ, Wehrmann HA, Varljen MD. *Groundwater* 1994;**32**:12.
95. Bachus DA, Ryna JN, Groher DM, MacFarlane JK, Gschwend PM. *Groundwater* 1993;**31**:466.
96. Puls RW, Clark DA, Bledsoe B, Powell RM, Paul C, Paul C. *J Hazard Waste Hazard Mater* 1992;**9**:149.
97. Powell RM, Puls RW. *J Contam Hydrol* 1993;**12**:51.
98. Gibs J, Brown GA, Turner KS, MacLeod CL, Jelinski JC, Koehnlein SA. *Groundwater* 1993;**31**:201.
99. Reilly TE, Gibs J. *Groundwater* 1993;**31**:805.
100. Zlotnik VA, Spalding RF, Exner ME, Burbach ME. *Water Sci Technol* 1993;**28**(3–5):409–13. [Diffuse pollution].
101. Kobrova M. *Chim Listy* 1973;**67**:762.
102. Johri KM, Hauda AC, Mehrer HC. *Anal Chim Acta* 1971;**57**:217.
103. Araki S, Suzuki S, Hobo T, Yoshida T, Yoshizaki K, Yamada M. *Jpn Anal* 1968;**17**:847.
104. Van der Winkel O, Mertens P, De Bearst G, Massart DL. *Anal Lett* 1972;**5**:567.
105. Nevoral V, Okac A. *Cslka Farm* 1968;**17**:478.
106. Toshio N. *Bull Chem Soc Jpn* 1969;**42**:3017.
107. Bermejo-Barrera P, Vazquez-Gonzalez JF, Bermejo-Martinez R. *Microchim Acta* 1986;**3/4**:259.
108. Vazquez-Gonzalez JF, Bermejo-Barrea P, Bermejo0Martinez F. *At Spectrosc* 1987;**8**:159.
109. Klose M. *Anal Chem* 1971;**254**:7.

110. Criaud A, Foullac C. *Anal Chim Acta* 1985;**167**:257.
111. Pepin D, Gardes A, Petit J. *Analysis* 1973;**2**:337.
112. Pulido C, Mareira de Almeida C. *Revta Port Quim* 1969;**11**:84.
113. Sontag G, Kerschbaumer M, Mainz G. *Wasser Abwasser* 1977;**10**:166.
114. Yamaoto M, Yasuda M, Yamamoto Y. *Anal Chem* 1985;**57**:1382.
115. Greene B, Mariel J, Tramsek G. *Microchim Acta* 1970;**1**:24.
116. Randato R, Van Lececki F, Owens L, Dani R. *J Anal At Spectrosc* 2000;**15**:341.
117. Balogh JJ, Andruch V, Kadar M, Billes F, Posta J, Szobova E. *Int J Environ Anal Chem* 2009;**89**:449.
118. Raddatz AL, Johnson TM, McLung TL. *Environ Sci Technol* 2011;**45**:502.
119. Yang D, Smitherman P, Hess CT, Cuthbertson CW, Marvinney RG, Smith AE, et al. *Environ Sci Technol* 2014;**48**:4298.

Chapter 5

Metals in Aqueous Precipitation

Chapter Outline

5.1 RAINWATER

The determination of trace elements in rainwater is becoming increasingly important. Rainfall has proved to be an exceptional form of deposition for trace-element input from the atmosphere to the terrestrial and aquatic environments. In addition to the input of the acid through rainwater, many heavy metals contribute to the hazards and destruction of forests, lakes, and coastal waters.

Until recently, studies of trace elements in rainwater have been neglected, largely due to problems of sample contamination and accurate analysis at the low trace-metal levels found in rainwater, particularly in marine areas. Only in the past few years have systematic investigations on the deposition of trace elements by rain and snow been started.

The element concentrations determined are normally present in the $\mu g\,L^{-1}$ to $ng\,L^{-1}$ range, and the analytical procedures applied prior to 1985 were usually based on inverse voltammetry and atomic absorption spectrometry as analytical principles that allow either single-element determination or simultaneous determination of three or four elements.

The growing need for information requires the analytical handling of large numbers of samples from systematic long-term investigations. Because of a favorable cost-benefit ratio, there is a need for further efficient trace analytical procedures that also allow multi-element determination.

Methods for the determination of metals in rainwater are reviewed in Table 5.1.

E. Journet et al.[45] have reported a quasi-online method of measurement of the oxidation states of iron coupled with a GFAAS analysis for the trace conditions found in atmospheric waters. The technique is based on the formation of a specific complex [Fe(Ferrozine)3]4 between Fe(III) and Ferrozine. Sep-Pak tC 18 solid-phase extraction cartridges are used to separate the [Fe(Ferrozene)3]4 and Fe(III), so as to limit the risk of redox evolution of the sample. The adaption

TABLE 5.1 Determination of Metals in Rainwater

Element	Technique	Detection Limit	Comments	References
Ag	Atomic absorption spectrometry	–	In rain	16
Ag	X-ray fluorescence spectroscopy	–	In rain	31
Ag	Atomic absorption spectrometry	$0.001\ \mu g\ L^{-1}$	In rain	27
Ag	Anodic stripping voltammetry	–	In rain	28
Ag	Stable isotope dilution methods	$1\ \mu g\ L^{-1}$	In rain	29
Al	In aluminum	–	Effect of filtration and centrifugation on determination of aluminum	1
Al	Spectrophotometry	$10\ \mu g\ L^{-1}$	–	2
Al	Chelating agent Chelex 100 for determination of aluminum loss	–	Four aluminum species identified	3
Al	Neutron activation analysis	$5\ \mu g\ L^{-1}$	In rain	20
As	X-ray fluorescence spectroscopy	–	In rain	15,31
Ba	X-ray fluorescence spectroscopy	–	In rain	15,31
Ca	Emission spectrometry at 422.7 nm	Low $ng\ L^{-1}$	In rain	19
Ca	X-ray fluorescence spectroscopy	–	In rain	15,31
Ca	Atomic absorption spectrometry	$10\ \mu g\ L^{-1}$	In rain	2
Ca	Atomic absorption spectrometry	–	In rain	20

Cd	X-ray fluorescence spectroscopy	–	In rain	15,31
Cd	Atomic adsorption spectrometry	–	In rain	16
Cd	Atomic adsorption spectrometry	$1\,\mu g\,L^{-1}$	In rain	17
Co	X-ray fluorescence spectroscopy	–	In rain	15,31
Cr	X-ray fluorescence spectroscopy	–	In rain	15
Cu	X-ray fluorescence spectroscopy	–	In rain	15,31
Fe	X-ray fluorescence spectroscopy	–	In rain	15,31
Fe	Atomic absorption spectrometry	$10\,\mu g\,L^{-1}$	In rain	2
Fe	Atomic absorption	–	In rain	20
Ga	X-ray fluorescence	–	In rain	31
Hg	Bioluminescence	–	In rain	24
Hg	Samples technique	–	In rain, snow	22
In	Atomic absorption spectroscopy	–	In rain	6
In	Neutron activation analysis	–	Preconcentration in ferricoxide – X-ray counting	21,22
K	Atomic absorption spectrometry	$5\,\mu g\,L^{-1}$	In rain	2,20
K	X-ray fluorescence spectroscopy	–	In rain	31
K	Ion chromatograph	–	In rain	26
Li	Atomic absorption spectrometry	$10\,\mu g\,L^{-1}$	In rain	17

Continued

TABLE 5.1 Determination of Metals in Rainwater—cont'd

Element	Technique	Detection Limit	Comments	References
Mg	Atomic absorption spectrometry	$10\,\mu g\,L^{-1}$	In rain	2,20
Mn	X-ray fluorescence spectroscopy	–	In rain	15,31
Mn	Atomic absorption spectroscopy	$10\,\mu g\,L^{-1}$	In rain	2
Mn	Neutron activation analysis	$0.5\,\mu g\,L^{-1}$	In rain	20
Mo	X-ray fluorescence spectroscopy	–	In rain	15
Multication analysis	Proton induced X-ray	–	In rain	32
Multication analysis	Particle induced X-ray emission spectrometry	–	In rain	33
Na	Atomic absorption spectrometry	$10\,\mu g\,L^{-1}$	In rain	2
Na	Atomic absorption spectrometry	–	In rain	20
Na	Neutron activation analysis	$0.5\,\mu g\,L^{-1}$	In rain	20
Na	Ion chromatography	–	In rain	26
NH_4	Membrane diffusion technique continuous spectrofluorimetry	–		4
NH_4	Laser spectrophotometry detection combined with indophenol spectrophotometry	–		5,14

NH_4	Optical sensor-based incorporation of an ammonium ion selective ionophore and hydrogen ion selective chromophore into plasticized PVC	–	In rain	6
NH_4	Ion chromatography	–	–	26
NH_4	Gas chromatography		Metal based on measurement of nitrogen liberated by quantitative oxidation of ammonia in alkaline solution	7
NH_4	Combined chemiluminescence combined with flow injection system	6.1×10^{-4} m of AM^{-3}	Method applied to rainwater and fog	8
NH_4	Ion chromatography	–	–	9
NH_4	Ultraviolet spectroscopy at 197.2 nm	$100\,\mu g\,L^{-1}$	Rainwater rendered alkaline with sodium hydroxide and released ammonia purged out with nitrogen, applied to rainwater	10
NH_4	Flow injection system with a gas diffusion membrane	$0.03\,\mu g\,L^{-1}$		11
NH_4	Ion selective electrode	–	In rain	12
NH_4	Spectrophotometric	–	In rain	13,18
Ni	X-ray fluorescence spectroscopy	–	In rain	15,31
Ni	Absorptive voltammetry with mercury film electrode	$20\,mg\,L^{-1}$	In rain	25

Continued

TABLE 5.1 Determination of Metals in Rainwater—cont'd

Element	Technique	Detection Limit	Comments	References
Pb	X-ray fluorescence spectroscopy	–	In rain	15,31
Pb	Atomic absorption spectroscopy	$5\,\mu g\,L^{-1}$	In rain	17,16
Pb	Spectrophotometry		Based on catalytic effect of divalent lead on the formation of a complex between Mn^{2+} and porphyrin 5, 10, 15, 20 tetra bis (4-sulphonato-phenyl) porphine	23
Pb, Cu	–	–	In rain	46
Pb, Od, Cr, Zn	–	–	In rain	47
Rb	X-ray fluorescence spectroscopy	–	In rain	31
Rh	X-ray fluorescence	–	In rain	15
Sb	X-ray fluorescence spectroscopy	–	In rain	31
Se	X-ray fluorescence spectroscopy	–	In rain	15,31
Se	X-ray fluorescence spectroscopy	–	In rain	31
Sr	X-ray fluorescence spectroscopy	–	In rain	15,31
Sr	Atomic absorption spectrometry	$10\,\mu g\,L^{-1}$	In rain	2
Ti	Atomic absorption spectrometry	$10\,\mu g\,L^{-1}$	In rain	2

Ti	X-ray fluorescence spectroscopy	–	In rain	31
Tl	Anodic stripping voltammetry	–	Also determines lead; 2–40 μg L^{-1} lead found in rain	30
Tl	X-ray fluorescence spectrometry	–	In rain	15
V	X-ray fluorescence	–	In rain	15
V	Neutron activation analysis	0.5 μg L^{-1}	In rain	20
V	X-ray fluorescence	–	In rain	31
Y	X-ray fluorescence spectroscopy	–	In rain	31
Zn	X-ray fluorescence spectroscopy	–	In rain	15,31
Zn	Atomic absorption spectrometry	10 μg L^{-1}	In rain	2
Zn	Atomic absorption spectrometry	–	In rain	20
Zr	X-ray fluorescence spectroscopy	–	In rain	31

to dilute aqueous media, via acidification of pH = 2 rainwater sample, and atmospheric interferences are discussed, and Fe(II) recovery in rainwater is determined. This method, coupled with a quasi-online sampling protocol, has been tested on rainfalls in Guadeloupe Island in the Caribbean (16°W) during a field campaign in May 2005. The results show that the method can satisfactorily be applied to the determination of Fe(II) and Fe(III) in atmospheric waters under in situ conditions.

5.2 SNOW AND ICE

For 20 years there has been a growing interest in the investigation of the occurrence of lead (and of several other heavy metals such as cadmium, copper, zinc, and mercury) in the well-preserved snow and ice layers deposited in the central areas of the Antarctic and Greenland ice sheets. This is indeed a unique way to reconstruct the past natural tropospheric flux of these highly toxic heavy metals on a global scale and to determine to what extent these fluxes are now influenced by human activities.

Such investigation has unfortunately proved to be very difficult because of the extremely low concentrations to be measured. As an illustration, lead concentrations in Holcome Antarctic ice have been shown to be as low as about 0.4 pg of lead g^{-1}. First of all, it is mandatory to decontaminate the snow or ice samples before final analysis; most available samples are more or less contaminated on the outside, regardless of the precautions taken to collect them cleanly in the field. Ultrasensitive analysis techniques must then be used. Due to the extremely low concentrations involved, ultraclean procedures are needed throughout the entire analytical process, from sample contamination to final analysis.

Based on these considerations, Bolshov et al.[39] have described a procedure for the measurement of lead in ancient Antarctic ice down to the sun-pg g^{-1} level by laser-excited atomic fluorescence spectrometry with electrothermal atomization. Detailed calibration of the spectrometer was successfully achieved down to the sub-pg g^{-1} level by using ultra-low-concentration lead standards. The ice-core samples, which had previously been mechanically decontaminated, were directly analyzed for lead by using very small volumes (20 μL only), without any preconcentration step or chemical treatment. The results are in very good agreement with those previously obtained for the same ice samples by isotope dilution mass spectrometry.

Further information on the analysis of metals in snow and ice is presented in Table 5.2.

TABLE 5.2 Determination of Metals in Snow and Ice

Element	Technique	Detection	Comments	References
Ag	Graphite furnace atomic absorption spectrometry	$0.05\ ng\ L^{-1}$	–	37
Ag	Anodic stripping voltammetry	10 pg	–	42
Ag	Atomic absorption spectrometry	$0.5\ \mu g\ kg^{-1}$	–	37
Ag	Atomic absorption	$12\ \mu g\ kg^{-1}$	–	43
Ag	Neutron activation analysis	1 ng	–	44
Ag	Neutron activation analysis	$0.5\ ng\ kg^{-1}$	–	40
Al	Aluminum atomic absorption spectrometry	–	–	34
As	Radio activation analysis	$0.4\ ng\ L^{-1}$	–	35
Ca	Atomic absorption spectrometry	–	–	34
Cd	Anodic stripping voltammetry	$5\ mg\ kg^{-1}$	–	36
Cd	Graphite furnace atomic absorption spectrometry	$0.05\ ng\ L^{-1}$	–	37
Cd	Radio activation analysis	$0.03\ ng\ L^{-1}$	–	35
Cu	Atomic absorption spectrometry	–	–	34
Cu	Flameless atomic absorption	–	–	34
Cu	Anodic stripping voltammetry	$20\ ng\ kg^{-1}$	–	36
Cu	Radio activation analysis	$0.1\ ng\ L^{-1}$	–	35

Continued

TABLE 5.2 Determination of Metals in Snow and Ice—cont'd

Element	Technique	Detection	Comments	References
Fe	Atomic absorption spectrometry	–	–	34
Hg	Radio activation analysis	0.3 ng L^{-1}	–	35
K	Isotope dilution spectrometry	1 mg kg^{-1}	–	40
Mn	Anodic stripping voltammetry	–	–	34
Mn	Radio activation analysis	0.01 ng L^{-1}	–	35
Ni	Atomic absorption spectrometry	–	–	34
Pb	Atomic absorption spectrometry	–	–	34,38
Pb	Anodic stripping voltammetry	–	–	36
Pb	Laser atomic fluorescence spectrometry	sub pg g^{-1}	–	39
Sb	Radio activation analysis	0.3 ng L^{-1}	–	35
Se	Radio activation analysis	5 ng kg^{-1}	–	41
Zn	Atomic absorption spectrometry	–	–	34
Zn	Anodic scanning voltammetry	–	–	36

REFERENCES

1. Royset O, Stanices AO, Ognal G, Sjotveit G. *Int J Environ Anal Chem* 1987;**29**:141.
2. Wagner GH, Steele KR. *Int Lab* September 1985;**92**.
3. Miller JR, Andelman JB. *Water Res* 1987;**21**:999.
4. T. Anoki, S. Uemura.
5. Munemori A. *Anal Chem* 1983;**55**:1620.
6. Strauss E, Favier JP, Bicanic D, Asselt KV, Lubbers M. *Analyst* 1991;**116**:77.
7. Jenkins RW, Cheek CH, Linnenbom VJ. *Anal Chem* 1966;**38**:1257.
8. Hu X, Takenaka N, Takasuna S, Kitano M, Bandow H, Maeda Y, et al. *Anal Chem* 1993;**65**:3489.
9. Small H, Stevens TS, Bauman WC. *Anal Chem* 1975;**47**:1301.
10. Vijan PN, Wood GR. *Anal Chem* 1981;**53**:1447.
11. Nakata R, Kamamura I, Sakashita H, Nitta A. *Anal Chim Acta* 1988;**208**:81.
12. Beckett MJ, Willson AL. *Water Res* 1974;**8**:333.
13. Harwood JE, Kuhn AL. *Water Res* 1970;**4**:805.
14. Tellow JA, Wilson AL. *Analyst (London)* 1964;**89**:453.
15. Stoggel RP, Pronge A. *Anal Chem* 1985;**57**:2880.
16. Rothenetti A. *Anal Chem* 1974;**46**:739.
17. Holroyd PM, Snodin DJ. *J Absorpt Public Analysts* 1972;**10**:110.
18. *Standard methods for the examination of water and wastewater*. Washington, DC: American Public Health Association; 1985.
19. Searle PL, Kenedy G. *Analyst (London)* 1972;**97**:457.
20. Shanina J, Mols JJ, Baeerel JH, Van der Stoot HA, Van Rapp Horst JG, Aswan W. *Int J Environ Anal Chem* 1979;**7**:161.
21. Bhatki KS, Dingle AN. *Radiochem Radioanalytical Lett* 1970;**3**:71.
22. Uchino E, Kogusa S, Konishi S, Nichimura M. *Environ Sci Technol* 1987;**21**:920.
23. Tabate M. *Analyst* 1987;**112**:141.
24. Selilanovy O, Burlage R, Barkay Hophead T. *Environ Microbiol* 1993;**59**:3083.
25. Braun H, Metzger M. *Zeitschrifr Für Anal Chem* 1984;**318**:321.
26. Xiang D, Chong Y. *Fenxi Ceshi Tongbao* 1987;**6**:15.
27. Ha J, Shi Guongpuxue S. *Fresenius Z Für Anal Chem* 1986;**6**:57.
28. Eisnei N, Mark HB. *J Electro Anal Chem* 1970;**24**:345.
29. Bickford ME, Silka IR, Shuster RD, Angino EE, Ragsdale CR. *Anal Chem* 1978;**50**:459.
30. Daneschwar RG, Zarapkar LR. *Analyst (London)* 1980;**105**:386.
31. Prange A, Kinoth J, Stobel RB, Bodekker H, Kramer K. *Anal Chim Acta* 1987;**195**:275.
32. Hansson HC, Erholm AKB, Ross B. *Environ Sci Technol* 1988;**22**:527.
33. Tanaka S, Darzi M, Winchester EV. *Environ Sci Technol* 1981;**15**:354.
34. Londslerger S, Jarvis RE, Aufreiter S, Van Bron JC. *Chemisphere* 1982;**11**:237.
35. Weiss HV, Bertin KK. *Anal Chim Acta* 1973;**65**:253.
36. Londy MF. *Anal Chim Acta* 1980;**121**:39.
37. Woodroft R, Culver BR, Shrader D, Super AB. *Anal Chem* 1973;**45**:230.
38. Wolf EW, Landy HF, Peel DA. *Anal Chem* 1981;**53**:1566.
39. Bolshov MA, Boutron CF, Zybin AV. 1989;**61**:1758.
40. Murozumi M, Nakamura G. *Jpn Anal* 1973;**22**:145.
41. Wass HV. *Anal Chim Acta* 1971;**56**:136.
42. Elsner N, Mark HB. *J Electro Anal Chem* 1970;**24**:345.
43. Warburton JA. *J Appl Meteorol* 1969;**8**:464.

44. Warburton JA, Young LG. *Anal Chem* 1972;**44**:2043.
45. Journet E, Desboeufs KV, Safikitis A, Verrault G, Colin JL. *Int J Environ Anal Chem* 2007;**87**:647.
46. Jo KW, Park JK. *Environ Sci Technol* 2010;**44**:9324.
47. Marbak P, Ayoka GA, Goonetillate A, Egodawatta P, Kokot S. *Environ Sci Technol* 2010;**44**:8904.

Chapter 6

Analysis of Metals in Sediments

Sampling Procedures

Chapter Outline

6.1 INTRODUCTION

Sampling procedures are extremely important in the analysis of sediments. It is essential to ensure that the composition of the portion of the sample being analyzed is representative of the material being analyzed. This fact is even more evident because the size of the portion of sample being analyzed in many modern methods of analysis is extremely small. It is therefore essential to ensure before the analysis is begun that correct statistically validated sampling procedures are used to ensure, as far as possible, that the portion of the sample being analyzed is representative of the bulk of material from which the sample was taken.

The collection and handling of samples prior to analysis has been discussed by various workers and organizations, including Smith and James,[1] Kratochvil et al.[2,3] Gy,[4] Woodget and Cooper,[5] Harrison,[6] Walton and Hoffman,[7] Laitinen et al.[8,9] Ingamells and Pitard,[10] Kratochvil and Taylor,[11] Kratochvil,[12] Wallace and Kratochvil,[13] the Ministry of Agriculture, Fisheries and Food,[14] and HMSO.[15]

Other bodies that have discussed sampling procedures include the American Society for Testing Materials, the U.S. Environmental Protection Agency, the American Public Health Association, and the British Standards Institution.[16]

The principal step of the sampling process is the taking of the sample. Here we intend to deal only with the risk of contaminating the sample during its collection, storage, and processing, since any subsequent separation is applied only after the sample has been brought into solution.

The "art" of the analytical chemist consists of knowing the history of the sample, and of choosing the simplest possible analytical procedure.

Ristenpart et al.[17] and Houba[18] have evaluated various sediment samplers. A sediment shovel was highly practical but limited because small particles tend to be lost when the shovel is lifted. A cryogenic sediment sampler was less convenient to use, but allowed the collection of nearly undisturbed samples. Houba described a different device for the automatic sub-sampling of sediments for proficiency testing.

Thoms[19] showed that freeze-sampling collects representative sediment samples, whereas grab-sampling introduces a bias in the textural composition of the 120 mesh fraction, due to washout and elutriation of finer fractions.

Rubio and Ure[20] have discussed the risks of the contamination of sediment samples using inappropriate materials, containers, and tools. They also discussed possible analyte loss during sample handling.

Wehrens et al.[21] have discussed a decision-support system for the sampling of aquatic sediments in lakes.

Vernet et al.[22] used three methods of fluvial sediment sampling to determine the validity and representativeness of the geochemical information obtained by each technique. Bottom sediments and trap sediments showed similar results for metals.

Fortunati et al.[23] have reviewed problems associated with techniques and strategies of soil sampling.

Droppo et al.[24] studied the effects of concentrating suspended sediment samples on the primary grain size distribution. The initial and resuspended size distributions were not significantly different using either cellulose or polycarbonate filters.

Truckenbrodt and Einax[25] have shown in an analysis of overall analytical error that, independent of grain size, sampling was the main source of variance in the determination of metals in river sediments. Two approaches are described to determine the number of samples required for representative sampling.

Ruiz et al.[26] have described a method for the determination of the distribution of metallic constituents in sediment particles in torrential rivers according to particle size. A rapid sampling method using passive sampling devices for soil contaminant characterization has been shown to provide a more thorough site assessment.[30]

6.2 SAMPLE HOMOGENEITY

The problem of homogeneity of a sample is closely related to the problem of the history of the sample.

Whereas the literature reports many specific sampling situations, there are few papers that consider the fundamental aspects of the sampling process and their implications for the general analytical process. The aspects that relate to the homogeneity of the sample have to be considered in the context of the nature of the analytical process. Also the nature of the analytical process is determined by the characteristics of the sample to be analyzed.

The homogeneity of the sample depends on the physical state of the material. Because of the natural diffusion processes, liquid and gaseous samples are much more homogeneous than solid ones. Solid samples are often heterogeneous, and have first to be homogenized by mechanical means (grinding, ball-milling, etc.) before specimens are selected. The lack of homogeneity of solid samples is the main factor that renders their processing difficult.

There are two principal methods employed in the analysis of sediments: destructive analysis and nondestructive analysis.

6.3 DESTRUCTIVE ANALYSIS

"Destroying" a sample means to bring it into a homogenous form as a solution, normally in an aqueous or a partially nonaqueous medium. There are two means of bringing solid sediment and sludges into solution, either by dissolving them or by decomposing or disintegrating them in dry form by means of fluxes. These supplementary operations not only increase the duration of the analysis proper, but also introduce the risk of contamination of the samples by reagents and working techniques.

To dissolve certain solid samples, acids or mixtures of acids may be used. However, besides dissolving the sample, the acids may interfere with the subsequent analysis by converting some components of the sample into extremely stable complexes or by creating volatile components that may be lost partly or even totally during the dissolution process.

Some wet dissolution/decomposition reagents, such as hydrofluoric or hydrochloric acid, may have strong competing action. Very often, the complex formed may prevent the proper determination from being performed, because it is kinetically or thermodynamically very stable. In many cases the dissolution/decomposition reagents are used to destroy an organic substrate. For example, the use of nitric-perchloric or nitric-sulfuric-perchloric acid mixtures is well known.

The art of the analytical chemist consists in choosing the most suitable dissolution/decomposition system for a given sample, so that the resulting solution contains the components in a form directly usable in the subsequent concentration and separation process.

When the aim of the analytical chemist is to determine trace components after a wet decomposition with water or acids, care must be taken to use clean vessels and pure reagents in order to minimize contamination risks. Trace analysis presuppose an appropriate sampling procedure and dedicated high-purity reagents. The water to be used must be purified by ion-exchange and then distilled, and stored in polythene vessels. The acids and other reagents used for decomposing samples must be of suitable purity, e.g., so-called "electronic" or "semiconductor" grade.

For samples that are virtually insoluble in water or acids, either cold or hot, so-called dry decomposition may be used. This system is more tedious than wet decomposition since it involves two independent operations, the decomposition proper and the succeeding dissolution of the product. For dry decomposition, various fluxes can be used, such as $Na_2CO_3+K_2CO_3$, Na_2CO_3+borax, or Na_2CO_3+S (Freiberger decomposition). To transfer the sample completely into solution, it must first be perfectly homogenized with the decomposition agents.

Unlike acids, which can now be obtained in a high degree of purity, solid reagents are often of insufficient purity for trace analysis. It is this aspect of trace analysis that has led to the development of some noncontaminant decomposition systems. The simplest way of achieving faster (and noncontaminating) decomposition has been to resort to an additional physical parameter, namely pressure, coupled with an adequate decomposition temperature. As will be discussed later, the use of high-pressure decomposition vessels requires much lower temperatures for decomposing a sample, than those necessary for dry decomposition at atmospheric pressure. The appearance of the high-pressure decomposition vessels (bombs) is a direct result of the availability of a chemically inert plastic, Teflon.

In analyzing solid samples, regardless of the chosen decomposition system, a preliminary and extremely important step is granulometry, which plays a decisive role when preparing solid samples for chemical analysis. Although many studies have been written on granulometry, these studies could also be considered as part of the general sampling process. Wolfson and Belyaev[27] reviewed current work in this field and discussed the role and importance of granulometry for the general sampling process. This work underlines the necessity of granulometric control of the composition of a sample during its preparation before chemical analysis. Vulfson and Belyaev[27] examined the modern methods of fine grinding and granulometric analysis, and attention was given to problems of the influence of the granulometric composition of the dispersed substance on the chemical analysis results and sampling errors.

A great number of separation processes are based on solvent extraction, especially since this is also a concentration technique. For these reasons, solvent extraction will be considered, both from the point of view of the sampling process and from that of the general analytical process. Solvent extraction is ultimately a process of partitioning between two immiscible solvents, and for its optimization, it is necessary to know first of all the operational parameters of the system.

The technique of solvent extraction has long been used in organic chemistry for concentrating and purifying some substances. In the case of organic compounds, the separation process is simple, in many cases being based only on differences in the solubility of the compounds in different solvents.

Many attempts at classifying solvent extraction systems have been made. Diamond and Tuck[28] have described a classification of the solutes that can be separated by solvent extraction.

6.3.1 Comminution of Samples

Various comminution devices are available from Fritsch for handling these types of samples. Grinding elements are offered in various noncontaminating materials such as corundum (Al_2O_3), agate (SiO_2), or zirconium oxide (ZrO_2).

6.3.2 Sieving Analysis of Samples

Having comminuted the sample, it may now be required to carry out a sieving analysis in order to obtain different size fractions for chemical analysis. Fritsch supply a range of devices for sieving analyzers.

6.3.3 Grinding of Samples

The air-dried sample is ground until the whole of the sample, excluding stones, passes through a 2-mm mesh sieve. There is a limited range of apparatus available for grinding, but the Rukuhia-type grinding machine is suitable (this is obtainable from D. Mackay, 85 East Road, Cambridge, CB1 1BY, United Kingdom). The apparatus consists of a number of cylinders into which the samples and metal pestles are placed. The cylinders, which have walls of 2-mm mesh perforated steel, are rotated horizontally by means of electrically driven rollers. As the cylinders rotate, the sample is ground by the pestle and falls through the mesh into a tray below.

Houba et al.[29] studied the influence of grinding procedures and demonstrated that the availability of some analytes is significantly influenced by the grinding process adopted.

6.3.4 Particle-Size Distribution Measurement

Complete particle-size analyses can require the use of various analysis technologies. A microscopic examination may be performed before the sieve analysis, which in turn can be followed by a sedimentation analysis or the recording and the evaluation of a diffraction pattern.

The working range of the analysis methods overlap and can be subdivided as shown in Table 6.1, which also details equipment suppliers.

6.3.5 Pressure Dissolution of Sediments

Pressure dissolution and digestion bombs have been used to dissolve samples for which wet digestion is unsuitable. In this technique, the sample is placed in a pressure dissolution vessel with a suitable mixture of acids, and the combination

TABLE 6.1 Suppliers and Working Ranges of Particle-Size Distribution Methods

Method	Particle Size Range	Equipment	Model
Dry sieving	63 μm–63 mm	Fritsch	Analysette 3, (20 μm–24 mm)
Wet sieving	20–200 μm	Fritsch	Analysette 18
Microsieving	5–100 μm	Fritsch	
Sedimentation in gravitation field	0.5–500 μm	Fritsch	Analysette 20
Laser diffraction	0.1–1100 μm	Fritsch	Analysette 22
Electrical zone sensing	0.4–1200 μm	Coulter	Model ZM, Coulter multisizer
Electron microscopy	0.5–100 μm	–	–
Photocorrelation spectroscopy	0.5–5 μm	–	–
Sedimentation in centrifugal field	0.5–10 μm	Fritsch	Analysette 21 (Anderson Pipette centrifuge)
Diffraction spectroscopy	1 μm–1 mm	–	–
Optical microscale	0.5μm–1 mm	–	–
Projection microscopy	0.5μm–1 mm	–	–
Image analysis systems	0.8–150 μm Down to 0.5 μm	Joyce-Leebl Leitz Karl Zeiss Cambridge Instruments	Magiscan and Magiscan P Autoscope P Videoplan II Quantimet 520

of temperature and pressure effects dissolution of the sample. This technique is particularly useful for the analysis of volatile elements that may be lost in an open digestion.[30]

6.3.6 Microwave Pressure Dissolution

More recently, microwave ovens have been used for sample dissolution. These are available from Park Instruments. The sample is sealed in a Teflon bottle or a specially designed microwave digestion vessel with a mixture of suitable acids. The high-frequency microwave, temperature (~100–250 °C), and increased pressure have a role to play in the success of this technique. An added advantage is the significant reduction in sample dissolution time.[31,32]

The list of applications has been expanded to include metals such as chromium, iron, nickel, manganese, beryllium, cadmium, copper, lead, vanadium, and zinc by using a quartz liner to eliminate interference from trace amounts of heavy metals leached from the bomb walls and electrodes.[33,34]

6.3.6.1 Digestion of Nonsaline Sediments

Various workers have applied microwave digestion to the determination of metals in nonsaline sediments.[35–41] Mahan et al.[39] used a microwave digestion technique in the sequential extraction of calcium, iron, chromium, manganese, lead, and zinc in nonsaline sediments.

Nieuwenholze et al.[40] analyzed six reference sediments after microwave aqua regia extraction. The results obtained showed close agreement with the reference values, and microwave extraction gave the same or slightly higher results than those obtained by conventional reflux extraction methods for 7 metals tested in 30 samples.

Elwaer and Belzile[41] have compared the use of a closed vessel microwave-assisted dissolution method and conventional hotplate digestion for the determination of selenium in lake sediments. A mixture of hydrochloric acid, nitric acid, and hydrofluoric acid with microwave digestion resulted in the best recoveries of selenium. Poor recoveries were obtained by hotplate digestion.

Several manufacturers supply microwave ovens and digestion bombs (Table 6.2). CFM Corporation states that its solid PTFE bombs are suitable for the digestion of sediments.

TABLE 6.2 Pressure Digestion Bombs

Supplier	Oven Part No.	Bomb Part No.
Acid Digester Types		
Parr Instruments	Not supplied	4781
CEM Corporation	MD581D	4782 Solid PTFE
Prolab	Microdigest 300 Microdigest A300	Solid PTFE
Oxygen Combustion Types		
Parr Instruments	Not supplied	1108

6.4 NONDESTRUCTIVE ANALYSIS OF SOLID SAMPLES

6.4.1 Introduction

There are methods that are capable of showing the distribution of elements on the surface of solid samples such as sediments. As such, they enable one to ascertain the homogeneity of distribution of elements on the surface of, and presumably within, the portion of sample analyzed.

We have at our disposal a large number of methods for analyzing solid materials without altering the sample in any way, all of which enable us to characterize them qualitatively, quantitatively, and sometimes structurally, by the direct action of a "reagent" upon a previously prepared surface of the sample. Although there are a number of techniques that involve the destruction of at least a fraction of the sample, either because of the width of the reagent beam used or the sensitivity of the determination (e.g., laser-source emission spectrometry or spark-source mass spectrometry). In this discussion we will refer mainly to those techniques for analysis of solid materials that maintain the sample almost intact after impact of the reagent. These are the so-called surface-analysis techniques, or more correctly, beam-analysis techniques.

As mentioned already, many surface-analysis techniques are available nowadays. In the opinion of some specialists in this field,[42,43] four of these are greater in importance: X-ray photoelectron spectrometry (ESCA), Auger electron spectrometry (AES), secondary ion mass spectrometry (SIMS), and low-energy ion-scattering spectrometry (ISS). I would add X-ray fluorescence spectroscopy to the list.

The importance of these surface-analysis techniques has resulted in the development of a range of highly automated instruments. In the effort to obtain multiple types of analytical data, a trend has occurred to build combined instruments, that is, devices that will permit measurements by several techniques in a single vacuum system. In this way, greater utilization of the complex instrumentation involved, and a more economic use of the functional parameters of the instruments, are ensured.

6.4.2 X-Ray Fluorescence Spectroscopy

X-ray fluorescence spectroscopy is extremely useful for determining the surface concentration and distribution of elements, arsenic, etc., in sediments. It was the first nondestructive technique for analyzing surfaces, and produced some remarkable results. Some advantages of nondestructive methods are no risk of loss of elements during sample-handling operations; the absence of contamination from reagents, etc.; and the avoidance of capital outlay on expensive instruments and highly trained staff.

A wide variety of X-ray fluorescence spectrometers may be used, depending on the nature and complexity of the sample and on the number of samples to be analyzed. To prove this and to indicate the substantial influence that the sample has on the choice of some measuring instruments, let us consider some of the main characteristics of some X-ray fluorescence instruments used today.[44]

6.4.3 Electron Probe X-Ray Microanalysis

Even though X-ray fluorescence is now widely used to analyze a large variety of samples, it does have some drawbacks. For instance, the X-ray beam used is wide, this is of no great use for analyzing tiny inclusions present in samples, and also does not allow point-by-point analysis on surfaces (scanning analysis). This was possible because the electron beam had a diameter of only about 1 μm. Although the small size of this beam permitted the analysis of some micro-inclusions in samples, and also multiple analyses by scanning, the main problem is that of the microhomogeneity and microtopography of the samples. Thus, whereas polishing the solid samples with a 30 to 100-μm grade abrasive is usually satisfactory for X-ray fluorescence spectrometry, a 0.25-μm grade abrasive or finer may be required for electron-probe microanalysis.

In principle, the difference between X-ray fluorescence spectrometry and electron-probe microanalysis lies in the fact the analytical information is provided, in the first case, by secondary fluorescence X-rays, and in the second by primary X-rays emitted as a result of the impact of the electron beam on the sample's electrons.

Owing to the small size of the electron beam on the one hand, and to the high sensitivity of the method on the other (a sensitivity that can go down to detection of 10^{-16}g), electron microanalysis has found application in many fields.

6.4.4 Auger Electron Spectrometry

Auger electron spectrometry (AES), reported by Auger in 1923,[45] is also a valuable technique for analyzing surfaces. The technique is somewhat similar to X-ray photoelectric spectroscopy, measuring electrons emitted from a surface as a result of electron bombardment. In both cases, the sampling depth is ~20A. Coupling this technique with scanning electron microscopy (SEM) produced a tandem (AES-SEM) technique that has proved extremely productive.

6.4.5 Secondary Ion Mass Spectrometry

In secondary ion mass spectrometry (SIMS), a primary ion beam bombards the surface and a mass spectrometer analyses the ions sputtered from the surface by the primary bombardment. This extremely sensitive technique provides both elemental and structural information.

REFERENCES

1. Smith R, James GV. *The sampling of bulk materials*. London: The Royal Society of Chemistry; 1981.
2. Kratochvil B, Wallace D, Taylor JK. *Anal Chem* 1984;**56**:113R.
3. Kratochvil B, Taylor, JK. *A survey of recent literature on sampling for chemical analysis*. NBS. Technical note 1153. Washington, DC: US Department of Commerce; January 1982.
4. Gy PM. *Sampling of particulate mixtures: theory and practice*. New York: Elsevier; 1979.
5. Woodget BW, Cooper D. *Samples and standards*. Basingstoke: Wiley; 1987.
6. Harrison TS. *Handbook of control of iron and steel production*. Chichester: Haward; 1979.
7. Walton WW, Hoffman JI. In: 1st ed. Kolthoff IM, Elving PJ, editors. *Treatise on analytical chemistry, part I*, vol. 1. New York: Wiley-Interscience; 1969. p. 67–97.
8. Laitinen HA. *Chemical analyses*. 1st ed. New York: McGraw-Hill; 1960.
9. Laitnen HA, Harris WE. *Chemical analysis*. 2nd ed. New York: McGraw-Hill; 1975.
10. Ingamells CO, Pitard FF. *Applied geochemical analysis*. New York: Wiley-Interscience; 1986.
11. Kratochvil B, Taylor JK. *Anal Chem* 1981;**58**:924A.
12. Kratochvil B. Samples for microanalysis: theories and strategies. In: *Paper presented at the 11th international symposium of microchemical techniques, Wiesbaden, 28th August–1st September 1989*; 1989.
13. Wallace D, Kratochvil B. *Anal Chem* 1987;**59**:226.
14. Ministry of Agriculture. *Fishers and food, the analysis of agricultural materials*. R.B. 427. London: HMSO; 1979.
15. HMSO. *Sampling and initial preparation of sewage and waterworks sludges, soils*. London: Sediments and Plant Materials; 1977.
16. *Sampling procedure*. London: British Standards Institution; 1971.
17. Ristenpart E, Gitzel R, Uhl M. *Water Sci Technol* 1992;**25**:63.
18. Houba VJG. *Fresenius J Anal Chem* 1994;**51**:131.
19. Thoms MC. *J Geochem Explor* 1994;**51**:131.
20. Rubio R, Ure AM. *Int J Environ Anal Chem* 1993;**51**:205.
21. Wehrens R, Van Hoof D, Buydens L, Kateman G, Vossen M, Mulder WH, et al. *Anal Chim Acta* 1992;**271**:11.
22. Vernet JP, Favarger PY, Span D, Martin C. *Trace Met Enviorn (1 Heavy Met Environ)* 1991;**1**:397.
23. Fortunati GL, Banfi C, Pasturenzi M. *Fresenius J Anal Chem* 1994;**348**:86.
24. Droppo IG, Krishnappan BG, Onley ED. *Environ Sci Technol* 1992;**26**:1655.
25. Truckenbrodt D, Einax J. *Fresenius J Anal Chem* 1995;**352**:437.
26. Ruiz R, Echeandia A, Romero F. *Fresenius J Anal Chem* 1991;**240**:223.
27. Wulfson EK, Beyaev Yu I. *Zhur Analt Khim* 1985;**40**:1364.
28. Diamond RM, Tuck DG. In: Colton FA, editor. *In organic chemistry*, Vol. 2. New York: Wiley Interscience; 1960. p. 109–92.
29. Houba VJG, Charden WJ, Roelsc K. *Commun Soil Sci Plant Analysis* 1993;**24**:1591.
30. Adrian Perkin Eluner WA. *At Absorpt Newsl* 1973;**10**:96.
31. Reverz R, Hasty E. Recovery study using an elevated pressure temperature microwave dissolution technique. In: *Paper presented at the Pittsberg conference and exposition on analytical chemistry and applied spectroscopy*. March 1997.
32. Nadkarni RA. *Anal Chem* 1984;**56**:2233.
33. Nadkarni RA. *Am Lab* 2nd August, 1981;**1981**(13).
34. Parr Manual 207M Parr Instrument Co., 53 Rd St. Moine, Illinois, 61265, 1974.

35. Li M, Barban R, Zucchi B, Martinotti W. *Water Air Soil Pollut* 1991;**57**:495.
36. Hewitt AD, Reynolds CM. *At Spectrosc* 1992;**11**:187.
37. Paudyn AM, Smith RG. *Can J Appl Spectrosc* 1992;**37**:94.
38. Hewitt AD, Reynolds CM. Report CRREL-SP-90–19 CETHA-TS-CR-90052 order No AD-a-226367 (Avail NTIS) 1990.
39. Mahan KI, Foderaro TH, Garza TL, Martinez M, Maroney GA, Trivisonno MR, et al. *Anal Chem* 1987;**59**:938.
40. Nieuwenholze J, Poley-Vos CH, Van den Akker AH, Van Delft WI. *Analyst* 1991;**116**:347.
41. Elwaer N, Belzile N. *Int J Environ Anal Chem* 1995;**61**:189.
42. Hercules DM. *Anal Chem* 1978;**50**:734A.
43. Hercules DM. *Anal Chem* 1986;**58**:1177A.
44. Jenkins R. *Anal Chem* 1984;**56**:1099A.
45. Auger P. *Compt Rend* 1923;**177**:169.

Chapter 7

Metals in Sediments

Chapter Outline

7.1 ALUMINUM

See Section 7.55.

7.2 ANTIMONY

7.2.1 Spectrophotometric Method

Abu-Hilal and Riley[1] have investigated a spectrophotometric method for the determination of antimony in sediments and clays.

In this procedure, 1 g of the finally ground sample is weighed into a polytetrafluoroethylene beaker and heated with 40% hydrofluoric acid. The beaker is covered with a PTFE lid and heated overnight on a boiling water bath. The residue is dissolved in 3 mL of 6 M hydrochloric acid and transferred quantitatively to a 1 L Erlenmeyer flask.

Antimony is then determined by a spectrophotometric method utilizing crystal violet in which the extract is treated with this chromogenic agent and the colored complex extracted with benzene. The benzene extract is evaluated spectrophotometrically at 610 nm.

7.2.2 Inductively Coupled Plasma Atomic Emission Spectrometry and Inductively Coupled Plasma Mass Spectrometry

Arrowsmith[2] have discussed a laser ablation inductively coupled plasma atomic emission spectrometric method for the determination of down to 0.2 $\mu g\,g^{-1}$ of antimony in sediments. No interference was exhibited in this method by 10 mg of acetate, arsenate, bromide, nitrate, cyanide, fluoride, iodide, nitrate, oxalate, phosphate, sulfate, thiocyanate, thiosulfate, or tartrate; or by 0.5 mg of mercury II or 50 μg of thallium.

7.2.3 Gas Chromatography

Cutter et al.[3] have described a selective hydride generation-gas chromatographic procedure using a photoionization detector for the determination of down to 3.3 $p\,mol\,L^{-1}$ of antimony III and antimony V (also arsenic III and arsenic V) in sediments.

7.2.4 Miscellaneous Techniques

Brannon and Patrick[4] give details of studies on the distribution and mobility of antimony in sediments from several sites in rivers, waterways, and coastal waters throughout the United States. Most of the naturally occurring and added antimony in the sediments was associated with relatively immobile iron and aluminum compounds.

7.3 ARSENIC

7.3.1 Spectrophotometry

A direct spectrophotometric procedure has been described for the determination of parts per billion of hydrochloric acid-releasable arsenic in river sediments.[5] In this method, the arsenic in the sediment sample is reduced by stannous chloride and zinc to arsine, which is then swept from the generator into silver diethyldithiocarbamate chromogenic reagent to be evaluated spectrophotometrically at 535 nm. Only inorganic arsenic is included in this determination. Organically bound arsenic is not determined unless the sample is oxidized.

Chromium, copper, nickel, mercury, antimony, or organo-compounds, arsenic did not interfere in this method.

7.3.2 Inductively Coupled Plasma Atomic Emission Spectrometry

Goulden et al.[6] have described a continuous-flow semiautomated system for the determination of arsenic (and selenium) in river sediments. By use of a fourfold preconcentration step, detection limits of $0.02\,\mu g\,L^{-1}$ were achieved for arsenic.

Brzezinska-Paudyn et al.[7,9] and Liversage et al.[8] compared results obtained in determinations of arsenic by conventional atomic emission spectrometry, flow injection/hydride generation inductively coupled plasma atomic emission spectrometry, graphite furnace atomic absorption spectrometry, and combined furnace flame atomic absorption and neutron activation analysis. Results obtained show that all these methods can be used for the determination of down to $5\,\mu g\,g^{-1}$ of arsenic in certified sediments.

For the determination of arsenic by conventional inductively coupled plasma atomic emission spectrometry, the samples were digested with mixtures of concentrated nitric acid, perchloric acid, and hydrofluoric acid in closed Teflon vessels.

7.3.3 Inductively Coupled Plasma Mass Spectrometry

Lasztity et al.[10] have reported an inductively coupled plasma mass spectrometric method for the determination of total arsenic in nonsaline sediments.

Cheam and Chau[12] used certified Great Lakes performance sediments for the determination of arsenic. Brannon and Patrick[11] reported on the transformation and fixation of arsenic(V) in anaerobic sediment, the long-term release of natural and added arsenic, and sediment properties that affected the mobilization of arsenic(V), arsenic(III), and organic arsenic.

Cutter[13] has described a selective hydride generation technique as the basis for the differential determination of total arsenic in oxidatively digested river sediment.

7.4 BARIUM

See Section 7.55.

7.5 BERYLLIUM

See Section 7.55.

7.6 BISMUTH

7.6.1 Atomic Absorption Spectrometry

Ebdon et al.[14] and Zhe-Ming et al.[15] have discussed the application of this technique. Zhe-Ming et al.[15] determined bismuth in amounts down to 1 $mg\,kg^{-1}$ in river sediments by electrothermal atomic absorption spectrometry with low-temperature atomization of argon/hydrogen (90:10 L^{-1}). Absorption was maximal at 850–950 °C. Interference effects from the matrix were reduced, and the sensitivity increased, by using trisodium phosphate as matrix modifier. The relative standard deviation was 35% for replicate determinations of 2.4 mg bismuth per kg river sediment.

7.7 CADMIUM

7.7.1 Atomic Absorption Spectrometry

Lum and Edgar[16] used a polarized Zeeman flame atomic absorption spectrometer at 228.8 nm to determine traces of cadmium in chemical extracts of river sediments. The detection limit was 0.1 mg in aqua reigias extracts digested in a PTFE bomb.

Recoveries were in the range of 80–102% at the 6–185 $mg\,L^{-1}$ cadmium level.

Sakata and Shimoda[17] have described a simple and rapid method in which 0.5-g sediment is digested for 1 h at 140 °C with a mixture of 10 mL hydrofluoric acid and 4 mL nitric acid and together with 1 mL perchloric acid in a Teflon-lined bomb prior to measurement by graphite furnace atomic absorption spectrometry at 228.8 nm. After digestion, 5 g of boric acid is added to the solution to dissolve precipitated metal fluorides. Sodium and potassium interference is overcome by the addition of 1% ammonium sulfate matrix modifier.

See also Section 7.55.

7.8 CALCIUM

See Section 7.55.

7.9 CERIUM

See Section 7.55.

7.10 CAESIUM

See Section 7.55.

7.11 CHROMIUM

7.11.1 Atomic Absorption Spectrometry

Pankow et al.[18] determined total chromium in river sediment. The samples were dried on a hotplate, approximately 1 g of the dry material was acid-washed with 25 mL of 1 mol L^{-1} nitric acid (to effect a removal of surface-bound chromium), and the acid wash was filtered through medium-speed filter paper. The filtrate was collected, and analyzed by atomic absorption spectrometry using the method of standard addition.

Chakraborty et al.[20] determined chromium in nonsaline sediments by microwave-assisted sample digestion followed by atomic absorption spectrometry without the use of any chemical modifier.

Scott[19] discussed the cause and control of chromium losses during nitric acid-perchloric acid oxidation of river sediments. A three-step sequential extraction scheme has been proposed for extracting chromium from sediments. This scheme employs (1) acetic acid, (2) hydroxylamine hydrochloride, and (3) ammonium acetate as extracting agents.[21] From the results obtained it is recommended that chromium content in steps 1 and 2 be measured by electrothermal atomic absorption spectrometry, and the chromium content of step 3 by flame atomic absorption spectrometry. Interfering effects when measuring chromium were circumvented by the use of 1% 8-hydroxyquinoline as a suppressor agent.[22]

See also Section 7.55.

7.12 COPPER

See Section 7.55.

7.13 GALLIUM

7.13.1 Atomic Absorption Spectrometry

Xiao-Quan et al.[23] used graphite furnace atomic absorption spectrometry with a nickel matrix modifier to determine μg kg^{-1} levels of gallium in perchloric acid agents of sediments.

7.14 GOLD

Xu and Schramel[24] have reviewed methods for the determination of gold in nonsaline sediments.

See also Section 7.55.

7.15 HAFNIUM

See Section 7.55.

7.16 INDIUM

See Section 7.55.

7.17 IRIDIUM

See Section 7.55.

7.18 IRON

See Section 7.55.

7.19 LANTHANIDES

Three techniques have been employed for the determination of rare earths in river sediments, namely neutron activation analysis, γ-ray spectrometry, and inductively coupled plasma mass spectrometry. The rare earths praseodymium, thulium, promethium, holmium, and erbium are not specifically mentioned in these references.

7.19.1 Neutron Activation Analysis

In the method described by Jundi,[146] 11 elements including La, Ce, Nd, Sm, Eu, Gd, Tb, Dy, Tm, Yb, and Lu were analyzed. Mean results varied from $0.52\,\mu g\,kg^{-1}$ for tamerium to $85\,\mu g\,kg^{-1}$ for cerium.

7.19.2 Inductively Coupled Plasma Mass Spectrometry

Shabani and Masuda[134] have described a method of sample introduction by online two-stage solvent extraction and back extraction to eliminate matrix interference and to enhance sensitivity in the determination of rare earth elements in sediments and rocks, by inductively coupled plasma mass spectrometry.

Two steps of extraction and back-extraction are linked together by multichannel pumping with the final back-extracts in aqueous solution being introduced to the inductively coupled plasma mass spectrometer. A mixture of 65% bis(2-ethylhexyl) hydrogen phosphate (HDEHP) and 35% 2-ethylhexyl dihydrogen phosphate (H_2MEHP) in heptane is used as the extracting agents, and octyl alcohol and nitric acid are used for the back-extraction. A preconcentration factor of up to 10 has been achieved.

7.19.3 Gamma-ray Spectrometry

Labresque et al.[152] determined 11 lanthanide elements, and also thorium and uranium in river sediments, employing a germanium detector for gamma ray spectrometry.

7.20 LEAD

7.20.1 Spectrometry

Savvin et al.[153] have discussed a spectrophotometric method for the determination of lead in nonsaline sediments.

7.20.2 Atomic Absorption Spectrometry

Various workers[25–32] have studied the application of this technique to the determination of lead in nonsaline sediments. Hinds et al.[31] investigated the application of low-pressure electrothermal atomic absorption spectrometry to the determination of lead in nonsaline sediments.

7.20.3 X-ray Fluorescence Spectroscopy

Wegrzynek and Holynska[33] have developed a method for the determination of lead in arsenic-containing sediments by energy dispersive X-ray fluorescence spectroscopy. Correction of arsenic interference is based on the use of an arsenic-free reference sample.

Koplitz et al.[34] have shown that in their X-ray fluorescence method for determining lead in sediments by using known masses of a sediment matrix made from all of the samples, there is no need for standard additions to each separate sample.

Lead has been determined in sediments by using a slurry sampling technique with lead nitrate and magnesium nitrate as a chemical modifier. Results were in good agreement with known concentrations of a standard reference material.[35]

7.21 LITHIUM

See Section 7.55.

7.22 MAGNESIUM

See Section 7.55.

7.23 MANGANESE

See Section 7.55.

7.24 MERCURY

7.24.1 Sample Digestion

The following procedure can be used to release bound mercury in solid sediment samples prior to analysis by suitable procedures such as atomic absorption spectrometry.

Pillay et al.[36] used a wet-washing procedure with sulfuric acid and perchloric acid to digest samples. The released mercury was precipitated as the sulfide. The precipitate was then re-digested using aqua regia.

Bretthaur et al.[37] described a method in which samples were ignited in a high-pressure oxygen-filled bomb. After ignition, the mercury was absorbed in a nitric acid solution.

Feldman[38] digested solid samples with potassium dichlormate, nitric acid, perchloric acid, and sulfuric acid. Bishop et al.[39] used aqua regia and potassium permanganate for digestion. Jacobs and Keeney oxidized sediment samples using aqua regia, potassium permanganate, and potassium persulfate.[40] The approved U.S. Environmental Protection Agency digestion procedure requires aqua regia and potassium permanganate as oxidants.[41]

The digestion procedures are slow and often hazardous because of the combination of strong oxidizing agents and high temperatures. In some of these methods, mercuric sulfide is not adequately recovered.

Various workers have used nitric sulfuric acid[42–44] and then potassium persulfate oxidation and stannous chloride reduction[45] to digest sediments prior to the determination of mercury.

Jurka and Carter[45] claim a relative standard deviation of 6% at the 20–30 mg kg^{-1} level for an automated cold vapor atomic absorption method, while Agemian and Chau[44] claim 14% at the 0.1 mg kg^{-1} level and 2% at the 2 mg kg^{-1} level.

Horvat et al.[46] compared distillation with alkaline digestion methods for the determination of mercury in nonsaline sediments by isothermal gas chromatography/cold vapor atomic fluorescence. The distillation approach produced results similar to those obtained by conventional digestion procedures, but with fewer matrix effects.

7.24.2 Miscellaneous Techniques

Kamburova[47] has reported a spectrophotometric method based on the formation of the mercury-triphenyltetrazolium chloride complex determination of mercury in nonsaline sediments.

7.24.3 Atomic Absorption Spectrometry

Various workers have studied the application of this technique to the determination of mercury in sediments.[48–55] Iskander et al.[43] applied flameless atomic absorption to a sulfuric acid–nitric acid digest of the sample following reduction with potassium permanganate, potassium persulfate, and stannous chloride. A detection limit of 1 part in 10^9 is claimed for this somewhat laborious method. Craig and Morton found a 2.2 $\mu g\,g^{-1}$ mean total mercury level in 136 samples of bottom deposits from the Mersey Estuary.

Jurka and Carter[48] have described an automated determination of down to 0.1 $\mu g\,L^{-1}$ mercury in river sediment samples in this method.

There was not significant interference due to sulfide in the solutions containing up to 10 mg sulfide L^{-1}. However, a negative interference was observed for both organic and inorganic standards containing 100 mg sulfide L^{-1}. This interference was overcome by ensuring that an excess of dichromate was present during the automated analysis.

This automated procedure was estimated to have a precision of 0.13–0.21 mg Hg kg^{-1} level, with standard decisions varying from 0.011 to 0.02 mg Hg kg^{-1}.

Banydopadhyay and Das[53] extracted mercury from nonsaline sediments with the liquid anion exchanger Aliquot-336 prior to determination by cold vapor atomic absorption spectrometry. A gold-coated graphite furnace atomic absorption spectrometer has been used to determine mercury in nonsaline sediments.[56]

Azzaria and Aftabi[55] showed that stepwise compared to continuous heating of nonsaline sediment samples before the determination by flameless atomic absorption spectrometry gives an increased resolution of the different phases of mercury.

7.24.4 Inductively Coupled Plasma Atomic Emission Spectrometry

Smith et al.[57] determined mercury in 1:1 hydrochloric acid:nitric acid digests of sediments in amounts down to 0.2 $ng\,L^{-1}$ (using a 200-mL sample) with isotope dilution inductively coupled plasma atomic emission spectrometry with a 201-Hg enriched spike.

Walker et al.[58] compared inductively coupled plasma mass spectrometry with inductively coupled plasma atomic emission spectrometry (ICPAES) as methods for the determination of mercury in Great Barrier Reef sediments. Typical instrument variabilities were 1 in 10^9 for inductively coupled plasma atomic emission spectrometry and 1 in 10^{12} for inductively coupled plasma mass spectrometry.

7.24.5 Inductively Coupled Plasma Mass Spectrometry

Using individual isotope measurements, Hintlemann and Wilken[63] measured mercury methylation rates in nonsaline sediments using inductively coupled plasma mass spectrometry.[59]

Pillay et al.[36] applied neutron activation analysis to the determination of 1.9–6.1 $mg\,kg^{-1}$ mercury in Lake Erie. The errors of the procedure were less than 15% at the 0.01 $mg\,kg^{-1}$ level and less than 5% at the 2 $mg\,kg^{-1}$ level. Robinson et al.[60] also used this technique to determine mercury in nonsaline sediments.

7.24.6 Anodic-Stripping Voltammetry

Mercury has been determined in acid-digested river sediment samples by differential-pulse anodic-stripping voltammetry.[61] Four types of working electrodes (glassy carbon and gold rotating-disk electrodes, and two types of gold film electrode, AuFe performed or in situ) were used and the analytical parameters of the procedures compared. The lowest limit of detection, 0.02 $\mu g\,L^{-1}$, was obtained with the gold rotating disk.

7.24.7 High-Performance Liquid Chromatography

Hintlemann and Wilken[63] used high-performance liquid chromatography to separate organomercury compounds. These were then converted to elemental mercury in a continuous-flow system and detected using atomic fluorescence.

7.24.8 Gas Chromatography

Emteborg et al.[64] have reported a method for determining mercury in nonsaline sediments that employed supercritical fluid extraction and gas chromatography coupled to microwave-induced plasma atomic-emission spectroscopy. Butyl magnesium chloride was used to derivatize mercury to butyl methylmercury for gas chromatographic determination.

7.24.9 Miscellaneous Techniques

In lakes and streams, mercury can collect in the bottom sediments, where it may remain for long periods of time. It is difficult to release the mercury from these matrices for analysis. Several investigators have liberated mercury from soil and sediment samples by the application of heat to the samples and the collection of the released mercury on gold surfaces. The mercury was then released from the gold by application of heat or by absorption in a solution containing oxidizing agents.[65,69,87]

Other techniques that have been used to determine mercury include isotope measurements[66] and gold film electrodes.[70]

Mudrock and Kokitich[65] determined mercury in lake sediments from the St Clair Lake using a gold film mercury analyzer. The mercury was extracted from the sediment by extraction with a mixture of nitric and hydrochloric acids (9:1 *v/v*). An accuracy of 0.02 $mg\,kg^{-1}$ was achieved.

Hintlemann et al.[66] have measured mercury methylation rates in sediment by individual isotope measurements using inductively coupled plasma mass spectrometry.

Hammer et al.[67] have pointed out that the Qu'Appelle River in Saskatoon is contaminated with mercury. Since the lake is eutrotrophic, with reduced oxygen in the hypolimnion during stratification, a study was carried out on the effects of low oxygen concentrations on release of mercury from lake sediments and subsequent bioaccumulation by aquatic plants (*Ceratophyllum dimersum*) and clams (*Anodonta grandis*).

Wilken and Hintlemann[71] have reviewed methods for the determination of mercury species in nonsaline sediments.

Cela et al.[72] attempted to correlate the behavior of inorganic mercury and methylmercury in sediments close to wastewater treatment systems and outfalls.

See also Section 7.55.

7.25 MOLYBDENUM

Chappaz et al.[154] measured the molybdenum isotope compositions (δ^{98}Mo) of well-dated sediment cores from two lakes in eastern Canada in an effort to distinguish between natural and anthropogenic contributions to these freshwater aquatic systems. Previously, Chappaz et al. ascribed pronounced twentieth-century molybdenum concentration enrichments in these lakes to anthropogenic inputs. δ^{98}Mo values in the deeper sediments (reflecting predominantly natural molybdenum sources) differ dramatically between the two lakes: -0.32 ± 0.17‰ for oxic Lake Tantare and $+0.64 \pm 0.09$‰ for oxic Lake Vose. Sediment layers previously identified as enriched in anthropogenic molybdenum, however, reveal significant δ^{98}Mo shifts of ± 0.3‰, resulting in isotopically heavier values of $+0.05 \pm 0.18$‰ in Lake Tantare and lighter values of $+0.31 \pm 0.03$‰ in Lake Vose. Chappaz et al.[154] argue that anthropogenic molybdenum modifies the isotope composition of the recent sediments, and they determine $\delta^{98}Mo_{anthropogenic}$ values of 0.1 ± 0.1‰ (Lake Vose) and 0.2 ± 0.2‰ (Lake Tantare). These calculated inputs are consistent with the δ^{98}Mo of molybdenite (MoS_2) likely delivered to the lakes via smelting of porphyry copper deposits (Lake Vose) or through combustion of coal and oil also containing molybdenum (Lake Tantare). These results confirm the utility of molybdenum isotopes as a promising fingerprint of human impacts and perhaps the specific sources of contamination. Importantly, the magnitudes of the anthropogenic inputs are large enough, relative to the natural molybdenum cycles in each lake, to have an impact on the microbiological communities.

See also Section 7.55.

7.26 NEPTUNIUM

Kim et al.[73] have demonstrated good agreement between methods for determining neptunium-237 in nonsaline sediments, based on inductively coupled plasma mass spectrometry, neutron activation analysis and α-spectrometry.

See also Section 7.55.

7.27 NICKEL

See Section 7.55.

7.28 OSMIUM

See Section 7.55.

7.29 PALLADIUM

See Section 7.55.

7.30 PLATINUM

See Section 7.55.

7.31 PLUTONIUM

7.31.1 Inductively Coupled Plasma Mass Spectrometry

Kershaw et al.[74] applied inductively coupled plasma mass spectrometry to the determination of the isotopic composition of plutonium in nonsaline sediments. These workers obtained good agreement of the measured isotope ratios by two mass spectrometric methods.

Kim et al.[75] determined the plutonium-240 plutonium-239 ratio in nonsaline sediments using the fission track method and inductively coupled plasma mass spectrometry.

7.31.2 Miscellaneous Techniques

Packed column chromatography has been used to determine various plutonium isotopes in non-saline sediments.[76] Various workers have discussed mass spectrometric and other methods for the determination of plutonium in non-saline sediments.[77–79]

7.32 POLONIUM

See Section 7.55.

7.33 POTASSIUM

See Section 7.55.

7.34 RADIUM

7.34.1 Thermal Ionization Mass Spectrometry

Cohen and O'Nions[80] have applied thermal ionization mass spectrometry to the determination of 226 radium in sediments and rocks in amounts down to

4×10^{-5} pg. The chemical separation techniques employed involved initial digestion of the sample, followed by co-precipitation of radium onto strontium (as $SrRaSO_4$), spiking of the sample with 228 radium, digestion with hot concentrated sulfuric acid, and then centrifuging. The washed precipitates are then counted. The abundance of 226 radium achieved by this method can be measured with 10^3 times the sensitivity achieved by conventional radioactive counting methods.

7.35 RHENIUM

See Section 7.55.

7.36 RUBIDIUM

See Section 7.55.

7.37 RUTHENIUM

See Section 7.55.

7.38 SCANDIUM

See Section 7.55.

7.39 SELENIUM

The fate of selenium in natural environments such as soil and sediments is affected by a variety of physical, chemical, and biological factors that are associated with changes in its oxidation state. Selenium can exist in four different oxidation states (-II, O, IV, and VI) and as a variety of organic compounds. The different chemical forms of selenium can control selenium solubility and availability to organisms. Selenate (Se(VI)) is the most oxidized form of selenium, is highly soluble in water, and is generally considered to be the most toxic form. Selenite (Se(IV)) occurs in oxic to suboxic environments and is less available to organisms because of its affinity to sorption sites of sediment and soil constituents. Under anoxic conditions, elemental selenium and selenide(-II) are the thermodynamically stable forms. Elemental selenium is relatively insoluble, and selenide(-II) precipitates as metal selenide(-II) of very low solubility. Organic selenium(-II) compounds such as selenomethionine and selenocystine can accumulate in soil and sediments or mineralize to inorganic selenium. Therefore, Se(VI), Se(IV), and organic selenium(-II) are the most important soluble forms of selenium in natural environments.

7.39.1 Spectrofluorimetry

Wiersma and Lee[81] determined selenium in lake sediments. The sample is digested with 4:1 concentrated nitric acid:6% perchloric acid, and the residue treated with 6M hydrochloric acid and then reduced with H_3PO_2. The fluorescence agent used was 2,3-diaminonaphthalene.

7.39.2 Hydride-Generation Atomic-Absorption Spectrometry

Hydride-generation atomic-absorption spectrometry is widely used to determine selenium in soil-sediment extracts because of its low detection limits. Speciation of selenium is determined by subdividing sample solutions into selective treatments. Selenite is determined by directly analyzing aliquots of samples without any treatments, or by analyzing samples acidified to pH 2 with concentrated hydrochloric acid or samples in 4–7 N hydrochloric acid solutions. Selenate plus Se(IV) are determined after reduction of Se(IV) to Se(IV) in 4–7 N hydrochloric acid at high temperatures (80–100 °C) and analysis for selenium to obtain Se(VI+IV) concentrations. Selenate is determined by the difference between a determination of Se(VI+IV) and a determination of Se(IV) in another subsample. Total selenium is determined by oxidizing all selenium species (organic Se(-II) and Se(IV)) to Se(VI) with hydrogen peroxide or persulfate, and then reducing Se(VI) to Se(IV) with 4–7 N hydrochloric acid at a high temperature (80–100 °C) and analyzing for total selenium in the samples.

Zhang et al.[82] developed a method to determine organic selenium(-II) in sediments.

In this method persulfate is used to oxidize organic selenium(-II) and manganese oxide is used as an indicator for oxidation completion. This method was used to determine selenium speciation in soil sediments and agricultural drainage water samples. Results showed that organic selenium(-II) can be quantitatively oxidized to selenite without changing the selenate concentration in the soil-sediment extract or the agricultural drainage water sample and then quantified by hydride-generation atomic-absorption spectrometry. Recoveries of spiked organic selenium(-II) and selenite were 96–105%.

Itoh et al.[83] and Cutter et al.[13,84] and also other workers[7,84–90] used hydride-generation atomic-absorption spectrometry techniques to distinguish selenite, total selenium, and organic selenium in nonsaline sediments.

7.39.3 Inductively Coupled Plasma Atomic Emission Spectrometry

This technique can determine selenium down to $0.03\,\mu g\,g^{-1}$ in the solution obtained following fusion of the sediment with solid sodium hydroxide in a zirconium crucible. A reference sample with a nominal selenium content of $0.4\,mg\,kg^{-1}$ gave a value of $0.49\,mg\,kg^{-1}$ by this method.[5]

7.39.4 Inductively Coupled Plasma Mass Spectrometry

Hydride-generation inductively coupled plasma mass spectrometry has been used to determine selenium in sediments.[91]

7.39.5 Cathodic Stripping Voltammetry

Square-wave cathodic stripping voltammetry has been used to determine selenium in sediments.[92]

7.39.6 High-Performance Liquid Chromatography

Selenium(IV) reacts selectively with various thiols in accord with the following equation to form selenotrisulfide involving an S–Se–S linkage which is generally unstable.

$$4RSH + H_2SeO_3 \rightarrow RSSeSR + 3H_2O$$

Nakagawa et al.[93] have reported selenotrisulfide formed from penicillamine (Pen) was exceptionally stable, and that the reduction proceeded in acid solution where most metal ions do not form chelates with penicillamine. These findings prompted Nakagawa et al. to apply the reaction to the selective determination of selenium(IV). UV absorption of penicillamine selenotrisulfide (PenSTS) enabled a parts-per-million level of selenium to be determined; the limit of detection could be lowered to a parts-per-billion level by conversion of PenSTS to a fluorophore. Thus, to develop a new method that allows assay of a low level of selenium with easy operation, Nakagawa et al.[93] set out to find the optimum reaction conditions for the quantitative formation of PenSTS and subsequent derivatization of PenSTS to a fluorophore using 7-fluoro-4-nitrobenz-2,1,3-oxadiazole (NBD-F), a labeling reagent for amino groups. The conditions for separation and detection of the fluorophase were also investigated.

7.39.7 Gas Chromatography

De Oliveira et al.[94] digested sediments with a mixture of nitric, perchloric, and sulfuric acids prior to the determination of selenium by a procedure involving reaction of selenium with 4-nitro-*o*-phenylene-diamine to produce a volatile product, which was determined in amounts down to $100\,\mu g\,kg^{-1}$ by electron-capture gas chromatography.[95]

7.39.8 Miscellaneous Techniques

Selenium has been directly determined in sediments by PIXE.[96] Haygarth et al.[97] have compared the use of fluorometry, hydride-generation atomic-absorption spectrometry, hydride-generation atomic-emission spectrometry, hydride-generation inductively coupled plasma mass spectrometry, and radiochemical neutron-activation analysis for the determination of selenium in sediments. For low concentration samples, hydride-generation inductively coupled plasma mass spectrometry performed best.

7.40 SILVER

7.40.1 Atomic Absorption Spectrometry

Lum and Edgar[16] carried out a five-part sequential extraction procedure on a 1-g dry weight Moira Lake sample, showing the distribution of different forms of silver in the sediment. Regardless of core depth, most of the silver is organically or sulfide bound or bound to the residual phase.

7.41 STRONTIUM

Stella et al.[98] and Ryabukhin et al.[99] have reviewed methods for the determination of radiostrontium in nonsaline sediments.

7.42 TANTALUM

See Section 7.55.

7.43 TECHNETIUM

Morita et al.[100] and Harvey et al.[101] have discussed the determination of technetium in nonsaline sediments.

7.44 TERBIUM

See Section 7.55.

7.45 THALLIUM

Lukaszewski and Zembrzuski[102] and Sagar[103] have reviewed methods for the determination of thallium in nonsaline sediments. See also Section 7.55.

7.46 THORIUM

7.46.1 Spectrofluorimetry

Mukhtar et al.[104] have described a laser fluorometric method for the determination of thorium (and uranium) in nonsaline sediments.

7.46.2 Inductively Coupled Plasma Mass Spectrometry

Toole et al.[105] and Shaw and Francois[106] determined thorium (and uranium) in nonsaline sediments by inductively coupled plasma mass spectrometry.

7.46.3 Miscellaneous Techniques

Various workers[107–109] have reviewed methods for the determination of thorium in nonsaline sediments. No sample preparation was required in this method.[110]

Parsa et al.[111] described a sequential radiochemical method for the determination of thorium (and uranium) in nonsaline sediments.

7.47 THULIUM

7.47.1 Neutron Activation Analysis

See Section 7.55.

7.48 TIN

7.48.1 Atomic Absorption Spectrometry

Dogan and Haerdi[112] applied their flameless atomic absorption method to the determination of down to 0.5 μg kg^{-1} tin in humus-rich lake sediments. Sample digestion was carried out using lumaton, a quaternary ammonium hydroxide, dissolved in isopropanol (available from H. Kurner D-6451 Neuberg, Germany).

Long-Zhu et al.[113] described a graphite furnace atomic absorption spectrometric method for the determination of down to 2.5 mg kg^{-1} of tin in river sediments.

Legret and Divet[114] have described a method for the determination of tin in sedimentary hydride-generation atomic-absorption spectrometry. Hydride generation was carried out in a nitric acid-tartaric acid solution of sediment. The effect of various acids and the matrix effects and interferences from other trace elements were studied. Several procedures for decomposing the samples were compared; the preferred procedure involved reflux with a mixture of nitric acid and hydrochloric acid.

7.48.2 Inductively Coupled Plasma Atomic Emission Spectrometry

Brzenzinska-Paudyn and Van Loon[115] used inductively coupled plasma atomic emission spectrometry-mass spectrometry to determine tin in digested river sediments and compared results obtained by graphite-furnace atomic-absorption spectrometry with a palladium/hydroxylamine matrix modifier. The inductively coupled plasma technique was more sensitive, achieving a detection limit of less than 1 pg of tin in the sample aliquot analyzed.

7.49 TITANIUM

See Section 7.55.

7.50 TUNGSTEN

7.50.1 Spectrophotometry

Quin and Brookes[116] determined tungsten in sediments by fusing the sample with potassium hydrogen sulfate and leaching the melt in 10M hydrochloric acid. A clear portion of the acid extract is heated with a solution of stannous chloride in 10M hydrochloric acid. A solution of dithiol in isoamyl acetate is added and the mixture is heated under precisely defined conditions so that a globule containing the tungsten dithiol complex is formed in >6h. The globule is dissolved in light petroleum (boiling range of 80–100°C) and the extinction of the solution is measured at 630nm. Down to 0.2ppm of tungsten can be determined, and Beer's law is obeyed for up to 300ppm. See also Section 7.55.

7.51 URANIUM

7.51.1 Spectrofluorimetry

Mukhtar et al.[104] have described a laser fluorometric method for the determination of uranium (and thorium) in nonsaline sediments.

7.51.2 Inductively Coupled Plasma Spectrometry

Toole et al.[105] and Shaw and Francois[117] determined uranium (and thorium) in nonsaline sediments by inductively coupled plasma mass spectrometry.

7.51.3 X-ray Fluorescence Spectroscopy

To determine uranium (and thorium) in nonsaline sediments, fluorescent X-rays were measured by the use of a germanium plasma detector and chemometric techniques.[118] No sample preparation was required in this method.

Bertine et al.[119] have discussed the determination of uranium in sediments and water utilizing the fission track technique. In this technique a weighed aliquot (50–100mg) of the powered sample is made into a pellet with sufficient cellulose (as binder). The pellet is placed in a high-purity aluminum capsule and covered by polycarbonate plastic film (Lexan: 10μm thick).

Stolikar et al.[155] evaluated extraction techniques utilizing high pH bicarbonate concentrations for their efficacy in determining the oxidation state of uranium in reduced sediments collected from Rifle, Colorado. Differences in dissolved concentrations between oxic and anoxic extractions have been proposed as a means to quantify the U(VI) and U(IV) content of sediments. An additional step was added to anoxic extractions using a strong anion exchange resin to separate dissolved U(IV) and U(VI). X-ray spectroscopy showed the U(IV) in the sediments was present as polymerized precipitants similar to uraninite and/or less-ordered U(IV), referred to a non-uraninite U(IV) species associated with biomass (NUSAB). Extractions of sediment containing both uraninite and NUSAB displayed higher dissolved uranium concentrations under oxic than anoxic conditions while extractions of sediment dominated by NUSAB resulted in identical dissolved uranium concentrations. Dissolved U(IV) was rapidly oxidized under anoxic conditions in all experiments. Uraninite reacted minimally under anoxic conditions, but thermodynamic calculations show that its propensity to oxidize is sensitive to solution chemistry and sediment mineralogy. A universal method for quantification of U(IV) and U(VI) in sediments has not yet been developed, but the chemical extractions, when combined with solid-phase characterization, have a narrow range of applicability for sediments without U(VI).

7.52 VANADIUM

7.52.1 Spectrophotometric Method

Miura et al.[56] described a method for the determination of vanadium using 2-(8-quinolylazo)-5-(dimethylamino) phenol by reversed-phase liquid chromatography-spectrophotometry. As a result of the addition of tetra-alkyl ammonium salts, the retention of the chelates is remarkably reduced. Tetrabutyl ammonium bromide permits rapid separation and sensitive spectrophotometric detection of the vanadium(V) chelate with 2-(8-quinolylazo)-5-(dimethylamino) phenol, making it possible to determine trace vanadium(V). When a 100 mm^3 aqueous sample was injected, sensitivity and precision were as follows: peak height calibration curves of vanadium (V) were linear up to 800 pg at 0.005 absorbance until full scale (AUFS) and up to 160 pg at 0.001 AUFS; the relative standard deviation for 10 determinations at 0.005 AUFS was 2.3% at a level of 320 pg of vanadium (V); the detection limit was 2.6 pg at 0.001 AUFS. Many cations including iron(III) and aluminum(III) do not interfere with the determination.

7.52.2 Differential Pulse Polarography

Hasche et al.[120] and Hasebe et al. determined traces of vanadium in digests of pond sediments using differential pulse polarography of the vanadium(IV)–pyrocatecheol complex.

7.53 ZINC

Kratchvil and Mamba[121] showed all the zinc and copper were released from nonsaline sediments within 7 min using a commercial microwave oven.

See also Section 7.55.

7.54 ZIRCONIUM

See Section 7.55.

7.55 MULTIPLE METALS

A Review of the analysis of mixtures of metals in nonsaline sediments is given in Table 7.1. Metals mentioned in Table 7.1 that are not included in Sections 7.1–7.54 include Al, Ba, Na, Si, and Nd.

TABLE 7.1 Review of Published Work on Multiple Metal Analysis in Sediments

Element	References
Atomic Absorption Spectrometry	
7 elements	122,123
Cr, Mn, Ni, Cu_2, Zn	124–127
Pb, Fe, Co, Cd	16
Cd, Ag, Fe, Mn, Cu, Zn, Al	128
Pb, Cd, Te	129
As, Se	13
Cd, Cu, As	130
Inductively Coupled Plasma Atomic Emission Spectrometry	
Cd, Cr, Fe, Mn, Zn, Al, Ba, Cd, K, Mg	27,131
Na, Si, Se, Ti, As, Sb, Bi, Se	11
Pb, Pt, In	133
Inductively Coupled Plasma Mass Spectrometry	
Cd, Sb, Zn, As, Sb	2,27,119
Lanthanides	134
Misc metals	135
Flow-Injection Analysis	
Cd, Ca, Pb	136
Neutron Activation Analysis	
Ca, Cr, Fe, Zn, Mn, Ni, Pb, As, Sb, Ba, Sr, Mg,Na, K, La, Nd	13,17,137–143
Sm, Gd, Tb, Dy[g], Tm, Yb, Lu, Eu, Pd, Os, Pt, Ir, In, Ru, Au, Ce, Mo, Sc, Se, U, Hg, Ag, W, Al, Ti, Hf, Ta, Th, V, Zr	146
Miscellaneous Elements Plasma Emission Spectrometry	
Fe, Mn, Zn, Cu, Ca, Ni, Pb, Al	144

Continued

TABLE 7.1 Review of Published Work on Multiple Metal Analysis in Sediments—cont'd

Element	References
Potent Metric Stripping Analysis	
Cu, Pb	145
X-ray Fluorescence Spectroscopy	
25 elements	147–149
X-ray Spectrometry	
Cr, Mn, Co, Ni, Ca, Zn, Pd, Fe, Pb, As	150
Rb, Sr, Hg	151
Gamma-ray Spectrometry	
Lanthanides, Th, U	152
Gas Chromatography	
As, Sb	153
Ab, Ca, U	62
Hg, Ps	68
P	132,148

REFERENCES

1. Abu-Hilal AH, Riley JP. *Anal Chim Acta* 1981;**131**:175.
2. Arrowsmith P. *Anal Chem* 1987;**59**:1437.
3. Cutter LS, Cutter GA, San Siego McGlove MLC. *Anal Chem* 1991;**63**:1138.
4. Brannon JM, Patrick WH. *Environ Pollut Ser B* 1985;**9**:107.
5. Sandhu AS. *Analyst* 1981;**106**:311.
6. Goulden PD, Anthony DHJ, Austen KD. *Anal Chem* 1981;**53**:2027.
7. Brzezinska-Paudyn A, Van Loon JC, Hancock R. *At Spectrosc* 1986;**7**:72.
8. Liversage BR, Van Loon JC, de Andrade JC. *Anal Chim Acta* 1984;**161**:275.
9. Brzezinska-Paudyn A, Balicka A, Van Loon JC. *Water Air Soil Pollut* 1983:323.
10. Lasztity A, Krushevska A, Kotrebai M, Barnes RM, Amarasiriwardena DJ. *Anal At Spectrosc* 1995;**10**:505.
11. Brannon JM, Patrick WH. *Sci Technol* 1987;**21**:450.
12. Cheam V, Chau ASY. *Analyst* 1984;**21**:450.
13. Cutter GA. Electric Power Research Institute, Palo Alto, California, Report EPRI-EA4641, vol. 1. Speciation of selenium and arsenic in natural waters and sediments, Arsenic Speciation; 1986.
14. Ebdon L, Hutton RC, Ottaway JM. *Anal Chim Acta* 1977;**95**:117.

15. Zhe-Ming N, Xio-Chun L, Heng-Bin H. *Anal Chim Acta* 1986;**186**:147.
16. Lum KR, Edgar DG. *Analyst* 1983;**108**:918.
17. Sakata M, Shimoda O. *Water Res* 1982;**16**:231.
18. Pankow JF, Leta DP, Lin JW, Ohl SE, Shum WP, Janner GE. *Sci Total Environ* 1977;**7**:17.
19. Scott K. *Analyst (London)* 1978;**103**:754.
20. Chakraborty R, Das AK, Cervera ML, De La Guardia M. *J Anal Atom Spectrosc* 1995;**10**:3536.
21. Sahuquillo A, Lopez-Sanchez JF, Rubio R, Rauret G, Hatjie V. *Fresenius J Anal Chem* 1995;**351**:197.
22. Sahuquillo A, Lopez-Sanchez JF, Rubio R, Rauret G. *Mikrochim Acta* 1995;**119**:251.
23. Xiao-Quan S, Zhi-Neng Y, Zhe-Ming N. *Anal Chem* 1985;**57**:857.
24. Xu L, Schramel D. *Z Für Anal Chim* 1992;**342**:179.
25. Chen TC, Hong A. *J Hazard Mater* 1995;**41**:147.
26. Lopez-Garcia L, Sanchez-Melos M, Hernandex-Cordoba M. *Anal Chim Acta* 1996;**328**:19.
27. Zaray G, Kantor T. *Spectrochim Acta Part B* 1995;**50B**:489.
28. Manceau A, Boisset MC, Sarret G, Hazemann JL, Mench M, Cabier P. *Environ Sci Technol* 1996;**30**:1540.
29. Hinds MW, Lastimer KE, Jackson KW. *J Anal Atom Spectrosc* 1991;**6**:473.
30. Hinds MW, Jackson KW. *At Spectrosc* 1991;**12**:109.
31. Hinds MW, Latimer KE, Jackson JW. *J Anal Atom Spectrosc* 1992;**7**:171.
32. Bufflap SE, Allen HE. *Water Res* 1995;**29**:2051.
33. Wegrzynek D, Holynska B. *Appl Radiat Isot* 1993;**44**:1101.
34. Koplitz LV, Urbanik J, Harris S, Mills D. *Environ Sci Technol* 1994;**28**:538.
35. Bermejo-Barrara P, Barciel-Alonso C, Aboal-Samaza M, Barmejo-Barrera A. *J Atom Spectrosc* 1994;**9**:469.
36. Pillay KJS, Thomas CC, Sondel JA, Hyche CM. *Anal Chem* 1971;**43**:1419.
37. Betthaur EW, Moghissi AA, Snynder SS, Matthews NW. *Anal Chem* 1974;**46**:445.
38. Feldman C. *Anal Chem* 1974;**46**:1606.
39. Bishop JN, Taylor LA, Nary BP. *The determination of mercury in environmental samples.* Canada: Ministry of the Environment; 1973.
40. Jacobs LW, Keeney DR. *Environ Sci Technol* 1976;**8**:267.
41. Environmental Protection Agency. *Methods for the analysis of water and wastes* Cincinnati, Ohio, p. 134. 1974.
42. Agemian H, Chau ASY. *Anal Chim Acta* 1974;**75**:297.
43. Iskander K, Syers JK, Jakobs L, Feeney D, Gilmour JT. *Analyst* 1972;**97**:388.
44. Agemian H, Chau ASY. *Analyst* 1976;**101**:91.
45. Jurka AM, Carter MJ. *Anal Chem* 1978;**50**:91.
46. Horvat M, Bloom NS, Liang L. *Anal Chim Acta* 1993;**281**:135.
47. Kamburova M. *Talanta* 1993;**40**:719.
48. Jurka AM, Carter MJ. *Anal Chem* 1978;**50**:92.
49. Kozuchowski J. *Anal Chim Acta* 1978;**99**:293.
50. Craig PJ, Morton SF. *Nature* 1976;**261**:126.
51. EWl-Awady AA, Miller RB, Carter MJ. *Anal Chem* 1976;**48**:110.
52. Abo-Rady MDK. *Fresenius Z Für Anal Chem* 1979;**299**:187.
53. Bandyopadhyay S, Das AK. *J Indian Chem Soc* 1989;**66**:427.
54. Lee HS, Jung KH, Lee DS. *Talanta* 1989;**36**:999.
55. Azzaria LM, Aftabi A. *Water Air Soil Pollut* 1991;**56**:203.
56. Miura J. *Anal Chem* 1990;**62**:1424.
57. Smith RG. *Anal Chem* 1993;**65**:2485.

58. Walker GS, Ridd MJ, Brunskill GJ. *Rapid Commun Mass Spectrom* 1996;**10**:96.
59. Hintlemann H, Evans RD. *Anal At Spectrosc* 1995;**10**:619.
60. Robinson L, Dyer FF, Combs DW, Wade W, Teasley NA, Carlton JE. *J Radioanal Nucl Chem* 1994;**179**:305.
61. Hatle M, Golimowski J, Orzechowka A. *Talanta* 1987;**34**:1001.
62. Tong G, Luo N, Watson DB, Brooks SC, Gu B. *Environ Sci Technol* 2013;**47**:5787.
63. Hintlemann H, Wilken RD. *Appl Organomet Chem* 1993;**7**:173.
64. Emteborg H, Bjoerklund E, Oedman F, Karlsson L, Mathiasson L, Fresh W, et al. *Analyst* 1996;**121**:19.
65. Mudrock A, Kokitich E. *Analyst* 1987;**112**:709.
66. Hintlemann H, Evans RD, Villeneuve JY. *Anal At Spectrom* 1995;**10**:619.
67. Hammer UR, Merkowsky AJ, Huang DM. *Arch Environ Contam Toxicol* 1988;**17**:257.
68. Azoury S, Trozezynski J, Cliffeleau JF, Cossa D, Nakhie K, Schmitt S, et al. *Environ Sci Technol* 2013;**47**:7101.
69. Anderson DH, Evans JH, Murphy JJ, White WW. *Anal Chem* 1971;**43**:1511.
70. Lexa J, Stalick K. *Talanta* 1989;**36**:843.
71. Wilken RD, Hintlemann H. *NATO ASI Series 1990-23 (Met Speciation Enviorn)*, ;**339**. 1990.
72. Cela R, Lorenzo RA, Rubi E, Botana A, Valino M, Casais C. *J Radioanal Nucl Chem* 1989;**130**:443.
73. Kim CK, Takaku A, Yamamoto M, Kawamura H, Shiraiski K, Igarashi Y, et al. *J Radioanal Nucl Chem* 1989;**132**:131.
74. Kershaw PJ, Sampson KE, McCarthy W, Scott DR. *Radioanal Nucl Chem* 1995;**198**:113.
75. Kim CK, Oura Y, Takaku NH, Igarashi V, Ikeda N. *J Radioanal Nucl Chem* 1989;**136**:353.
76. Jia G, Testa C, Desideri D, Mell MA. *Anal Chim Acta* 1989;**220**:103.
77. Green LW, Miller FC, Sparling JA, Joshi SR. *J Am Soc Mass Spectrom* 1991;**2**:240.
78. Holgye Z. *J Radioanal Nucl Chem* 1991;**149**:275.
79. Barei-Funel G, Dalmasso J, Ardisson G. *J Radioanal Nucl Chem* 1992;**156**:83.
80. Cohen AS, O'Nions RK. *Anal Chem* 1991;**63**:2705.
81. Wiersma JH, Lee GF. *Environ Sci Technol* 1971;**5**:1203.
82. Zhang Y, Moore JN, Frankenberger WT. *Environ Sci Technol* 1999;**33**:1652.
83. Itoh K, Chikum M, Tanaka H. *Fresenius Z Für Anal Chim* 1988;**330**:600.
84. Cutters GA. Electric Power Research Institute, Palo Alta, California, Report EPRI-EA4641 Speciation of selenium and arsenic in natural waters and sediments; 1986.
85. Joshi SR. *Appl Radiat Isot* 1989;**40**:691.
86. Livens FR, Singleton DL. *Analyst* 1989;**114**:1097.
87. Hwang JD, Huxley HP, Diomiguardi JP, Vaughn W. *J Appl Spectrom* 1990;**44**:491.
88. Williams LR, Leggett RW, Espergren ML, Little CA. *Environ Monit Assess* 1989;**12**:83.
89. Hafez AF, Moharram BM, El-Khatib AM, Abel-Naby A. *Isotopenpraxis* 1991;**27**:185.
90. Cruvinel PE, Floccini RG. *Nucl Instrum Metal Phys Res Sect B* 1993;**375**:415.
91. McCurdy EJ, Nange JD, Haygarth PM. *Sci Total Environ* 1993;**135**:131.
92. Rojas CL, Maroto SB, Valanta D. *Fresenius J Anal Chem* 1994;**348**:775.
93. Nakagawa T, Aoyama E, Hasegawa N, Kobayashi N, Tanaka H. *Anal Chem* 1989;**61**:233.
94. De Olivera E, Laren JWN, Berman SS. *Anal Chem* 1983;**55**:2047.
95. Siu KWM, Berman SS. *Anal Chem* 1983;**55**:1603.
96. Cruvinel PE, Flocchini RG. *Nucl Instrum Meth Phys Res Sect B* 1993;**B75**:415.
97. Haygarth PM, Rowland AP, Sturup S, Jones KC. *Analyst* 1993;**118**:1303.
98. Stella R, Ganzerli UMT, Maggi L. *J Radioanal Nucl Chem* 1992;**161**:413.
99. Ryabukhin VA, Volynet MP, Myasoedoo BF, Radionova IM, Tuzova AM. *Fresenius J Anal Chem* 1991;**341**:636.

100. Morita S, Kim CK, Takaku Y, Seki R, Ikeda N. *Appl Radiat Isot* 1991;**42**:531.
101. Harvey BR, Williams KJ, Lovatt MG, Ibbett RD. *J Radioanal Nucl Chem* 1992;**158**:417.
102. Lukazewski Z, Zembrzuski W. *Talanta* 1992;**39**:221.
103. Sagar M. *Mikrochim Acta* 1992;**106**:241.
104. Mukhtar OM, Ghodes A, Khangi FA. *Radiochim Acta* 1991;**54**:201.
105. Toole J, McKay K, Baxter M. *Anal Chim Acta* 1990;**245**:83.
106. Shaw TJ, Francois R. *Geochim Cosmochim Acta* 1991;**55**:2075.
107. Shuktomova II, Kochan IG. *J Radioanal Chem* 1989;**129**:245.
108. Lazo EN. Report 1988 DOE (Department of Environment)/OR/0033–T424 Order No DE89010612 Avail NTIS 350 pp.
109. Noey KC, Liedle SD, Hickey CR, Doane RW. In: *Proceedings on symposium on waste management,* vol. 615; 1989.
110. Lazo EN, Doeeier GS, Bervan BA. *Health Phys* 1991;**6**:231.
111. Parsa B. *J Radioanal Nucl Chem* 1992;**157**:65.
112. Dogan S, Haerdi W. *Int J Environ Anal Chem* 1980;**8**:249.
113. Long-Zhu J. *At Spectrosc* 1984;**5**:91.
114. Legret M, Divet L. *Anal Chim Acta* 1986;**189**:313.
115. Brzenzinska-Paudyn A, Van Loon JC. *Fresenius Z Für Anal Chem* 1988;**331**:707.
116. Quin BF, Brooks RR. *Anal Chim Acta* 1972;**58**:301.
117. Shaw TJ, Francois A. *Geochim Cosmochim Acta* 1991;**55**:2075.
118. Lazo EN, Hossier GS, Bervan BA. *Health Phys* 1991;**6**:231.
119. Bertine KK, Chan LA, Tueckian KK. *Geochim Cosmochim Acta* 1970;**34**:641.
120. Hasche K, Kakizaki T, Tochida H. *Fresenius Z Für Anal Chim* 1985;**322**:486.
121. Kratchvil B, Mamba S. *Can J Chem* 1990;**68**:360.
122. Helinke PA, Schomberg PJ, Iskanda IK. *Environ Sci Technol* 1977;**11**:984.
123. Agemian H, Chau ASY. *Anal Chim Acta* 1975;**80**:61.
124. Zink-Neilsen I. *Vatten* 1977;**1**:14.
125. Sinex SA, Cantillo AW, Helz GR. *Anal Chem* 1980;**52**:2342.
126. Legret M, Demare D, Marchnadise P, Robbe D. *Anal Chim Acta* 1983;**149**:107.
127. Legret M, Divet L, Demare D. *Anal Chim Acta* 1985;**175**:203.
128. Agemian H, Chau ASY. *Analyst* 1976;**101**:761.
129. Garcia-Lopez I, Sanchez-Merlos M, Hernandez-Cordoba M. *Anal Chim Acta* 1996;**328**:19.
130. Welk B, Bleo N, Montiel A. *Environ Technol Lett* 1983;**4**:223.
131. Que-Hee SC. *Anal Chem* 1988;**60**:1022.
132. Ding S, Jiam F, Xu D, Sun Q, Zhang L, Fan C, Zhang C. *Environ Sci Technol* 2011;**45**:9680.
133. Colodner DC, Boyle EA, Edmond JM. *Anal Chem* 1997;**65**:1419.
134. Shabani MB, Masuda A. *Anal Chem* 1991;**63**:2099.
135. Rachid MA. *Chem Geol* 1974;**13**:175.
136. Ma R, Mol WW, Adams F. *Anal Chim Acta* 1994;**285**:33.
137. Nadkarni RA, Morrison GH. *Anal Chim Acta* 1975;**99**:133.
138. Slavic I, Draskovic R, Tasovac T. *Radiosavljevic* 1973;**9**:87.
139. Anders UW. *Anal Chem* 1972;**44**:1930.
140. Lieser K, Llamano W, Heuss H, Neitzert V. *J Radioanal Nucl Chem* 1977;**77**:717.
141. Ackermann E. *Dtsch Gweässer Kundliche Mittelugen* 1977;**21**:53.
142. Bart G, Von Gunten HR. *Int J Environ Anal Chem* 1979;**6**:25.
143. Bonifort R, Madaro M, Moauro A. *J Radioanal Nucl Chem* 1984;**84**:441.
144. Welte B, Bleo N, Montiel A. *Environ Technol Lett* 1983;**4**:223.
145. Madsen PP, Drabach I, Sorenson J. *Anal Chim Acta* 1983;**151**:479.
146. Al-Jundi J, Mamao C, Earwabec LG, Randl K, West J. *Anal Proc Lond* 1993;**30**:153.

147. Hellman HZ. *Fresenius Z Für Anal Chem* 1973;**263**:14.
148. Zhu Y, Wu F, He Z, Gao J, Qu X, Xie F, et al. *Environ Sci Technol* 2013;**47**:7679.
149. Prange A, Knoth J, Stossel RP, Baddaber H, Kramer K. *Anal Chim Acta* 1987;**195**:275.
150. Lichtfuss R, Brummer G. *Chem Geol* 1978;**21**:51.
151. Schneider B, Weiler K. *Environ Technol Lett* 1984;**5**:245.
152. Labresque JJ, Rosale PA, Meijas G. *Appl Spectrosc* 1986;**40**:1232.
153. Savvin SB, Petrova TU. *Fresenius Z Für Anal Chem* 1991;**340**:217.
154. Chappaz A, Lyons TU, Gordon GV, Anhar AD. *Environ Sci Technol* 2012;**48**:10934.
155. Stolikar PL, Campbell KM, Fox PM, Singer DM, Kavaiani N, Carey M, et al. *Environ Sci Technol* 2013;**47**:9225.

Chapter 8

Determination of Metals in Soils

Chapter Outline

8.1 ALUMINUM

8.1.1 Spectrophotometric Methods

An early spectrophotometric method[1] for determination of aluminum in soil involves the use of a Technicon sample changer, proportioning pump, and automatic colorimeter. The method is based on the measurement of the rate of color development in the reaction between aluminum and xylenol orange in ethanolic media. The calibration graph is rectilinear up to $2.7\,mg\,L^{-1}$ aluminum and the coefficient of variation is 4.5%.

8.1.2 Flow-Injection Analysis

Flow-injection analysis has been used to determine aluminum in soil. Reis et al.[2] studied the spectrophotometric determination of aluminum in soil using merging zones and sequential addition of pulsed reagents.

Tecator et al.[3] have described a flow-injection method for the determination of $0.5–100\,mg\,L^{-1}$ aluminum in 0.1 M potassium chloride extracts of soils, in which the acidified soil extract is injected into a carrier stream that has the same composition as the sample matrix (i.e., 0.1 M KCl) and merged with a masking solution for iron phenanthroline monohydrate and subsequently with the color reagent for aluminum (pyrocatechol violet) and a buffer (R3 aqueous hexamethylene tetramine). The colored complex formed between aluminum and pyrocatethol violet is measured at 585 nm. Repeatability is 1% RSD.

In addition to the above method, Tecator also describe a flow-injection analysis for determination $0.5–5\,mg\,L^{-1}$ aluminum in soil extracts based on the measurement of the chromazural–aluminum complex at 570 nm.[4,5]

8.1.3 Atomic Absorption Spectrometry

Ross et al.[35] analyzed samples of soil leachates from soil pore water from field porous cup lysimeters for aluminum by atomic absorption spectrometry under instrumental conditions employing either uncoated graphite tubes and wall atomization or employing a graphite-furnace pyrolitically coated platform and tubes.

8.1.4 Inductively Coupled Plasma Atomic-Emission Spectrometry

The determination of aluminum is discussed in Section 8.59.

8.1.5 Emission Spectrometry

The determination of aluminum is discussed in Section 8.59.

8.1.6 Photon Activation Analysis

The determination of aluminum is discussed in Section 8.59.

8.1.7 Miscellaneous Techniques

Mitrovic et al.[6] and Kozuk et al.[7] have carried out aluminum speciation studies on soil extracts. Various workers[231–233] have discussed the determination of aluminum in soils.

Using isotachoelectrophoresis, Schmidt and co-workers[234] were able to differentiate aluminum III and aluminum species in soil leachates.

8.2 AMERICIUM

8.2.1 Alpha-Spectrometry

Sill et al.[8] have discussed an α-spectrometric method for the determination of americium and other alpha-emitting nucleids including curium and californium in potassium fluoride-pyrosulfate extracts of soils. This method, discussed further in Section 8.59,[126] used α-spectrometry to determine americium in soil with a chemical recovery of 60–70%.

8.2.2 Miscellaneous Techniques

Joshi[235] and Livens and Singleton[236] have discussed methods for the determination of americium-241 in soils.

8.3 AMMONIUM

8.3.1 Spectrophotometric Method

Keay and Menage[9] have described an automated method for the determination of ammonium and nitrate in 2M potassium chloride extracts of soil. In this method, a sample of soil (2g) is shaken for 1h with 2 N-potassium chloride (10mL), and the filtrate is distilled, in the Auto-Analyzer, with a 0.25% suspension of magnesium oxide; the ammonia evolved is absorbed in 0.1 N-hydrochloric acid and determined spectrophotometrically at 625nm by the indophenol method. The sum of ammonium plus nitrate is determined similarly, but with addition of 4.5% titanous chloride solution before distillation; this reduced nitrate but not nitrite.

Waughman[14] have described a microdiffusion method for the determination of ammonium and nitrate in soils. Nitrate in the sample solution is reduced to ammonia by titanous sulfate and the ammonia is then released from the solution and diffused and absorbed onto a nylon square impregnated with dilute sulfuric acid. The nylon is then put into a solution that colors quantitatively when ammonia is present, and a spectrophotometer is used to measure the color.

8.3.2 Gasometric Method

Alder et al.[10] describe a method for determining low levels of ammonium ions in solution in which the ammonium ion is oxidized with sodium hypobromite in alkaline medium; the evolved nitrogen is passed into an argon plasma.

$$2NH_3 + 3NaBr = 3NaBr + 3H_2O + N_2$$

The nitrogen-hydrogen emission intensity produced in the plasma at 336 nm is monitored. A practical detection limit of 0.1 μg nitrogen per mL for 5 mL aqueous sample solutions was obtained. The method has been applied to the determination of the exchangeable ammonium content of soil samples.

A three-way tap allowed the system to be flushed free of air before the injector gas was introduced to the plasma. Addition of hypobromite reagent solution to the sample was achieved by rotation of the glass bulb in the sidearm through 180°.

In an application of the method, the content of ammonium nitrate was determined in the sample. Soil samples supplied by the Macaulay Institute for Soil Research (Aberdeen) were first air dried at 25 °C and sieved to 2 mm.

The ammonium ion was extracted by the method described by Bremner.[11] A 10-g sample of each of the soils was shaken with 40 mL of neutral 2 M potassium chloride solution for 1 h. After the extracts had settled, 5-mL aliquots were removed from the clear supernatant liquid and analyzed for ammonium nitrate. The values obtained are all within the expected range for exchangeable ammonium in soils. No interference was observed in the case of Na^+, K^+, Ca^{2+}, Mg^{2+}, Cl^{1-}, Na^{3-}, and SO_4^{2-}.

8.3.3 Flow-Injection Analysis

Tecator Ltd[12,69–73] described a flow-injection analysis method for the determination of 0.2–1.4 $mg L^{-1}$ of ammonia nitrogen in soil samples extractable by 2 M potassium chloride. The soil suspension in 2 M potassium chloride is centrifuged and filtered and introduced into the flow-injection system for analysis of ammonia (and nitrate) one parameter at a time. Ammonia is determined by the gas diffusion principle in which a PTFE membrane is mounted in the gas diffusion cell.

8.3.4 Ammonia-Selective Electrode

HMSO (UK)[13] published a method for the determination of ammonia nitrate and nitrite in potassium chloride extracts of soil extracts. An aliquot of the extract is made alkaline and the released ammonia, originating from ammonium ion, is determined either with an ammonia-selective probe or after removal by distillation by titration.

8.4 ANTIMONY

8.4.1 Atomic Absorption Spectrometry Inductively Coupled Plasma Atomic Spectrometry

Neutron activation analysis, spark source mass spectrometry, and carbon activation analysis have been applied to the determination of antimony in soils. See Section 8.59.

Chikhaikar et al.[15,237] have discussed the speciation of antimony in soil extract and soils. Asami et al.[238] have reviewed methods for the determination of antimony in soils.

8.5 ARSENIC

Arsenic occurs naturally in the earth's crust, but a considerable amount of arsenic is added to the human environment through its uses in wood preservatives, sheep dips, fly paper, arsenical soaps, rat poison, glass additives, dye pigment for calico prints, wallpaper, lead shot, and pesticides. Estimated production of organoarsenical herbicides, such as monosodium methanearsenate, disodium methanearsenate, and hydroxydimethylarsine oxide (cacodylic acid), in the United States exceeds 10.7×10^8 kg.[9,10,16,17] Generally, soils contain about 5.0 ppm of arsenic, but soils with a known history or arsenic application average about 165 ppm.[18] In some places, such as Buns, Switzerland and the Wiatapu Valley, New Zealand, the arsenic level in the soil may reach 10^4 ppm[19]; a substantial portion of arsenic in soil and soil-like material (sediment clays, sand, etc.) is expected to be found in soluble form and probably can be dislodged easily by the action of water moving through the soil. Soluble forms of arsenic are relatively more mobile in the environment and pose a greater potential for containing both groundwater and surface water. Soluble forms of arsenic from soil and soil-like material are likely to enter a bioconversion chain through their initial uptake by vegetation.

8.5.1 Spectrophotometric Methods

The limitations of the Gutzeit method for determining arsenic are well known. The spectrophotometric molybdenum blue or silver diethyldithiocarbamate procedures tend to suffer from poor precision and accuracy, as shown in collaborative studies.[20,21] Sandhu[22] have described a spectrophotometric method for the direct determination of hydrochloric acid-releasable inorganic arsenic in soils and sediments. The method provides reliable data on the quantitative recovery of arsenic (V) from soil clay, sand, and sediment samples. The method is simple, reliable, and relatively rapid; 24 samples can be analyzed in about 1 h. It does not require elaborate equipment and can be routinely used for the quantitative determination of arsenic in soil and soil-like material.

Merry and Zarcinas[29] have described a silver diethyldithiocarbamate method for the determination of arsenic and antimony in soil. The method involved the addition of sodium tetrahydroborate to an acid-digested sample that has been treated with hydroxylammonium chloride to prevent formation of insoluble antimony compounds. The generated arsine and stibine react with a solution of silver diethyldithiocarbamate in pyridine in a gas washtub. Absorbance is measured twice at wavelengths of 600 and 504 nm.

8.5.2 Atomic Absorption Spectrometry

Atomic absorption spectrometry has been applied extensively to the determination of arsenic in soils.[23–31,35,44] Forehand et al.[23] described a method involving reduction to AsIII with stannous chloride-potassium iodide prior to measurement at 193.7 nm for the determination of down to 0.8 $mg\,L^{-1}$ arsenic in sandy soils.

Arsenic recoveries between 85% and 90% were obtained for naturally occurring arsenic in sandy soils containing 91–98% sand, with the recoveries reducing as the sand content decreased from 90% to 23%.

Recoveries of arsenic from sandy soils (90% sand) spiked with sodium meta-arsenite, sodium meta-arsenate, arsenic trioxide, and arsenic pentoxide ranged between 86% and 92%.

The determination of arsenic by atomic absorption spectrometry with thermal atomization and with hydride generation using sodium borohydride has been described by Thompson and Thomerson,[24] and it was evident that this method could be modified for the analysis of soil. Thompson and Thorseby[25] have described a method for the determination of down to 0.001 $mg\,L^{-1}$ of arsenic in soil by hydride generation and atomic absorption spectrophotometry using electrothermal atomization. Soils are decomposed by leaching with a mixture of nitric and sulfuric acids or fusion with pyrosulfate. The resultant acidic sample solution was made to react with sodium borohydride and the liberated arsenic hydride swept into an electrically heated tube mounted on the optical axis of a simple atomic absorption apparatus.

Values obtained by atomic absorption spectrometry are higher than those obtained by the molybdenum blue method,[115,116] and this is believed to reflect the greater inherent accuracy of the former method.

To avoid problems previously encountered with flame atomic absorption spectrometry of arsenic and also with flameless methods, such as that in which the elements is converted to arsine, Ohta and Suzuki[27] proposed an alternative method based on electrothermal ionization with a metal microtube atomizer. Effective atomization can be achieved by the addition of thiourea to the arsenic solution or by preliminary extraction of the arsenic–thionalide complex. The second method is recommended for soil samples so as to avoid interference due to the presence of trace elements.

Haring et al.[28] determined arsenic and antimony by a combination of hydride generation and atomic absorption spectrometry. These workers found that compared to the spectrophotometric technique, the atomic absorption spectrophotometric technique with a heated quartz cell suffered from interferences by other hydride forming elements.

The antimony absorbance signal (2.5 $\mu g L^{-1}$ Sb) was found to decrease by more than 10% if the amount of selenium present in the sample is eight times higher than that of the antimony. Arsenic and bismuth interfere with the antimony determination only when present at many times the concentration of antimony. Formation of arsine from a solution of 2.5 $\mu g L^{-1}$ arsenic is suppressed at a fivefold concentration of both selenium and antimony.

Many of these low recoveries could be avoided by the inclusion of potassium iodide into the reaction mixture.

Jiminez de Blas et al.[31] have reported a method for the determination of total arsenic in soils based on hydride-generation atomic absorption spectrometry and flow-injection analysis. The method gave good recoveries and had a detection limit below 1 $\mu g L^{-1}$ for an injection volume of 160 μL.

A UK standard method also discusses the determination or arsenic in soil by atomic absorption spectrometry.[30] The determination of arsenic in soils by atomic absorption spectrometry is also discussed in Section 8.59.

8.5.3 Inductively Coupled Plasma Atomic-Emission Spectrometry

Hydride generation inductively coupled plasma atomic-emission spectrometry has been used to determine arsenic in soils. This technique was found to greatly reduce sample preparation time.[240] See also Section 8.59.

8.5.4 Inductively Coupled Plasma Mass Spectrometry

Lasztity et al.[239] have reported on inductively coupled plasma mass spectrometric methods for the determination of arsenic in soils. See also Section 8.59.

8.5.5 Neutron Activation Analysis and Photon Activation Analysis

The determination of arsenic is discussed in Section 8.59.

8.5.6 Miscellaneous Techniques

Agemian and Bedak[32] have described a semiautomated method for the determination of total arsenic in soils. Chappell et al.[34] have described an inexpensive but effective method for the quantitative determination of arsenic species in contaminated soils. Chappell found that the extraction efficiency varied with the ratio of soil to acid and with the concentration of the acid. Rurikova and Beno[241] accomplished speciation of arsenic III and arsenic V in soils by cathodic stripping voltammetry. Wenclawiak and Krab[33] used reactive supercritical fluid extraction in speciation studies of inorganic and organic arsenic in soils. In this method, derivatization with thioglycolic acid methyl ester was performed in supercritical carbon dioxide. Various other workers[242–244] have discussed the determination of arsenic in soils.

8.6 BARIUM

Inductively coupled plasma atomic-emission spectrometry, emission spectrometry, and stable isotope dilution have been applied to the determination of barium, and are discussed in Section 8.59.

8.7 BERYLLIUM

Plasma emission spectrometry has been applied to the determination of beryllium in soils. See Section 8.59.

8.8 BISMUTH

Atomic absorption spectrometry and inductively coupled plasma atomic emission spectrometry have been applied to the determination of bismuth, and are discussed in Section 8.59.

Asami et al.[238] have reviewed methods for the determination of bismuth in soils.

8.9 BORON

8.9.1 Spectrophotometric Methods

Spectrophotometric methods have been used have been used to determine water-soluble boron in soils. In one method,[36] the soil is extracted with boiling water, and then converted to fluoroborate, which is evaluated spectrophotometrically as the methylene blue complex.

Aznarez et al.[37] have described a spectrophotometric method using curcumin as chromophore for the determination of boron in soil. Boron is extracted from the soil into methyl isobutyl ketone with 2-methylpentane-2. In this method, 0.2–1 g of finely ground soil is digested with 5 mL concentrated nitric-perchloric acid (3 + 1) in a PTFE-lined pressure vessel for 2 h at 150 °C. The filtrate is neutralized with 5 M sodium hydroxide and diluted to 100 mL with hydrochloric acid 1 + 1. This solution is triple-extracted with 10 mL of methyl isobutyl ketone to remove iron interference. This solution is then extracted with 10 mL of 2-methylpentane-2,4-diol, and this extract is dried over anhydrous sodium sulfate. The development of the color is carried out in the organic phase used for extraction by the addition of curcumin in glacial acetic acid and phosphoric acid as dehydrating agent. Spectrophotometric evaluation is carried out at 510 nm.

The interference of iron at concentrations higher than 7×10^{-5} M can be eliminated as the chloro-complex by extraction with methyl isobutyl ketone. The total elimination of Fe(III) was not necessary, as the phosphoric acid masked the residual Fe(III) in the boric acid–curcumin reaction.

Recovery of boron was generally in the range of 97.2–104%.

In a further spectrophotometric method[38,39] for water-soluble boron in soil, boron is extracted from soil with boiling water. Borate in the extract is converted to fluoroborate by the action of orthophosphoric acid and sodium fluoride. The concentration of fluoroborate is measured spectrophotometrically as the blue complex formed with methylene blue and which is extracted into 1,2-dichlorethane. Nitrates and nitrites interfere; they are removed by reduction with zinc powder and orthophosphoric acid.

8.9.2 Inductively Coupled Plasma Atomic-Emission Spectrometry

Inductively coupled plasma optical emission spectrometry has been applied to the determination of boron in soil extracts in amounts down to $0.05\,mg\,L^{-1}$.[40]

To extract boron from the sample, a 50-g sample of air-dried soil was boiled under reflux with 100 mL of water for 10 min. The extract was filtered through an 18.5-cm Whatman no. 3 filter paper and an aliquot of the filtrate transferred to a silica beaker. The solution was taken to dryness and the residue was oxidized twice with 10 mL of 6% hydrogen peroxide solution. The residue was then diluted to 50 mL.

Boron has a very simple ICPAES spectrum, with the sensitive double at 249.7 and 249.8 nm being the only useful analytical lines. Between 0.4 and $0.7\,mg\,L^{-1}$, boron was found in soil extracts. Zarcinas and Cartwright[41] studied the acid dissolution of boron from soils prior to determination by inductively coupled plasma atomic-emission spectrometry. The application of this technique in the multimetal analysis of boron is also discussed in Section 8.59.

8.9.3 Molecular Absorption Spectrometry

Aznarez et al.[37] have described a method based on the molecular fluorescence of boron with dibenzoylmethane. The sampling first digested with nitric acid and perchloric acid. A methyl isobutylketone extract of this solution was used to determined boron.

The relative fluorescence intensity of the boron complex is measured at 400 nm with excitation of 390 nm and quinine sulfate as reference. The calibration graph was linear in the range of 0.5–5 μg of boron in aqueous solution (20–200 $\mu g\,L^{-1}$ of boron in the final solution to be measured). The detection limit and precision were 1 $\mu g\,L^{-1}$ of boron and 3% for 10 replicate determination of 1.2 μg of boron, respectively. Interference by foreign ions is minimal.

8.10 CADMIUM

Cadmium is readily taken up by most plants. The occurrence of cadmium in motor oils, car tires, phosphorus fertilizers, and as an impurity in zinc compounds explains its accumulation in soils; the cadmium contents of soils in nonpolluted areas are below 1 ppm, but values as high as 50 ppm can be found.[45]

8.10.1 Atomic Absorption Spectrometry

The determination of cadmium by graphite-furnace atomic-absorption spectrometry is especially difficult because cadmium is a volatile element, and matrix constituents cannot be removed by charring without a loss of cadmium. The use of selective volatilization often makes it possible to obtain a cadmium peak before the background has risen to such a high value that it interferes with the cadmium measurement. Another unrecognized source of interference is char loss resulting from the salt matrix. Although uncoated graphite tubes can be used for the determination of cadmium because of its volatility, some workers have found that pyrolitically coated tubes give better results when cadmium is determined in the presence of high contents of alkali and alkaline-earth elements.[113] Many studies of the determination of cadmium in soil extracts have been reported, but a chelation-extraction step has always been used prior to determination by graphite-furnace atomic-absorption spectrometry in order to reduce matrix interferences and to improve detection limits.[104,114–118,149]

Atomic absorption spectrometry with[19] or without[20–22] preliminary solvent extraction of metal has been applied extensively to the determination of cadmium in soils.[42–44] In an early standard official method,[42] the sieved soil sample is digested with hot nitric perchloric acids then the filtered extract dissolved in hydrochloric acid. The extract is evaluated at 228.8 nm using a cadmium hollow cathode lamp with a spectral width of 0.6 nm.

Acetic acid extractable cadmium in soil[43] is determined on a 0.5 M acetic extract. This is removed from an acid solution as its pyrrolidine dithiocarbamate complex by extraction into chloroform. The chloroform is removed by evaporation and the organic matter destroyed by wet oxidation. This residue is dissolved in hydrochloric acid and cadmium determined by atomic absorption spectrometry utilizing the 228.8-nm emission.

Berrow and Stein[75] have described a procedure based on digestion with aqua regia followed by atomic absorption spectrometry for the determination of cadmium (also iron and zinc), in concentrated nitric acid, hydrochloric acid extracts in soils.

Baucells et al.[44] applied graphite-furnace atomic-absorption spectrometry to the determination of cadmium in soils with a precision of 0.4% at the 69-μL^{-1} cadmium level. The loss of cadmium during the charring cycle was high, preventing the use of any char in the atomization process in order to remove the organic matrix or minimize interference effects. Baucells[44] used a procedure described by Henn[119] in which cadmium is converted to heteropolymolybdate anion with a central metal atom. In this method, the absorbance remained constant when a temperature of 1000 °C was exceeded, and 1200 °C was chosen for good reproducibility. At this temperature, the background absorbance was zero; the matrix was atomized at higher temperatures (>1600 °C).

The low atomization temperature used and the presence of heavy metals in the samples required a subsequent clean-up step at a high temperature (3000 °C) to avoid erroneous readings.

An interference study carried out with calcium, magnesium, sodium, potassium, iron, and aluminum (as nitrates) at two concentration levels, 10 and 100 $\mu g\,mL^{-1}$, showed that iron and calcium are most important interferences.

The determination of cadmium by atomic absorption spectrometry is also discussed in Section 8.59.

8.10.2 Inductively Coupled Plasma Atomic-Emission Spectrometry and Inductively Coupled Plasma Spectrometry

The application of inductively coupled plasma atomic-emission spectrometry and graphite-furnace atomic-absorption spectrometry to the determination of cadmium (and molybdenum) in soils has been discussed by Baucells et al.[44] and others.[107,120,121] Baucells et al. chose the 228.802-nm cadmium line because it is well resolved from the 228.763-nm iron line with the spectrometer used in this work.

Inductively coupled plasma atomic emission spectrometry has proved to be an excellent technique for the direct analysis of soil extracts because it is precise, accurate, and not time-consuming, the level of matrix interference being very low. Of course, the graphite-furnace technique yields better detection limits than the inductively coupled plasma procedure.

Other applications of inductively coupled plasma atomic-emission spectrometry have been discussed.[107,120,121] Further discussion of the determination of cadmium in multi-element analyses by this technique of soils is given in Section 8.59 (inductively coupled plasma atomic-emission spectrometry).

8.10.3 Miscellaneous Techniques

Differential rubber anodic stripping voltammetry, X-ray fluorescence spectrometry, and emission spectrometry have been applied to the determination of cadmium in multimetal samples; see Section 8.59.

Lewin and Beckett[46] have shown that cadmium added to soils treated with sewage will quickly divide between a number of different forms of combination, from some of which it can become available to plants during crop growth. These workers investigated reagents that can extract cadmium and make it available for analysis. Acidified fluoride and EDTA were effective extractants.

Christensen and Lun[47] developed a speciation procedure using a cation-exchange resin (Chelex 100) in a sequential batch/column/batch system for determining free divalent cadmium and cadmium complexes of various stabilities at the cadmium concentrations typically found in landfill leachates (less than $100\,\mu g\,L^{-1}$). Results obtained on standardization solutions containing cadmium and on two actual leachates are included. The leachates had only a small percentage of free divalent cadmium and a large percentage of labile complexes.

Turner et al.[48] discussed the limitations of atomic absorption in trace metals on soils owing to inadequate control of composition and pH of the equilibrium solution. Use of chelating resins is suggested to establish and maintain constant pH and metal activity in a solution of constant ionic strength and composition.

Roberts et al.[49] have discussed the simultaneous extraction and concentration of cadmium and zinc from soil extracts. Extractions were conducted with calcium chloride adjusted to various pH values between 3 and 11. The simultaneous recovery of cadmium and zinc was essentially quantitative over the pH range 4 to 7, with values ranging from 92% to 102%. An extraction of pH 4.5 was adopted. Adequate recoveries were obtained when the procedure was applied to spiked soils.

Carlosena et al.[50] and Hirsch and Banin[245] have conducted studies on the speciation of cadmium in soil. Feng and Barratt[246] showed that microwave dissolution of soil and dust samples with nitric-hydrofluoric acid gave recoveries of cadmium (and lead) of over 90% in 30 min digestion. Various other workers[247–250] have reviewed methods for the determination of cadmium in soils.

8.11 CAESIUM

Caesium contamination on soil is one system for which spectroscopic information would be of great interest.[251,252] The 134 and 137 isotopes decay by γ emission and are formed in high-fission yield. The 137 isotope has a moderately long half-life (30 years). The caesium isotopes comprise one of the lasting health problems from the Chernobyl accident.[253,254] Caesium can be highly mobile in some environments, and geochemically it has many of the same characteristics as potassium as a consequence of similar ionic radii in solution.[252] Hence, there is motivation for understanding the interaction of Cs^+ with naturally occurring mineral surfaces at the molecular level.

Caesium sorption has been extensively investigated, primarily by using sequential extractions together with γ-spectroscopy (for radioisotopes) or atomic absorption for detection.[251,253–256] This approach has been applied to the study of caesium contamination in soils.[257] Caesium was shown to prefer the mineral soil horizons in high organic soils.[258] From these studies, it has been possible to infer mechanistic details: Caesium will tenaciously adhere to adsorption sites and can be supplanted only by K^+ and NH_4^+. It appears to prefer surface “defects,” which have been termed frayed edge and wedge sites.[259–261] However, understanding of caesium soil systems would benefit from direct spectroscope information.

8.11.1 Imaging Time-of-Flight Secondary-Ion-Mass Spectrometry

Groenewald et al.[263] used an imaging time-of-flight secondary-ion-mass spectrometer (SIMS)[262] for characterization of soil particles that had been exposed to caesium iodide solutions. SIMS is well suited to the analysis of Cs^+ because it readily forms gas-phase secondary ions. The SIMS instrument utilizes microfocused primary ion guns, achieving spatial resolutions of less than 1 μm. The ion optics transmit the secondary ions through three electrostatic sectors to a channel-plate detector, such that the spatial information is preserved. They also employed scanning electron microscopy/energy-dispersive X-ray spectroscopy (SEM/EDS). The results showed that Cs^+ could be detected and imaged on the surface of the soil particles readily at concentrations down to 160 ppm, which corresponds to 0.04 monolayer. Imaging revealed that most of the soil surface consisted of aluminosilicate material.

8.12 CALCIUM

8.12.1 Spectrophotometric Method

Qin Xing-chu et al.[51] have described a spectrophotometric method using chlorophosphonazo 30 mA for the determination of exchangeable calcium in soils. A 1 M ammonium acetate extract of the soil is treated with triethanolamine quinolin-8-ol masking agents and the chromogenic reagent and evaluated spectrophotometrically at 630 nm. Recoveries of 99.3–102.7% were obtained from calcium-spiked soil samples. Magnesium, iron, aluminum, manganese, copper, zinc, tungsten, chromium, molybdenum, and lead did not interfere at the levels normally present in soils.

Soil-cation exchangeable capacity is not only an index for evaluating the nutrient and water retention ability of the soil, but also is an important basis for amelioration of soil and for applying rationally, fertilizer. Exchangeable cations absorbed by soil colloid include K^+, Na^+, Ca^{2+}, Mg^{2+}, $A1^{3+}$; and H^+, K^+, Na^+, Ca^{2+}, and Mg^{2+} are exchangeable bases. $A1^{3+}$ and H^+ are exchangeable acids and the sum of these ions is known as the cation exchangeable capacity. Exchangeable Cu^{2+}, Zn^{2+}, and Mn^{2+} are present at negligible concentrations.

According to the pH balance method recommended by Jackson,[52] acetic acid can be used to extract the total amount of exchangeable metal cations from the acidic soil, and the pH change measured carefully. This is then compared with the pH calibration graph of acetic acid so as to obtain the decrease in H^+ in solution. However, this method requires a very precise acidity measurement (generally an accuracy of ±0.01 pH unit when the pH value is within 2.3–2.8). To overcome this difficulty, Qin Xing-chu and Zhu Ying-quan[53] developed the bromophenol blue spectrophotometric method for the determination of exchangeable calcium in soils. The pH range of bromophenol blue is 3.0–4.6, and its acidic and basic forms are yellow and blue, respectively. Comparison of the absorption spectra of the acidic and basic forms of bromophenol blue measured against water over the 420 to 660-nm range shows that the absorption spectrum of the basic form has a maximum at 580nm, whereas the absorption of the acidic form is almost negligible at this wavelength.

A 98% recovery of added calcium is obtained by this procedure. Relative standard deviations are below 1.4% at the 6.5m equivalent calcium per 100g soil level. Satisfactory agreement was obtained in a comparison of analysis performed on a variety of soils by this method and the pH balance method described by Jackson.[52]

8.12.2 Miscellaneous Techniques

Inductively coupled plasma atomic-emission spectrometry, stable isotope dilution, and photon activation analysis have been applied to the determination of calcium in soil. See also Section 8.59.

8.13 CALIFORNIUM

The determination of californium by alpha-spectrometry is discussed in Section 8.59.

8.14 CERIUM

The determination of cerium in multi-element analysis is discussed in Section 8.59.

8.15 CHROMIUM

8.15.1 Spectrophotometric Methods

Qi and Zhu[54] investigated a highly sensitive method for the determination of chromium in soils. In this method, chromium VI is reacted with *o*-nitrophenyl-fluorone in the presence of cetyltrimethyl ammonium bromide to form a purplish-red complex at pH 4.7 to 6.6 by heating at 50°C for 10min. The wavelength of maximal absorbance was 582nm, and the molar absorptive was 111,000 $L\,mol^{-1}\,cm^{-1}$. Interference due to copper(II), iron(III), and aluminum(III) was eliminated by the addition of a masking reagent containing potassium fluoride *trans*-1,2-diaminocyclohexanetetra-acetic acid and potassium sodium tartrate. This method was more sensitive than the diphenyl-carbazone method.

Fodor and Fischer[59] have investigated problems of chromium speciation in soils. When employing spectrophotometric detection, only a method based on the diphenyl-carbazide reaction was found suitable for chromium speciation analysis.

8.15.2 Atomic Absorbance Spectrometry

Smith and Lloyd[55] determined chromium VI in soil by a method based on complexation with sodium diethyl-dithiocarbamate in pH 4 buffered medium followed by extraction of the complex with methyl-isobutyl ketone and analysis of the extract by atomic absorption spectrometry.[56] Levels of chromium V between 90 and 176 mgL^{-1} were found in pastureland on which numerous cattle fatalities had occurred.

Chakraborty et al.[264] determined chromium in soils by microwave assisted sample digestion followed by atomic absorption spectrometry without the use of a chemical modifier.

The determination of chromium by atomic absorption spectrometry is also discussed in Section 8.59.

8.15.3 Miscellaneous Techniques

Inductively coupled plasma, atomic emission spectrometry, X-ray fluorescence spectroscopy, emission spectrometry, spark-source spectrometry, neutron activation analysis, and photon activation analysis have all been applied to the determination of chromium in soils, as discussed in Section 8.59.

The sequential extraction of chromium from soils has been studied.[57] A three-step sequential extraction scheme has been proposed using acetic acid, hydroxylamine hydrochloride, and ammonium acetate as extracting agents. Steps 1 and 2 were measured by electrothermal atomic absorption spectrometry. Step 3 was measured by flame atomic absorption spectrometry.

Interfering effects when measuring chromium in soils were circumvented by the use of a 1% 8-hydroxyquinoline suppressor agent.[58]

Fodor and Fischer[59] have investigated chromium speciation in soils.

Prokisch et al.[60] have described a simple method for determining chromium speciation in soils. Separation of different chromium species was accomplished by the use of acidic-activated aluminum oxide. Polarographic methods have been applied in speciation studies on chromium VI in soil extracts.[61] Milacic et al.[265] have reviewed methods for the determination of chromium VI in soils.

8.16 COBALT

8.16.1 Spectrophotometric Method

A method based on measurement of the ammonium pyrollidine-dithiocarbamate complex at 240.7 nm has been described for the determination of 0.5 M acetic acid[62] and nitric-perchloric-acid-soluble cobalt in soils.[63]

8.16.2 Miscellaneous Techniques

Atomic absorption spectrometry, inductively coupled plasma atomic emission spectrometry, neutron activation analysis, photon activation analysis, differential pulse anodic scanning voltammetry, X-ray fluorescence spectroscopy, and emission spectrometry have been applied to the determination of cobalt in soil. See Section 8.59.

8.17 COPPER

8.17.1 Atomic Absorption Spectrometry

Atomic absorption spectrometry has been used to determine 0.05 M ammoniacal ethylene diamine tetracetic acid-extractable[64] and nitric-perchloric acid-soluble copper in soils.[65] The determination of copper by atomic absorption spectrometry is also discussed in Section 8.59.

8.17.2 Neutron Activation Analysis

Fast neutron activation analysis has been studied[230] as a screening technique for copper (and zinc) in waste soils. Experiments were conducted in a sealed tube neutron generator and a germanium γ-ray detector.

8.17.3 Miscellaneous Techniques

Atomic absorption spectrometry inductively coupled plasma atomic emission spectrometry, differential pulse anodic stripping, voltammetry X-ray, and fluorescence spectroscopy have all been applied to the determination of copper in soils; see Section 8.59.

Mesuere et al.[266] and Gerringa et al.[267] have reviewed methods for the determination of copper in soils. Residual copper II complexes have been determined in soil by electron spin resonance spectroscopy.

8.18 CURIUM

Alpha spectroscopy has been applied to the determination of curium in soils, as discussed in Section 8.59.

8.19 EUROPIUM

8.19.1 Neutron Activation Analysis

The determination of europium in soil is discussed in Section 8.59.

8.20 HAFNIUM

8.20.1 Neutron Activation Analysis

The determination of hafnium in soil is discussed in Section 8.59.

8.21 INDIUM

8.21.1 Atomic Absorption Spectrometry

The determination of indium in soil is discussed in Section 8.59.

8.22 IRIDIUM

8.22.1 Neutron Activation Analysis

Stefanov and Daieve[67] have described a neutron activation analysis procedure for the determination of down to 30 ng of iridium in soil. Short activation times were used to avoid activation of other trace impurities.

The determination of iridium by atomic absorption spectrometry is discussed in Section 8.59.

8.23 IRON

8.23.1 Spectrophotometric Method

Jayman et al.[68] have pointed out that the iron and the aluminum complexes of phenanthroline exhibit identical absorption characteristics. Attempts to mask aluminum to facilitate the determination of iron were unsuccessful. These workers have described an alternate procedure in which aluminum and phosphates are separated from iron and then the iron can be simply determined without interferences.

A nitric acid-perchloric acid digest of the soil sample is reacted with phenanthroline prior to spectrophotometric evaluation at 490 nm.

8.23.2 Miscellaneous Techniques

Atomic absorption spectrometry, individually coupled plasma atomic emission spectrometry, neutron activation analysis, X-ray fluorescence spectroscopy, and emission spectrometry have all been applied to the determination of iron in metals, as discussed in Section 8.59.

8.24 LANTHANUM

8.24.1 Neutron Activation Analysis

The determination of lanthanum in soil is discussed in Section 8.59.

8.25 LEAD

Most of the lead in soil exits in sparingly soluble forms. When 2784 ppm of lead nitrite were added to soil, it was found that after three days the soluble lead content was only 17 ppm.[69] It is to be expected that all ions in nature will accumulate as their less soluble compounds, such as oxides, carbonates, silicates, and sulfates, the relative proportions of each depending on the nature of the soil and on solubility.

Several acids and acid mixtures have been used for the digestion of soil samples prior to the analysis of lead, including nitric-acid–perchloric acid (1+1),[70] hydrochloric acid,[71] perchloric acid,[72] nitric acid–hydrofluoric acid (1+1),[73,74] and aqua regia.[75]

8.25.1 Spectrophotometric Method

A spectrophotometric method employing ammonium pyrolidine-diethyl-dithiocarbamate[76] as chromogenic reagent has been applied to the determination of 0.5 M acetic extractable lead in soils. Savvin et al.[268] have discussed a spectrophotometric method for the determination of lead in soils.

8.25.2 Atomic Absorption Spectrometry

Standard official methods have been described for the determination of nitric–perchloric acid-soluble lead in soil.[77] Atomic absorption measurements were performed by measuring at 217-nm emission from a lead hollow cathode lamp at a special hand width of 1.0 nm. Tills and Alloway[80] investigated the speciation of lead in soil solution using a fractionation scheme, ion exchange chromatography, and graphite-furnace atomic-absorption spectrophotometry. The sources of contamination of soils from various sites and its chemical properties are described. The percentages of lead in cationic, anionic, neutral, and less polar organic complexes were determined and discussed with respect to organic matter content and pH. No direct relationship was established between total lead content of the soil and total lead content in soil solution.

Using a palladium-magnesium nitrate mixture as chemical modifier, Hinds and Jackson[269] effectively delayed the atomization of lead until atomic absorption spectrometer furnace conditions were nearly isothermal. This technique was used to determine lead in soil slurries. Zhang et al.[270] investigated the application of low-pressure electrothermal atomic absorption spectrometry to the determination of lead in soils.

Hinds et al.[271,272] investigated the application of slurry electrothermal atomic absorption spectrometry to the determination of lead in soils. Hinds and Jackson also investigated the application of vortex-mixing slurry graphite-furnace atomic-absorption spectrometry to the determination of lead in soils. The determination of lead is also discussed in Section 8.59.

8.25.3 Atomic Fluorescence Spectroscopy

Rigin and Rigina[81] determined lead in soil using flameless atomic fluorescence spectroscopy on an extract of the sample that had been preconcentrated by electrolysis on a silanized graphite rod. The limit of detection was 15 pg lead, and the relative standard deviation was 0.4.

8.25.4 Inductively Coupled Plasma Atomic-Emission Spectrometry and Inductively Coupled Plasma Mass Spectrometry

The determination of lead is discussed in Section 8.59.

8.25.5 Anodic Stripping Voltammetry

Somer and Aydin[78] determined the lead content of soil adjacent to roads in Turkey using anodic stripping voltammetry. These workers found that aqua regia was the most suitable acid for extracting lead from roadside soil. The lead salt that may be trapped in the silicate crystal lattice of the soil was brought into solution by extracting the soil in acid overnight. To avoid the possibility of the presence of undissolved lead salts even after digestion, EDTA was added to the digested sample in order to ensure quantitative dissolution of lead. The lead content of this solution was determined by anodic stripping voltammetry and platinum at 6.0 mv S^{-1}. Calibration was achieved by standard additions of 10 to -5 M lead nitrite solution.

Using this procedure, the lead content of samples of soil adjacent to some Ankara roads was determined. As expected, the lead content of the soil increased with increasing traffic volume.

Differential pulse-scanning voltammetry has been applied to the determination of lead in soils.[273] The determination of lead in soil by this technique is also discussed in Section 8.59.

8.25.6 Polarography

Sakharov[79] determined lead in soil polarography by digesting the sample with sodium carbonate followed by dissolution in hydrochloric acid. He found that when hydrochloric, sulfuric, or nitric acids were used as a digestion medium instead of sodium carbonate, no lead could be detected in the resulting solution.

8.25.7 X-ray Fluorescence Spectroscopy

Wegrzynek and Holynska[274] have developed a method for the determination of lead in arsenic-containing soils by energy-dispersive X-ray fluorescence spectroscopy. Correction for arsenic interference was based on the use of an arsenic-free reference sample. The determination of lead is also discussed in Section 8.59.

8.25.8 Miscellaneous Techniques

Several investigations have reviewed the determination of lead in soils.[201,208–211] Lead has been determined in soil by using a slurry sampling technique with lead nitrate and magnesium nitrate as chemical modifiers.[275] Results were in good agreement with known concentrations of a standard reference material. Feng and Barratt[246] showed that microwave dissolution of soil and dust samples with nitric-hydrofluoric acid gave recoveries of lead (and cadmium) of over 90% in 30 min digestion.

Chen and Hong[208] found that 5-carboxy methyl-L-cysteine was especially effective for the chelating extraction of lead from contaminated soils. The chelator could be recovered and reused over consecutive runs with no loss in performance. Atomic absorptions spectrometry, inductively coupled plasma, atomic emission spectrometer, photoactivation analysis, emission spectrometry, and anodic scanning voltammetry have been applied to the determination of lead in multimetal mixtures, as discussed in Section 8.59.

8.26 MAGNESIUM

8.26.1 Atomic Absorption Spectrometry

This technique used to determine magnesium in 1 M ammonium nitrate extracts of soil.[82] The determination of magnesium in soils by this technique is also discussed in Section 8.59.

8.26.2 Miscellaneous Techniques

Inductively coupled plasma, atomic emission spectrometry, and photon activation analysis have also been applied to the determination of magnesium in soils, as reported in Section 8.59.

8.27 MANGANESE

8.27.1 Spectrophotometric Method

Alekseeva and Davydova[83] determined micro amounts of manganese II in sulfuric and hydrofluoric extracts of clays by a kinetic spectrophotometric method involving the oxidation of *o*-dianisidine by potassium periodate. The reaction between *o*-dianisidine and potassium periodate is catalyzed by manganese II. Spectrophotometric measurements were conducted at 460 nm.

8.27.2 Atomic Absorption Spectroscopy

Atomic absorption spectrometry utilizing the 403-nm emission has been employed to determine exchangeable and easily reduced manganese in M ammonium acetate pH 7 extracts of soils.[84] Atomic absorption spectrometry has been extensively used in the multimetal methods for the determination of manganese in soils; see Section 8.59.

8.27.3 Miscellaneous Techniques

Inductively coupled plasma, atomic emission spectrometry, and X-ray fluorescence spectroscopy have all been applied to the determination of manganese, as discussed in Section 8.59.

8.28 MERCURY

8.28.1 Spectrophotometric Method

Kimura and Miller[85] have shown that at room temperature, reduction with tin(II) and aeration are suitable for quantitative separation of microgram quantities of mercury(II) from sulfuric and nitric acid extracts of soil over wide ranges of concentrations. Mercury is concentrated during the separation and is determined by a direct photometric dithizone procedure. The standard deviation for a single determination of 0.05 μg mercury was in the 0–0.5 μg range.

Kamburova[276] reported a spectrophotometric method based on the formation of the mercury-triphenyltetrazolium chloride complex for the determination of mercury in soils.

8.28.2 Atomic Absorption Spectrometry

Cold vapor (or flameless) atomic absorption spectrometry is the method of choice for the determination of mercury in soils.[86–97] Ure and Shand[88] investigated various procedures for the digestion of soil samples prior to analysis by cold-vapor atomic-absorption spectrometry. They found good agreement between two digestion methods, involving digestion of the soil with the mixture of nitric and sulfuric acids and digestion with potassium permanganate and oxygen flask combustion over acid potassium permanganate solution.

Agemian and Chau[90] have reported an improved digestion method for the extraction of mercury from soils and clays in which the sample is digested at 60 °C with sulfuric acid–nitric acid (2 + 1) containing a trace amount of hydrochloric acid, and subsequently oxidized with permanganate and persulfate solutions prior to atomic absorption spectrometry. With this procedure, mercury is successfully recovered from organic matter and resistant inorganic forms such as mercury(II) sulfide. Unlike digestion with aqua regia, this procedure is simple and safe, and is applicable to the digestion of a large number of samples simultaneously and is sensitive and capable of determining down to 0.2 $\mu g\,L^{-1}$ of mercury. The method can be adapted to the automated cold vapor and flame atomic absorption techniques and is therefore ideal for routine monitoring.

Kuwae et al.[91] used high-frequency induction heating of the sample for the rapid release of mercury from soil samples prior to absorption in 0.5% potassium permanganate in 0.5 M sulfuric acid and cold vapor atomic absorption spectrometry. Mercury recoveries of 98–99% were obtained following a 5-min induction heating period, which is considerably more rapid than wet digestion procedures.

Nicholson[92] has described a rapid thermal decomposition technique for the determination of mercury in soils (and sediments). The method is applicable to samples rich in organic matter. In this method, the sample is heated in a nickel boat at 650 °C and the products collected on gold. The gold is then heated to

750 °C to release mercury, which is swept into a cold-vapor atomic-absorption spectrophotometric-cell gold film. In general, results obtained by this method were in the 0.6 to 0.33 $\mu g\,kg^{-1}$ range, some 14% higher than those obtained by wet oxidation.[93]

Other workers[94] have used flameless atomic-absorption spectrometry or atomic fluorescence spectrometry to determine mercury in amounts down to 10 $\mu g\,kg^{-1}$ in soil.

Cold vapor atomic absorption spectrometry and atomic fluorescence spectrometry (253-nm emission) have been applied to the determination of down to 0.01 $mg\,kg^{-1}$ of mercury in soils and sediments.[94]

Sakamoto et al.[277] have shown that the differential determination of different forms of mercury in soil can be accomplished by successive extraction and cold-vapor atomic absorption spectrometry.

Azzaria and Aftabi[278] showed that stepwise, compared to continuous heating of soil samples before determination of mercury by flameless atomic absorption spectrometry gives increased resolution of the different phases of mercury. A gold-coated graphite-furnace atomic-absorption spectrometry has been used to determine mercury in soils.[279]

Bandyopadhyay and Das[280] extracted mercury from soils with the liquid anion-exchanger Aliquat-336 prior to determination by cold-vapor atomic-absorption spectrometry. The application of atomic absorption spectrometry to the determination of mercury in soils is also discussed in Section 8.59.

8.28.3 Anodic Stripping Voltammetry

Voltammetric methods have been used to determine mercury in soil composts. The amount of mercury leaching from composts was shown to be very low.[281]

8.28.4 Miscellaneous Techniques

A study by Rasemann demonstrated by example of a nonuniformly contaminated site, to what extent mercury concentrations depend on the method of handling soil samples between sampling and chemical analysis.[98] Sample pretreatment contributed substantially to the variance in results and was of the same order as the contribution from sample inhomogeneity. Welz et al.[283] and Baxter et al.[284] have conducted speciation studies on mercury in soils. Lexa and Stulik[285] employed a gold-film electrode modified by a film of tri-n-octylphosphine oxide in a PVC matrix to determine mercury in soils. Concentrations of mercury as low as 0.02 ppm were determined.

Cherian and Gupta[286] have described a simple field test for the determination of mercury in soil. Saouter et al.[287] showed that the use of hydrogen peroxide as an oxidizing agent for organics in soils can result in loss of mercury. This is because hydrogen peroxide can act as a reducing agent for mercury compounds. Neutron activation analysis has been used to determine mercurium soils.[282]

8.29 MOLYBDENUM

8.29.1 Spectrophotometric Methods

Various chromogenic reagents have been used for the determination of molybdenum in soil. These include toluene 3.4 dithiol[99,100] and iron thiocyanate.[101,102] The complexes formed with toluene 3.4 dithiol are extracted with iso-amyl acetate prior to spectrophotometric evaluation at 680 nm. Down to 0.1 µg, molybdenum can be determined in the soil sample. The iron thiocyanate method determined total molybdenum[101] or ammonium oxalate-oxalic acid soluble molybdenum[102] in soils utilizing the 470-nm absorption of the complex.

8.29.2 Atomic Absorption Spectrometry

Earlier atomic absorption methods[104–106] from the determination of molybdenum in soils employed a preliminary solvent extraction step to improve sensitivity in view of the low concentrations of molybdenum occurring in most soils.[40,107] Baucells et al.[44] developed a graphite-furnace atomic-absorption procedure that was capable of determining down to 8.4 pg of molybdenum in a soil matrix solution with a precision of 4% for 100 $\mu g L^{-1}$ molybdenum. These workers showed that a char temperature of 1500 °C and an atomization temperature of 2400 °C are optimum for molybdenum. Under these conditions, the background absorbance is 0.015. However, the use of a char temperature of 700 °C in part prevents attack of the graphite, gives a better precision, and the background absorbance in the atomization step is only 0.030.

The most important interferences in this method were given by aluminum, iron, and magnesium.

Five replicate determinations of molybdenum in a siliceous soil sample gave a precision of 17.1% with a mean concentration of 35 $\mu g L^{-1}$.

The application of atomic absorption spectrometry to the determination of molybdenum is also discussed in Section 8.59.

8.29.3 Inductively Coupled Plasma Atomic-Emission Spectrometry

Atomic emission spectrometry is not sufficiently sensitive to determine molybdenum at the levels at which it occurs in soils. Due to its greater intrinsic sensitivity, inductively coupled plasma atomic-emission spectrometry is capable of achieving the required sensitivity.

Manzoori et al.[40] utilized inductively coupled plasma optical-emission spectrometry to determine down to 0.01 $mg L^{-1}$ molybdenum in 1 M ammonium acetate extracts of soils. In this method, a 50 g sample of soil was shaken overnight with 800 mL of neutral 1 M ammonium acetate. The whole extract was filtered through an 18.5 cm Whatman no. 540 filter paper and the residue dissolved in dilute nitric acid.

The most sensitive molybdenum line (379.8 nm) was used. There is some interference from the iron line at 379.8 nm. Between 0.2 and 3.1 mg L^{-1}, molybdenum was found in soil extracts, equivalent to 1–15.5 mg kg^{-1} in soils.

Thompson and Zao[103] carried out rapid determination of 0.4–40.7 mg kg^{-1} molybdenum in soils by solvent extraction, followed by inductively coupled plasma atomic emission spectrometry. In soils, molybdenum is completely oxidized and largely associated as the MnO_4 ion with iron III oxide materials, although it may also be associated with organic matter.

The molybdenum is solubilized by treatment of the sample with 6 M hydrochloric acid in capped tubes at 120 °C. It is then extracted from the same medium into heptane-2-1, in which it is determined directly by inductively coupled plasma atomic emission spectrometry using the 281–332 nm molybdenum line. Extraction of iron, which can otherwise interfere with the determination, is minimized by the presence of a reducing agent.

This procedure is capable of providing a detection limit of about 0.06 μg g^{-1} of molybdenum, and a coefficient of variation of 3% at concentrations well above the detection limit. Baucells et al.[44] applied ICPAES to the determination of very low levels of molybdenum (and cadmium) in soils. Of the most sensitive molybdenum lines, 202.030 nm proved to be an excellent analytical line; there are two iron lines, at 201.99 and 202.074 nm, but they are free from interference. Other molybdenum-sensitive lines are subject to interference from iron, chromium, and vanadium. These results agree with those of Maessen et al.[108]

The application of inductively coupled plasma atomic absorption spectrometry to the determination of molybdenum in soils is also discussed in Section 8.59.

8.29.4 Miscellaneous Techniques

The determination of molybdenum in soil is of interest because molybdenum is necessary for normal crop growth, but an excess in forage has a toxic effect on ruminants. The absorption of molybdenum by plants is influenced by other soil components, especially extractable iron, pH, and organic matter. The average abundance of molybdenum in soils is about 2 ppm, but deficient soils can have much less than 1 ppm.[112]

Jiao et al.[288] and Rowbottom[289] have reviewed methods for the determination of molybdenum in soils.

Neutron activation analysis and emission spectrometry[112–114] have also been applied to the determination of molybdenum; see Section 8.59.

8.30 NEPTUNIUM

8.30.1 Miscellaneous Techniques

Kim et al.[290] have demonstrated good agreement between methods for determining neptunium-237 in soils, based on inductively coupled plasma mass spectrometry, neutron activation analysis, and alpha-spectrometry.

8.31 NICKEL

8.31.1 Spectrophotometric Method

A spectrophotometric method has been used to determine 0.5 M acetic acid extractable nickel in soil.[122]

8.31.2 Atomic Absorption Spectrometry

This technique has been applied to the determination of nitric-perchloric acid-soluble nickel in soils.[123] The determination of nickel by atomic absorption spectrometry is also discussed in Section 8.59.

8.31.3 Miscellaneous Techniques

Inductively coupled plasma atomic emission spectrometry, spark-source mass spectrometry, differential-pulse anodic-stripping voltammetry, X-ray fluorescence spectrometry, photon activation analysis, and emission spectrometry have all been applied to the determination of molybdenum in multimetal mixtures, as discussed in Section 8.59.

8.32 PALLADIUM

8.32.1 X-ray Absorption Fine-Structure Analysis

Manceau et al.[211] studied the application of X-ray absorption fine-structure analysis (EXAFS) to the speciation and quantification of the various trace metals in solid materials. Palladium was studied in particular.

8.33 PLUTONIUM

8.33.1 Inductively Coupled Plasma Mass Spectrometry

Kim et al.[291] determined the plutonium-240 to plutonium-239 ratio in soils using the fission track method and inductively coupled plasma mass spectrometry.

8.33.2 Gas Chromatography

Packed column gas chromatography has been used to determine various plutonium isotopes in soils.[292]

8.33.3 Alpha-Spectrometry

Dienstbach and Bachmann[125] have determined plutonium in amounts down to 20 fCiPug^{-1} soil in sandy soils by an automated method based on gas chromatographic separation and alpha-spectrometry. In this procedure, the sample is decomposed completely by hydrogen fluoride. The hydrogen fluoride is evaporated and the residue is chlorinated. Plutonium is separated from the sample by volatilization and separation of the chlorides in the gas phase. The plutonium is deposited on a glass disk by condensation of volatilized plutonium chloride. The concentration of plutonium is then determined by alpha-spectroscopy.

Sekine et al.[126] used alpha-spectrometry to determine plutonium (and americium) in soil. The chemical recovery of plutonium was 51–99% and averaged 81%, while for americium the recovery was 60–70%. The method is coupled with the liquid–liquid extraction stage taking about two days less than would the ion exchange method. A complete analysis takes about 1 week.

Talvitie[124] has described a radiochemical method for the determination of plutonium in soil based on chromatography on an anion exchange resin of a 9 M hydrochloric acid extract of the sample. Following clean-up, plutonium is desorbed by reductive elution with 1.2 M hydrochloric acid 30% hydrogen peroxide (50:1) at pH 2 followed by alpha-particle counting. The lowest detectable activity for 1000 m counts was 0.02 pCi of ^{239}PU, which is sufficient to detect global nuclear contamination in 1 g of soil.

Various workers[293,294,380] have discussed mass spectrometric and other methods for the determination of plutonium in soils. Plutonium in soils has been quantified using plutonium-238 as a yield trace.[295] Hollenbach et al.[216] used flow-injection preconcentration for the determination of ^{230}Th, ^{234}U, ^{239}Pu, and ^{240}Pu in soils. Detection limits were improved by a factor of about 20, and greater freedom from interferences was observed with the flow-injection system compared to direct aspiration.

8.34 POLONIUM

8.34.1 Miscellaneous Techniques

Various workers[296,297] have reviewed methods for the determination of polonium-210 in soils.

8.35 POTASSIUM

8.35.1 Flame Photometry

Potassium has been determined in 1 M ammonium nitrate extracts of soil by flame photometry.[127]

8.35.2 Inductively Coupled Plasma Atomic-Emission Spectrometry

This technique has been used to determine potassium in soils, as discussed in Section 8.59.

8.35.3 Stable Isotope Dilution

This technique has been used to determine potassium, as discussed in Section 8.59.

8.36 RADIUM

8.36.1 Miscellaneous Techniques

Various workers have discussed methods for the determination of radium-226 in soils.[298,299]

8.37 RUBIDIUM

8.37.1 Spark-Source Mass Spectrometry

This technique has been used to determine rubidium, as discussed in Section 8.59.

8.37.2 Stable Isotope Dilution

This technique has been used to determine rubidium, as discussed in Section 8.59.

8.38 SCANDIUM

8.38.1 Neutron Activation Analysis

This technique has been used to determine scandium, as discussed in Section 8.59.

8.39 SELENIUM

The fate of selenium in natural environments such as soil and sediments is affected by a variety of physical, chemical, and biological factors that are associated with changes in its oxidation state. Selenium can exist in four different oxidation states (-II, 0, IV and VI) and as a variety of organic compounds. The different chemical forms of selenium can control selenium solubility and availability to organisms. Selenate (Se(VI)) is the most oxidized form of selenium, is highly soluble in water, and is generally considered to be the most toxic form. Selenite (Se(VI), Se(IV)) occurs in oxic to suboxic environments and is less available to organisms because of its affinity to sorption sites of sediment and soil constituents. Under anoxic conditions, elemental selenium and selenide(-II) are the thermodynamically stable forms. Elemental selenium is relatively insoluble, and selenide(-II) precipitates as metal selenides(-II) of very low solubility. Organic selenium(-II) compounds such as selensomethionine and selenocystine can accumulate in soil and sediments or mineralize to inorganic selenium. Therefore, Se(VI) and organic selenium(-II) are the most important soluble forms of selenium in natural environments.

8.39.1 Spectrofluorimetric Methods

A widely used method for the routine determination of selenium in soils is based on the reaction between selenium and 2,3-diaminonaphthalene to form a fluorescent piazoselenol product.[142–147] Complete destruction of organic matter is necessary in order to avoid the possibility of fluorescent interference from these amounts of residual material, and is achieved by treatment of the sample with hot acidic oxidizing mixtures. Excessive temperature or prolonged heating bring about the loss of selenium by volatilization and considerable ingenuity is required to devise methods that will satisfy these conflicting requirements. These considerations and problems associated with the purity of the 2,3-diaminonaphthalene reagent are now tending to preclude recommendation of fluorometric methods.

8.39.2 Atomic Absorption Spectrometry

Hydride generation methods are finding increasing favor for the determination of selenium in soils and sediments. This method consists of measuring the atomic absorption of selenium hydride formed as a result of reduction of selenium and its compounds with different reducing mixtures such as sodium borohydride or, occasionally, zinc-stannous chloride-potassium iodide. Hydride generation techniques are about three orders of magnitude more sensitive for determining selenium than are classical flame ionization techniques, a detection limit of $0.2\,ng\,g^{-1}$ being achievable. They have an additional advantage of separating selenium from the matrix before atomization, thus avoiding interferences inherent to the conventional atomic absorption technique. Practical working ranges for selenium are 3–250 $\mu g\,L^{-1}$, up to 250 $\mu g\,L^{-1}$, and up to 0.12 $\mu g\,L^{-1}$ respectively, for flame atomic absorption, atomic absorption, and vapor generation methods.[128–135]

The most intense resonance line of selenium (196.03 nm) corresponds to a range near to the vacuum ultraviolet. Moreover, the most frequently applied air acetylene flame absorbs about 55% of radiation intensity of the light source. When using electrodeless discharge lamps and an air-acetylene flame, appreciably lower detection limits can be achieved by application of a deuterium lamp for background correction. The argon-hydrogen flame is often used for augmentation of sensitivity, but it increases interferences, too. Extraction has also been attempted[136] as a means of improving sensitivity, but in selenium determination a re-extraction to a water solution is necessary.

The addition of nickel enhances significantly the sensitivity for selenium by about 30% and allows higher ashing temperatures (1000 °C) without losses.[137–139] Other elements capable of forming selenides (i.e., barium, copper, iron, magnesium, and zinc) did not interfere and arsenic interference was minimized. A detection limit of 10–12 $\mu g\,kg^{-1}$ selenium has been achieved using a graphite electrothermal furnace and background correction with a deuterium lamp.[140]

A method has been reported[31] for determining total arsenic (and selenium) in soils based on atomic-absorption spectrometry and flow-injection analysis. The method exhibits good recoveries and detection limits below 1 $\mu g\,L^{-1}$ for an injection volume of 160 µL.

Some drawbacks for the speciation of selenium using hydride-generation atomic-absorption spectrometry have been found by some researchers. Thus Se(IV) is recovered poorly from many samples after a reduction with 6N hydrochloric acid 100 °C. The addition of ammonium persulfate increased the recovery of Se(VI). However, part of the organic Se(-II) was included in the value reported for Se(VI) due to the oxidation of organic Se(-II) by persulfate.

To overcome these drawbacks, Zhang et al.[300] developed a method to determine organic selenium(-II) in soils and sediments. In this method, persulfate is used to oxidize organic selenium(-II) and manganese oxide is used as an indicator for oxidation completion. This method was used to determine selenium speciation in soil-sediments and agricultural drainage water samples collected from the western United States. Results showed that organic selenium can be quantitatively oxidized to selenite without changing the selenate concentration in the soil-sediment extract and agricultural drainage water and then quantified by hydride generation atomic absorption spectrometry. Recoveries of spiked organic selenium and selenite were 96–105% in the soil-sediment extracts and 96–103% in the agricultural water.

The application of atomic absorption spectrometry to the determination of selenium in soils is also discussed in Section 8.59.

8.39.3 Nondispersive Atomic Fluorescence Spectrometry

Azad et al.[141] determined selenium or tellurium in amounts down to 10 $ng\,mL^{-1}$ in nitric acid digests of soil by nondispersive atomic fluorescence spectrometry using an argon-hydrogen flame and the hydride-generation technique utilizing sodium tetrahydroborate(III). Recoveries were in the range of 94–107%.

8.39.4 Inductively Coupled Plasma Atomic-Emission Spectrometry

Pahlavanpour et al.[148] determined traces of selenium in nitric acid digested soils by the introduction of hydrogen selenide into an inductively coupled plasma source (in emission spectrometry).

The soils digest containing the selenium is reduced with sodium tetrahydroborate(III) and the hydrogen selenide formed is swept into an inductively coupled plasma source for determination by emission spectrometry. Calibrations are linear from the detection limit (1 ng mL^{-1}) to about 1000 ng mL^{-1} of selenium. The detection limit obtained is comparable to most of the values reported for selenium by atomic absorption-hydride systems, and the linear calibration range is much greater.

With few exceptions, the results given by fluorometry, chromatography, and neutron activation analysis compare well with those obtained by the method described above.

The application of hydride-generation inductively coupled plasma mass spectrometry to determine selenium in soils has also been discussed by McCurdy et al.[301] Selenium is discussed further in Section 8.59.

8.39.5 Neutron Activation Analysis

Neutron activation analysis has been used to determine selenium in soil.[149–154] The determination of selenium in soil by neutron activation analysis is also discussed in Section 8.59.

8.39.6 Miscellaneous Techniques

Agemian and Bedek[155] have described a semiautomated method for the determination of total selenium in soils.

Dong et al.[179] used mixtures of phosphoric acid, nitric acid, and hydrogen peroxide in the digestion of soils prior to the determination of selenium.

Bem et al.[156] reviewed methods for the determination of selenium in soil. These methods include neutron activation analysis, atomic absorption spectrometry, gas chromatography, and spectrophotometric methods. Square-wave cathodic-stripping voltammetry has been used to determine selenium in soils.[302] Selenium has been directly determined in soils by PIXE.[303]

8.40 SILICON

8.40.1 Atomic Absorption Spectrometry

Pellenberg et al.[157] analyzed soils and river sediment for silicon content by nitrous oxide-acetylene flame atomic absorption spectrophotometry. They showed that total carbon and total carbohydrates both correlate well with silicon.

8.40.2 Inductively Coupled Plasma Atomic-Emission Spectrometry

Que-Hee and Boyle[158] analyzed soils for total silicon using Parr bomb digestion with hydrofluoric-nitric-perchloric acids followed by inductively coupled plasma atomic absorption spectrometry. The determination of silicon by this technique is also discussed in Section 8.59.

8.41 SILVER

8.41.1 Atomic Absorption Spectrometry

The determination of silver by this technique is discussed in Section 8.59.

8.42 SODIUM

8.42.1 Atomic Absorption Spectrometry

Sodium has been determined in 1 M ammonium extracts of soil by this method.[159]

8.42.2 Inductively Coupled Plasma Atomic-Emission Spectrometry

Sodium has been determined by this technique, as discussed in Section 8.59.

8.43 STRONTIUM

8.43.1 Beta Spectrometry

Martin[160] described a beta-counting technique for the determination of $1 \times 10^{-1}\,Ci\,g^{-1}$ strontium 89 and 90 in soils. In this method, fusion of the soil with potassium fluoride and potassium pyrosulfate converts strontium to the ionic state. After dissolving the fusion cake, strontium sulfate is co-precipitated with lead and calcium, which are preferentially dissolved into EDTA and discarded. Subsequently, strontium sulfate is dissolved in EDTA and hydrolytic elements are separated on ferric hydroxide. Strontium sulfate is precipitated from EDTA at a lower pH, dissolved in DTPA, and set aside for ^{90}Y ingrowth. After ingrowth, the strontium sulfate is re-precipitated and the ^{90}Y is extracted from the supernate into bis(2-ethylhexyl) phosphoric acid. When both ^{89}Sr and ^{90}Sr are being determined, barium chromate is separated from a DTPA solution of strontium. Strontium sulfate is precipitated from the barium chromate supernate and counted for ^{89}Sr plus ^{90}Sr. Yttrium-90 is counted on yttrium oxalate to determine ^{90}Sr, and ^{89}Sr is determined by difference.

8.43.2 Miscellaneous Techniques

Akcay et al.[304] have shown that extraction of total strontium using an ultrasonic extraction procedure was not as good as was achieved by using conventional extraction methods. Stella et al.[305] and Ryabukhin et al.[306] have discussed the determination of radiostrontium in soils.

Photon activation analysis, inductively coupled plasma atomic emission spectrometry, and stable isotope dilution methods have been used in the determination of strontium in multimetal samples, as discussed in Section 8.59.

8.44 TAMERIUM

8.44.1 Emission Spectrometry

The determination of tamerium is discussed in Section 8.59.

8.44.2 Neutron Activation Analysis

The determination of tamerium is discussed in Section 8.59.

8.45 TANTALUM

8.45.1 Neutron Activation Analysis

The determination of tantalum is discussed in Section 8.59.

8.46 TECHNETIUM

8.46.1 Inductively Coupled Plasma Mass Spectrometry

Tagami and Uchida[307] have reported a simple method for the determination of technetium-99 in soil by inductively coupled plasma mass spectrometry.

8.46.2 Miscellaneous Techniques

Morita et al.[308] and Harvey et al.[309] have reviewed methods for the determination of technetium in soils.

8.47 TELLURIUM

8.47.1 Emission Spectrometry

The determination of tellurium is discussed in Section 8.59.

8.48 TERBIUM

8.48.1 Emission Spectrometry

The determination of terbium is discussed in Section 8.59.

8.48.2 Neutron Activation Analysis

The determination of terbium is discussed in Section 8.59.

8.49 THALLIUM

8.49.1 Anodic Stripping Voltammetry

Opydo[311] used anodic stripping voltammetry to determine thallium in soil extracts in the presence of a large excess of lead.

8.49.2 Miscellaneous Techniques

Atomic absorption spectrometry[310] has been used to determine thallium in soil, as discussed in Section 8.59.

Chikhalikar et al.[15] have studied the speciation of thallium (and antimony) in soil. Luckaszewski and Zembrzuski[312] and Sagar[313] have discussed the determination of thallium in soils.

8.50 THORIUM

8.50.1 Inductively Coupled Plasma Mass Spectrometry

Toole et al.[314] and Shaw and Francois[315] determined thorium (and uranium) in soils by inductively coupled plasma mass spectrometry.

8.50.2 Miscellaneous Techniques

Parsa[316] described a sequential radiochemical method for the determination of thorium (and uranium) in soils. Mukhtar et al.[317] have described a laser fluorometric method for the determination of thorium (and uranium) in soils. Stream digestion has been employed in the preparation of soil samples for the determination of thorium (and uranium).[318] To determine thorium (and uranium) in soils, fluorescent X-rays were measured by the use of germanium planar detector and chemometric techniques. No sample preparation was required in this method.[319] Various other workers have been reported for the determination of thorium (and uranium) in soils.[320,321]

Neutron activation analysis has been applied in the determination of thorium in multimetal mixtures, as discussed in Section 8.59.

8.51 TIN

8.51.1 Spark-Source Mass Spectrometry

The determination of tin in soil by this technique is discussed in Section 8.59.

8.51.2 Miscellaneous Techniques

Li et al.[322] have reviewed methods for the determination of tin in soils.

8.52 TITANIUM

8.52.1 Spectrophotometric Method

Abbasi[161] described a spectrophotometric method using N-*p*-methoxyphenyl-2-furohydroxamine acid chromogenic reagent for the determination of down to 0.007 $mg\,L^{-1}$ titanium in soils. The soil was brought into solution by alkali fusion.[125] The extract was acidified to $pH \approx 0$ with nitric acid, boiled for 5 min and filtered through a 0.45-mm membrane filter. The acidity of the extract was adjusted to 10–12 M in hydrochloric acid prior to color development and spectrophotometric evaluation at 385 nm. About 60 mg kg^{-1} titanium was found in a soil sample.

8.52.2 Miscellaneous Techniques

Photon activation analysis has also been used to determine titanium in multimetal mixtures; see Section 8.59.

8.53 TUNGSTEN

8.53.1 Spectrophotometric Method

Quinn and Brooks[162] have described a rapid method for the determination of down to 0.01 ppm tungsten in soils. The sample is fused with potassium hydrogen sulfate and the melt leached with 10 M hydrochloric acid, and then heated with stannous chloride. This solution is heated with a solution of dithiol in isoamyl acetate, and then dissolved in petroleum ether prior to spectrophotometric evaluation at 630 nm.

Tungsten has been determined by this technique, as discussed in Section 8.59.

8.54 URANIUM

8.54.1 Spectrophotometric Method

Spectrophotometry at 655 nm using arsenazo III as chromogenic reagent has been used to determine down to 0.01 mg uranium in nitric acid extracts of soils.[164]

8.54.2 Inductively Coupled Plasma Mass Spectrometry

Inductively coupled plasma mass spectrometry has been used for the analysis of uranium. However, the technique surfers from spectral interferences.

Inductively coupled plasma mass spectrometry has superior limits of detection over optical methods. Also, this technique has an order of magnitude better detection limit than that obtained for the conventional fluorometric method. Uranium has many stable and unstable isotopes, but ^{238}U has the largest percentage abundance (99.274%). ^{238}U is free from interference from other elements and it is therefore possible to detect lower concentrations.

Boomer and Powell[323] have developed an analytical technique using inductively coupled plasma mass spectrometry to estimate the concentration of uranium in a variety of environmental samples including soil. The lower limit for quantitation is 0.1 ng mL^{-1}. Calibration is linear from the low limit to 100 ng mL^{-1}. Precision, accuracy, and a quality-control protocol were established. Results are compared with those obtained by the conventional fluorometric method.

In this method the soil sample is digested with concentrated nitric acid.

The *m/z* 238 ion was monitored for 2 s with five replicates of this measurement carried out for each determination.

Toole et al.[314] and Shaw and Francois[315] determined uranium (and thorium) in soils by inductively coupled plasma mass spectrometry.

Neutron activation analysis and photon activation analysis have been used to determine uranium in multimetal mixtures, as discussed in Section 8.59.

8.54.3 Miscellaneous Techniques

Nass et al.[165] used a delayed neutron counting technique to determine down to 50 ng of uranium-235 in soils. Other workers have reported the determination of uranium (and thorium) in soils.[317,320,321,324] Steam digestion has been employed in the preparation of soil samples for the determination of uranium (and thorium).[318] Parsa[316] described a sequential radiochemical method for the determination of uranium (and thorium) in soils.

Two methods involving dissolution in hydrogen chloride gas and microwave dissolution have been compared for the remote dissolution of uranium in soil.[325] Mukhtar et al.[317] have described a laser fluorometric method for the determination of uranium (and thorium) in soils. To determine uranium (and thorium) in soils, fluorescent X-rays were measured by the use of a germanium planar detector and chromometric techniques.[319] No sample preparation was required in this method.

8.55 VANADIUM

8.55.1 Spectrophotometric Method

Abbasi[163] described a spectrophotometric method employing the violet-colored N(pN,N-dimethylanilo-3-methoxy-2-naphtho) hydroxamic acid vanadium V complex for the determination of down to 0.05 μg vanadium in soils. In this method, an extract of the soil was rendered 8 N in hydrochloric acid, then 0.1 mL 0.001 M potassium permanganate added to convert vanadium to the pentavalent state. The chromogenic reagent is then added prior to spectrophotometric evaluation.

8.55.2 Miscellaneous Techniques

Emission spectrometry has been used to determine vanadium in multimetal mixtures, as discussed in Section 8.59.

8.56 YTTRIUM

8.56.1 Emission Spectrometry

Yttrium has been determined in soils by this method, as discussed in Section 8.59.

8.57 ZINC

8.57.1 Atomic Absorption Spectrometry

Atomic absorption spectrometry has been used to determine 0.5 M acetic and extractable[166] and nitric-perchloric-acid-soluble[167] zinc in soils. The application of this technique to the determination of zinc is also discussed in Section 8.59.

8.57.2 Miscellaneous Techniques

External-beam photon-induced X-ray emission spectrometry has been used to determine total zinc in soils.[168]

Roberts et al.[49] have discussed the simultaneous extraction and concentration of zinc and cadmium from calcium chloride soil extracts. Adequate recoveries of zinc were obtained when the pH of the extractant was adjusted to the range 4 to 7.

Fast neutron activation analysis has been studied as a screening technique for zinc (and copper) in waste soils.[230] Experiments were conducted in a sealed-tube neutron generator and a geranium X-ray detector.

Other techniques that have been used to determine zinc in multimetal analysis are inductively coupled plasma atomic-emission spectrometry, inductively coupled plasma mass spectrometry, differential-pulse anodic-scanning voltammetry, X-ray fluorescence spectroscopy, electron probe microanalysis, photon activation analysis, and emission spectrometry, as discussed in Section 8.59.

8.58 ZIRCONIUM

8.58.1 Emission Spectrometry

This technique has been applied to the determination of zirconium in soils, as discussed in Section 8.59.

8.58.2 Photon Activation Analysis

The determination of zirconium is discussed in Section 8.59.

8.59 MULTIPLE METALS

A review of methods for the analysis of mixtures of metals in soils is given in Table 8.1. Methods of extracting metals from soils prior to analysis are reviewed in Table 8.2.

TABLE 8.1 Review of Published Work on Multiple Metal Analysis in Soils

Section	Metals	LD	References
Atomic Absorption Spectrometry			
8.59.1	Cd, Cr, Cu, Ni, Pb, Zn, Se, As, Hg, Mo	$mg\,kg^{-1}$	169–182
8.59.1	Cd, Cu, Co, Ni, Pb, Zn, Ir, Ag, Sg, Hg, Bi, Tl, Mo	0.1–1	184
8.59.1	Pb, Cd, Tl		209
8.59.1	As, Sb, Bi, Se		28
8.59.1	Pb, Cd, Tl		218
Emission Spectrometry and Inductively Coupled Plasma Atomic Emission Spectrometry			
8.59.2	General discussion		10,185–190
	Cr, Fe, Ni, Cu, Co		
8.59.2	Cd, Zn, Pb, Mn		115,229
8.59.2	Cd, Cr, Fe, Mn, Zn, Al		158
	Ba, Ca, Ki, Mg, Na, Si, Sr, Ti, V		210
8.59.2	Mo, Co, B	0.01–0.05 $mg\,kg^{-1}$	40
	Various metals		215
8.59.2	As, Sb, Bi, Se	0.1 $mg\,kg^{-1}$	149,191,192
8.59.2	17 elements		174,196
Neutron Activation Analysis			
8.59.3	Cr, Fe, Co, Sc, As, Se, Mo, Sb, La, Ce, Eu, Hg, Ta, Th, U	–	193,230
8.59.3	As, Se	–	151
8.59.3	Lanthanides	–	326
8.59.3	Medium-lived radionuclide	–	335
Photon Activation Analysis			
8.59.4	Cr, Fe, Co, Pb, Ni, Zn, Al, Mg, Ca, Ti, As, Sr, Zr, Sb, U	–	193
Emission Spectrometry			
8.59.5	Cr, Fe, Co, Pb, Ni, Zn, Al, Ba, Be, Mo, Se, V, W, Yt, Zr	–	194,327,380

TABLE 8.1 Review of Published Work on Multiple Metal Analysis in Soils—cont'd

Section	Metals	LD	References
Spark-Source Mass Spectrometry			
8.59.6	Cr, NI, Rb, Sn, Sb, Cs	–	195
Differential-Pulse Anodic-Stripping Voltammetry			
8.59.7	Cu, Pb, Co, Ni, Cd, Zn	–	173,192,197,198
8.59.7	Zn, Pb, Cu	–	329,330
X-ray Fluorescence Spectroscopy			
8.59.8	Heavy metals	–	199
8.59.8	Miscellaneous metals	–	200
Stable-Isotope Dilution Mass Spectrometry			
8.59.9	K, Rb, Cs, Ca, Sr, Ba	–	201
Electron Probe Microanalysis			
8.59.10	Zn, Cu	–	202
Alpha-Spectrometry			
8.59.11	Am, Cf, Cm, Ra, Cf	–	78
	Inter-comparison study	–	2,18
Miscellaneous Methods			
8.59.12	Comparison of GLC, fluorimetry hydride generation method, As, Se	–	203
8.59.12	Ion chromatography Mg, Ca, Mn	–	204
8.59.12	Electron spin-resonance spectroscopy Cu	–	66
8.59.12	Extraction of copper from humic acid complexes		178
8.59.12	Laser-induced breakdown spectroscopy, miscellaneous metals	–	212
8.59.12	EXAFS spectroscopy, miscellaneous metals	–	211
8.59.12	Nuclear track detectors, alpha-contaminated soils	–	213

Continued

TABLE 8.1 Review of Published Work on Multiple Metal Analysis in Soils—cont'd

Section	Metals	LD	References
8.59.12	Soil sampling, radon flux measurement	–	214
8.59.12	Selective extraction chromatography and ion-exchange resins, actinides	–	215
8.59.12	Flow inspection preconcentration ^{237}Th, ^{234}U, ^{259}Pu, ^{260}Pu	–	216
8.59.12	Sampling technique prior alpha-spectrometry	–	217
8.59.12	Sampling techniques trace elements, S	–	218
8.59.12	Sampling techniques prior to AAS	–	219
8.59.17	Review methods on alkalis in earth	–	328
8.59.18	Speciation studies of trace metal	–	331,332
8.59.19	Computerized digestion	–	226,228
	Procedures, heavy metals, As	–	324
8.59.20	Radionuclides	–	333
8.59.12	X-ray measurements radionuclides	–	334
8.59.12	Passive sampling device radon flux	–	214
8.60	Fe, As		382
8.60	Al, Cu, Fe, Zn		383
8.60	Pb, Cd		384
8.60	As, Cr, Ni, V		183
8.60	As, Cr, V		381

TABLE 8.2 Methods of Extracting Metals from Soil Prior to Chemical Analysis

Extraction Procedures	Metals Discussed	References
Ammonium acetate	Heavy metal	336
EDTA	Cu, Zn, Fe, Ma	337
EDTA	Miscellaneous metals	338
Acid extraction	Cd, Ce, Cu, Ni, Pb, Zn, Mo, Hg, As, Se, B	182
Hydrofluoric acid extraction	Miscellaneous metals	227
Hydrofluoric acid–nitric acid	Al, Ti, Cr, Cu, Cd, Fe, Mn, Ni, Pb, Zn	223,224
Extraction with ammonium methylene carbodithioate	As, Cd, Tl	339
Extraction as xanthate complexes	Co, Ni, Pb, Bi, Ir	340
Microwave-assisted acid extraction	Various metals	108–111,220,226,227, 264,325,331–352
Microwave-assisted acid extraction	Zn, Cu	345
Sequential extraction		356
Microwave-assisted acid extraction	Heavy metals	348
Microwave-assisted extraction with nitric–hydrofluoric acids	Pb, Cd	246
Microwave-assisted extraction with hydrogen chloride gas	U	325
Comparison of different extraction methods	Heavy metals, As	252
Comparison of different extraction methods	Miscellaneous metals	220,253
Ultrasonic extraction	Sr	219,304,353–379
Sequential extraction with carbamate dissolving extractions, acid-reducing extractants, cation-exchange extractants, strong acid extractants	Miscellaneous metals	207,221,222,225, 226,229
Carbamate-dissolving extractions, acid-reducing extractants, cation-exchange extractants, strong acid extractants	Pb, Zn, Cu, NI	174

Continued

TABLE 8.2 Methods of Extracting Metals from Soil Prior to Chemical Analysis—cont'd

Extraction Procedures	Metals Discussed	References
Carbamate-dissolving extractions, acid-reducing extractants, cation-exchange extractants, strong acid extractants	Au, Pb, Cd, Zn	205
Carbamate-dissolving extractions, acid-reducing extractants, cation-exchange extractants, strong acid extractants	Cd, Cu, Ni, Zn	206
Sequential extraction with carbamate-dissolving extractants, acid-reducing extractants, cation-exchange extractants, strong acid extractants	Ca, Mg, heavy metals	376
Sequential extraction with carbamate-dissolving extractants. acid-reducing extractants, cation-exchange extractants, strong acid extractants	Se, Al, V	207
Comparison of extraction methods	Miscellaneous metals	378

REFERENCES

1. Dodson A, Jennings VJ. *Talanta* 1972;**19**:801.
2. Reis BF, Bergamin FH, Zegatto EAG, Krug FJ. *Anal Chim Acta* 1979;**107**:309.
3. Tecator Ltd, Application Note ASN 78–31/85. *Determination of aluminium in soil by flow injection analysis*; 1985.
4. Tecator Ltd, Application Note ASTN 10/84. *Determination of aluminium by flow injection analysis*; 1984.
5. Zolter D. *Diplomarbeit*. West Germany: Inst Für Anong Chemie, University of Gottingen; 1982.
6. Mitrovic B, Milacic R, Pihlar B. *Analyst (London)* 1996;**121**:627.
7. Kozuk N, Milacic R, Gorenc B. *Ann Chim* 1996;**86**:99.
8. Sill CW, Puphal KW, Hindman FD. *Anal Chem* 1974;**46**:1725.
9. Keay J, Menage PMA. *Analyst (London)* 1970;**95**:379.
10. Alder JF, Gunn AM, Kirkbright GF. *Anal Chim Acta* 1977;**92**:43.
11. Bremner JM. *Monogram Am Soc Argon* 1965:9.
12. Tecator Ltd, Applications Note ASN 65–32/83. *Determination of ammonia nitrogen in soil samples extractable by 2 M KCl using flow injection analysis*; 1983.
13. HMSO, London. *The analysis of agricultural materials, RB 427*. 2nd ed. 1979, ISBN: 011240 352 2. Method 60, Ammonium Nitrate and Nitrite Nitrogen Potassium Chloride Extractable in Moist Soil.

14. Waughman GJ. *Environ Res* 1981;**26**:529.
15. Chikhalikar S, Sharma K, Patel KS. *Commun Soil Sci Plant Anal* 1995;**26**:621.
16. Agrawal YK, Patke SK. *Int J Environ Anal Chem* 1980;**8**:157.
17. Lawless EW, Von Rumber R, Ferguson RL. *Technical studies Dept T 500–72*. Washington, DC: Environmental Protection Agency; 1972.
18. Woolson EA, Axley JJ, Kearney PC. *Soil Sci Soc Am Proc* 1971;**35**:101.
19. Shroeder HA, Balassa JJ. *J Chronic Disord* 1966;**19**:1.
20. Analytical Sub Comm. *Analyst (London)* 1960;**86**:679.
21. Analytical Methods Committee. *Analyst (London)* 1975;**100**:54.
22. Sandhu SS. *Analyst (London)* 1981;**106**:311.
23. Forehand TJ, Dupey AE, Tai H. *Anal Chem* 1976;**48**:999.
24. Thompson KC, Thomerson DR. *Analyst (London)* 1974;**99**:595.
25. Thompson AJ, Thoresby PA. *Analyst (London)* 1977;**102**:9.
26. Wanchange RD. *At Absorpt Newsl Perkin-Elmer* 1976;**15**:64.
27. Ohta K, Suzuki M. *Talanta* 1978;**25**:160.
28. Haring BJA, van Delft W, Bom CM. *Fresenius Z Für Anal Chem* 1982;**310**:217.
29. Merry RH, Zarcinas BA. *Analyst (London)* 1980;**109**:998.
30. a. HMSO (London). *Selenium and arsenic in sludges, soils and related materials*, 1985. A note on the use of hydride generator kits; 1987.
    b. Jiminez de Blas O, Mateos NR, Sanchez AG. *J Am Assoc Anal Chem (AOAC) Int* 1996;**79**:764.
31. Jiminez de Blas O, Mateos NR, Sanchez AG. *J Am Assoc Anal Chem (AOAC) Int* 1996;**76**:764.
32. Agemian H, Bedak E. *Anal Chim Acta* 1980;**119**:323.
33. Wenclawiak BW, Krab M. *J Anal Chem* 1995;**351**:134.
34. Chappell J, Chiswell B, Olszowy H. *Talanta* 1995;**42**:323.
35. Ross DS, Bartlett RJ, Magdoff FR. *At Spectrosc* 1986;**7**:158.
36. HMSO. *The analysis of agricultural materials, RB 427*. 2nd ed. 1979, ISBN: 011240 352 2. Method 8, Baron, Water Soluble in Soil.
37. Aznarez J, Bonilla A, Vidal JC. *Analyst (London)* 1983;**108**:368.
38. Ducret L. *Anal Chim Acta* 1957;**17**:213.
39. Pasztor L, Bode JD, Fernando Q. *Anal Chem* 1960;**32**:277.
40. Manzoori JL. *Talanta* 1980;**27**:682.
41. Zarcinas BA, Cartwright MB. *Analyst (London)* 1987;**112**:1107.
42. HMSO, London. *The analysis of agricultural materials, RB 427*. 2nd ed. 1979, ISBN: 011240 352 2. Method 11, Cadmium Nitric Perchloric Acid Soluble in Soil.
43. HMSO, London. *The analysis of agricultural materials, RB 427*. 2nd ed. 1979, ISBN: 011240 352 2. Method 10, Cadmium Acetic Acid Extractable in Soil.
44. Baucells M, Lacort G, Roura M. *Analyst (London)* 1985;**110**:1423.
45. Bolt GH, Bruggenwert MGM, editors. *Soil chemistry part A*. Amsterdam: Elsevier; 1978.
46. Lewin VH, Beckett OHT. *Effl Waste Treat J* 1980;**20**:162.
47. Christensen TH, Lun XZ. *Water Res* 1989;**23**:73.
48. Turner MA, Hendrickson LL, Corey RB. *Soil Sci Soc Am J* 1984;**48**:763.
49. Roberts AHC, Turner MA, Syers JK. *Analyst (London)* 1976;**101**:574.
50. Carlosena A, Prada D, Andrade JM, Lopez P, Muniategui B. *Fresenius J Anal Chem* 1996;**355**:289.
51. Xing-chu Qin, Yu-sheng Zhang, Ying-quan Zhu. *Analyst (London)* 1983;**108**:754.
52. Jackson YL. In: *Soils chemical analysis*. Eaglewood Cliffs, NJ: Prentice-Hall; 1958.
53. Xing-chu Qin, Ying-quan Zhu. *Analyst (London)* 1985;**110**:185.

54. Qi WB, Zhu LZ. *Talanta* 1986;**33**:694.
55. Smith GH, Lloyd OL. *Chemistry in Britain* February 1986. p. 139.
56. Evans R. Personal Communications. Dundee: Dept of City Analyst; 1985.
57. Sahuquillo A, Lopez-Sanchez JF, Rubio R, Rauret G, Hatje V. *Fresenius J Anal Chem* 1995;**351**:197.
58. Sahuquillo A, Lopez-Sanchez JF, Rubio R, Rauret G. *Mikrochim Acta* 1995;**119**:251.
59. Fodor P, Fischer I. *Fresenius Z Für Anal Chem* 1995;**351**:454.
60. Prokisch J, Kovacs TS, Gyroi Z, Loch J. *Commun Soil Sci* 1995;**26**:2051.
61. Florez-Velez LM, Gutierrez-Ruiz HE, Reyes-Salas O, Cram-Heydrich S, Baesze-Reys A. *Int J Environ Anal Chem* 1995;**61**:177.
62. HMSO, London. *The analysis of agricultural materials, RB 427*. 2nd ed. 1979, ISBN: 011240 352 2. Method 22, Cobalt Extractable in Soil.
63. HMSO, London. *The analysis of agricultural materials, RB 427*. 2nd ed. 1979, ISBN: 011240 352 2. Method 23, Cobalt, Nitric-Perchloric Soluble in Soil.
64. HMSO, London. *The analysis of agricultural materials, RB 427*. 2nd ed. 1979, ISBN: 011240 352 2. Method 26, Copper, EDTA Extractable in Soil.
65. HMSO, London. *The analysis of agricultural materials, RB 427*. 2nd ed. 1979, ISBN: 011240 352 2. Method 27, Copper, Nitric-Perchloric Soluble, in Soil.
66. Senesi N, Sposito G. *Soil Sci Soc Am J* 1984;**48**:1247.
67. Stefanov G, Daieve L. *Isotopenpraxis* 1972;**8**:146.
68. Jayman TCZ, Sivasubramanian S, Wijeedasa MA. *Analyst (London)* 1975;**100**:716.
69. Keaton CM. *Soil Sci* 1937;**43**:40.
70. Agrawal YK, Raj KPS, Desai S, Patel SG, Merh SS. *Int J Environ Sci* 1980;**14**:313.
71. Chow TJ. *Nature (London)* 1970;**225**:295.
72. Motto HL, Daines RH, Chilko DM, Motto CK. *Environ Sci Technol* 1970;**4**:231.
73. Low KS, Lee CK, Arshad MM. *Pertanika* 1979;**2**:105.
74. Ward NI, Reeves RD, Brooks RR. *Environ Pollut* 1974;**6**:149.
75. Berrow ML, Stein WM. *Analyst (London)* 1983;**108**:277.
76. HMSO, London. *The analysis of agricultural materials, RB 427*. 2nd ed. 1979, ISBN: 011240 352 2. Method 43, Lead Extractable in Soil.
77. HMSO, London. *The analysis of agricultural materials, RB 427*. 2nd ed. 1979, ISBN: 011240 352 2. Method 44, Lead Nitric Perchloric Soluble in Soil.
78. Somer G, Aydin H. *Analyst (London)* 1985;**110**:631.
79. Sakharov AA. *Pockvovedenie* 1967;**1**:107.
80. Tills AR, Alloway BJ. *Environ Technol Lett* 1983;**4**:529.
81. Rigin VI, Rigina II. *Zhur Anal Khim* 1979;**34**:1121.
82. HMSO, London. *The Analyst of agricultural materials, RB 427*. 2nd ed. 1979, ISBN: 011240 352 2. Method 46, Magnesium, Extractable in Soil.
83. Alekseeva II, Davydova ZP. *Zhur Anal Khim* 1971;**26**:1786.
84. HMSO, London. *The analysis of agricultural materials, RB 427*. 2nd ed. 1979, ISBN: 011240 352 2. Method 48, Manganese Exchangeable and Easily Reducible in Soil.
85. Kimura Y, Miller VL. *Anal Chim Acta* 1962;**27**:325.
86. Hatch A, Ott WL. *Anal Chem* 1968;**40**:2085.
87. Hoggins FE, Brooks RRJ. *J Assoc Off Anal Chem* 1973;**56**:1306.
88. Ure AM, Shand CA. *Anal Chim Acta* 1974;**72**:63.
89. Floyd M, Sommers LE. *J Environ Qual* 1975;**4**:323.
90. Agemian H, Chau ASY. *Analyst (London)* 1976;**101**:91.
91. Kuwae Y, Hasegawa T, Shono T. *Anal Chim Acta* 1976;**84**:185.
92. Nicholson RA. *Analyst (London)* 1977;**102**:399.

93. Head PC, Nicholson RA. *Analyst (London)* 1973;**98**:53.
94. HMSO. *Methods for the examination of water and associated materials, mercury in waters effluents soils and sediments etc*; 1987. Methods 40453 pp. 39.
95. Lutze RG. *Analyst (London)* 1979;**104**:979.
96. Grantham PL. *Lab Pract* 1978;**27**:294.
97. HMSO, London. *The analysis of agricultural materials, RB 427*. 2nd ed. 1979, ISBN: 011240 352 2. Method 86, Mercury in Soil and Plant Material.
98. Rasemann W, Seltmann U, Hempel M. *Fresenius J Anal Chem* 1995;**351**:632.
99. Wenger R, Hagel O. *Mitt Geb Leb U Hyg* 1971;**61**:1.
100. Micham P, Maksvytis A, Barkus B. *Anal Chem* 1972;**44**:2102.
101. HMSO, London. *The analysis of agricultural materials, RB 427*. 2nd ed. 1979, ISBN: 011240 352 2. Method 51, Molybdenum Total in Soil.
102. HMSO, London. *The analysis of agricultural materials, RB 427*. 2nd ed. 1979, ISBN: 011240 352 2. Method 50, Molybdenum Extractable in Soil.
103. Thompson M, Zao L. *Analyst (London)* 1985;**110**:229.
104. Kim CH, Owens CM, Smythe LE. *Talanta* 1974;**21**:445.
105. Kim CH, Alexander PW, Smythe LE. *Talanta* 1976;**23**:229.
106. Ni ZM, Chin LC, Wu TH. *Huan K'o Hsueh* 1979;**6**:25. Analytical Abstracts, 1981, 40, 5H65.
107. Munter RC, Grande RA. In: Barnes, editor. *Developments in atomic plasma spectrochemical analysis*. London: Heyden; 1981. p. 653–72.
108. Maessen FJML, Balke J, de Beer JLM. *Spectrochim Acta Part B* 1982;**37**:517.
109. David DJ. *Prog Anal At Spectrosc* 1978;**1**:225.
110. Jones JB. *J Assoc Off Anal Chem* 1975;**58**:764.
111. Mitchell RL, Scott RO. *J Soc Chem Ind* 1974;**66**:330.
112. Reisenauer MH. In: *Methods of soul analysis, part 2*. Madison, WI: American Society of Agronomy; 1965. p. 1054–7.
113. Sperling KR. *Z Für Anal Chem* 1977;**283**:30.
114. Ure AM, Mitchell MC. *Anal Chim Acta* 1976;**87**:283.
115. Iu KI, Pulford ID, Duncan HJ. *Anal Chim Acta* 1979;**106**:319.
116. Crompton TR. [Unpublished work].
117. Ure AM, Hernandez MP, Mitchell MC. *Anal Chim Acta* 1978;**96**:37.
118. Pedersen B, Willems M, Storgaard JS. *Analyst (London)* 1980;**105**:119.
119. Henn EL. In: *Flameless atomic absorption analysis—an update*. Philadelphia: ASTM STP, 618 American Society for Testing and Materials; 1977. p. 54–64.
120. Dahlquist RL, Knoll JW. *Appl Spectrosc* 1978;**32**:1.
121. Jones JB. *Commun Soil Sci Plant Anal* 1977;**8**:340.
122. HMSO, London. *The analysis of agricultural materials, RB 427*. 2nd ed. 1979, ISBN: 011240 352 2. Method 53, Nickel Extractable in Soil.
123. HMSO, London. *The analysis of agricultural materials, RB 427*. 2nd ed. 1979, ISBN: 011240 352 2. Method 54, Nickel Nitric Perchloric Soluble in Soil.
124. Talvitie NA. *Anal Chem* 1971;**43**:1827.
125. Dienstbach E, Bachmann K. *Anal Chem* 1980;**52**:620.
126. Sekine K, Imai T, Kasai A. *Talanta* 1987;**34**:567.
127. HMSO, London. *The analysis of agricultural materials, RB 427*. 2nd ed. 1979, ISBN: 011240 352 2. Method 68, Ammonium Acetate Extractable Potassium in Soil.
128. Weltz B. *Atomic absorption spectrometry*. New York: Verlag Chemie; 1976.
129. Pinta M. *Atomic spectrometry applications to chemical analysis*. PWN Warsaw; 1977.
130. Shrenk WG. *Modern, analytical chemistry analytical atomic spectroscopy*. New York: Plenum Press; 1975.

131. Brodie KG. *Int Lab* July/August 1979;**40**.
132. Laurakis V, Barry E, Golembeski T. *Talanta* 1975;**22**:547.
133. Ohta K, Suzuki M. *Talanta* 1975;**22**:465.
134. Thompson KC. *Analyst (London)* 1975;**100**:307.
135. HMSO, London. *Selenium and arsenic in sludges, soils and related materials*, 1985. A note on the use of hydride generation kits; 1987.
136. Chambers JC, McClellan D. *Anal Chem* 1976;**48**:2061.
137. Inhat M. *Anal Chim Acta* 1976;**82**:292.
138. Henn EL. *Anal Chem* 1975;**47**:428.
139. Ishizaka. *Talanta* 1978;**25**:167.
140. Montaser A, Mehrabzadeh AA. *Anal Chem* 1978;**50**:1697.
141. Azad J, Kirkbright GG, Snook RD. *Analyst (London)* 1979;**104**:232.
142. Parker CA, Harvey LG. *Analyst (London)* 1962;**87**:558.
143. Jolloway WH, Cary EE. *Anal Chem* 1964;**36**:1359.
144. Watkinson JH. *Anal Chem* 1960;**32**:98.
145. Watkinson JH. *Anal Chem* 1966;**38**:92.
146. Ewan RC, Baumann CA, Pope AL. *J Agric Food Chem* 1968;**16**:212.
147. Hall RJ, Gupta PL. *Analyst (London)* 1969;**94**:292.
148. Pahlavanpour B, Pullen JH, Thompson M. *Analyst (London)* 1980;**105**:274.
149. Thompson M, Pahlavanpour B, Walton SL, Kirkbright GF. *Analyst (London)* 1978;**103**:568.
150. Zmijewska W, Semkow T. *Chem Anal (Warsaw)* 1978;**23**:583.
151. Kronborg OJ, Steinnes E. *Analyst (London)* 1975;**100**:835.
152. Mignosin EP, Roelandts I. *Chem Geol* 1975;**16**:137.
153. Van der Klugt N, Poelstra P, Zwemmer E. *J Radio Anal Chem* 1977;**35**:109.
154. Baedecker PA, Rowe JJ, Steinnes E. *J Radio Anal Chem* 1977;**40**:115.
155. Agemian H, Bedek E. *Anal Chem* 1980;**119**:394.
156. Bem EM. *Environ Health Perspect* 1981;**37**:183.
157. Pellenberg R. *Environ Health Perspect* 1981;**10**:267.
158. Que-Hee SG, Boyle JR. *Anal Chem* 1988;**60**:1033.
159. HMSO, London. *The analysis of agricultural materials, RB 427*. 2nd ed. 1979, ISBN: 011240 352 2. Method 72, Sodium Extractable, in Soil.
160. Martin DB. *Anal Chem* 1979;**51**:1968.
161. Abbasi SA. *Int J Environ Anal Chem* 1982;**11**:1.
162. Quinn BF, Brooks RR. *Anal Chim Acta* 1972;**58**:301.
163. Abbasi SA. *Int J Environ Stud* 1981;**18**:51.
164. Prister BS, Zubach SS. *Radiokhimiya* 1968;**10**:743.
165. Nass HW, Molinski VJ, Kramer HH. *Mater Res Stand* 1972;**12**:24.
166. HMSO, London. *The analysis of agricultural materials, RB 427*. 2nd ed. 1979, ISBN: 011240 352 2. Method 82, Zinc Extractable, in Soil.
167. HMSO, London. *The analysis of agricultural materials, RB 427*. 2nd ed. 1979, ISBN: 011240 352 2. Method 83, Zinc, Nitric Perchloric Soluble, in Soil.
168. Abdullah M, Zaman MB, Kaliquzzaman M, Khan AH. *Anal Chim Acta* 1980;**118**:175.
169. Chao TT, Sanzolone RF. *J Res US Geol Surv* 1973;**1**:681.
170. Weitz A, Fuchs G, Bachmann K. *Fresenius Z Für Anal Chem* 1982;**313**:38.
171. HMSO, London, Department of the Environment/National Water Council Standing Technical Committee of Analysts. *Methods for the examination of water and associated materials—determination of extractable metals in soil sewage sludge treated soils and associated materials*; 1983. 17 pp. (22 BC ENV).

172. Hinds MW, Jackson KW, Newman AP. *Analyst (London)* 1985;**110**:947.
173. Edmonds TE, Guogang P, West TS. *Anal Chim Acta* 1980;**120**:41.
174. Kheboian C, Bauer CF. *Anal Chem* 1987;**59**:1417.
175. Bowman WS, Fayer GH, Sutarno B, McKeague JA, Kodama H. *Geostand Newsl* 1979;**3**:109.
176. Jackson KW, Newman AP. *Analyst (London)* 1983;**108**:261.
177. Chester R, Hughes M. *Chem Geol* 1967;**2**:249.
178. Chowdbury AN, Bose BB. *Geochem Explor* 1971;**11**:410.
179. Dong A, Rendig VV, Barau RG, Besga GS. *Anal Chem* 1987;**59**:2728.
180. Stupar J, Ajlec R. *Analyst (London)* 1982;**107**:144.
181. Davi RP, Carlton-Smith CH, Stark JH, Campbell JA. *Environ Pollut* 1988;**49**:99.
182. Davis RP, Carlton-Smith CH. *Pollut Control* 1983;**82**:290.
183. Mayes WM, Jarvis AP, Burke IT, Walton M, Feigl V, Klelerez O, et al. *Environ Sci Technol* 2011;**45**:5147.
184. Eidecker R, Jackwerth E. *Fresenius Z Fur Anal Chem* 1987;**469**:1987.
185. Schramel P, Xu LQ, Hasse S. *Fresenius Z Für Anal Chem* 1982;**313**:213.
186. Greenfield S, Jones H, McGeachin HD, Smith PB. *Anal Chim Acta* 1975;**74**:225.
187. Butler CC, Kniseley RN, Fassel VA. *Anal Chem* 1975;**47**:825.
188. Fassel VA, Peterson FN, Abercrombie FN, Kniseley RN. *Anal Chem* 1976;**48**:516.
189. Gunn AM, Kirkbright GF, Openeim LN. *Anal Chem* 1977;**49**:1492.
190. Watson E, Russel GM, Bales S. *South Africa National Institute of metal Report No 1815*; April 15, 1976.
191. Pahlavanpour B, Thompson M, Thorne L. *Analyst (London)* 1980;**105**:756.
192. Goulden PD, Anthony DJH, Keith D. *Anal Chem* 1981;**53**:2027.
193. Randle K, Hartman EH. *J Radioanal Nucl Chem* 1985;**90**:309.
194. Ure AM, Ewen GJ, Mitchell MC. *Anal Chim Acta* 1980;**118**:1.
195. Ure AM, Bacon JR. *Analyst (London)* 1978;**103**:807.
196. Kanda Y, Taira M. *Anal Chim Acta* 1988;**207**:269.
197. Reddy SJ, Valenta P, Nurnberg HW. *Fresenius Z Für Anal Chem* 1982;**313**:390.
198. Meyer A, de la Chevallerie-Haaf U, Henze G. *Fresenius Z Für Anal Chem* 1987;**328**:565.
199. Muntau H, Crossman G, Schramel P, Gallorni M, Orvini E. *Fresenius Z Für Anal Chem* 1987;**326**:634.
200. Levinson A, Pablo L. *J Geochem Explor* 1975;**4**:339.
201. Hinkley T. *Nature (London)* 1979;**277**:444.
202. Lee FY, Kittrick JA. *Soil Sci Am J* 1984;**48**:548.
203. HMSO, London. *Method for the examination of waters and associated materials, selenium in waters*, 1984. Selenium and Arsenic in Sludges Soils and Related Materials, 1985, a Note on Hydride Generator Kits, 1985, 42 pp. (40548), 1987.
204. Jan D, Schwedt G. *Fresenius Z Für Anal Chem* 1985;**320**:121.
205. Slavek J, Wold J, Pickering WF. *Talanta* 1982;**29**:743.
206. Hickey MG, Kittrick JA. *J Environ Qual* 1984;**13**:372.
207. Bin Xiao-Quan S. *Anal Chem* 1993;**65**:802.
208. Chen TC, Hong A. *Hazard Mater* 1995;**41**:147.
209. Lopez-Garcia L, Sanchez-Merlos M, Hernandez-Cordoba M. *Anal Chim Acta* 1996; **328**:19.
210. Zaray G, Kantor T. *Spectrochim Acta Part B* 1995;**50B**:489.
211. Manceau A, Boisset MC, Sarret G, Hazemann JL, Mench M, Cambier P, et al. *Environ Sci Technol* 1996;**30**:1540.

212. Yamamoto KY, Cremers DA, Ferris MJ, Foster LE. *Appl Spectrosc* 1996;**50**:222.
213. Espinosa G, Silva R. *J Radioanal Nucl Chem* 1995;**194**:207.
214. Burnett WC, Cable PH, Chanton JP. *J Radioanal Nucl Chem* 1995;**193**:281.
215. Smith LL, Crain YS, Yeager JS, Horwitz RP, Diamond H, Chiarizla R. *J Radioanal Nucl Chem* 1995;**194**:151.
216. Hollenbach M, Grohs J, Kraft M, Mamich S. ASTM Special Technical Publication STP 11291. *Applications of inductively coupled plasma mass spectrometry to radionucleide determination*; 1996.
217. Lettner H, Andrasi A, Hubmer AK, Lavranich E, Steger F, Zombori P. *Nucl Instr Meth Phys Res Sect A* 1996;**369**:547.
218. Merimether JR, Burns SF, Thompson RH, Beck JN. *Health Phys* 1995;**69**:406.
219. Klemm W, Bombach G. *Fresenius J Anal Chem* 1995;**353**:12.
220. Torres P, Ballesteros E, Luque de Castro MD. *Anal Chim Acta* 1995;**308**:371.
221. Ure AM, Davidson CM, Thomas RP. *Tech Instrum Anal Chem* 1995;**17**:505.
222. Magni E, Guisto T, Frache R. *Anal Proc* 1995;**32**:267.
223. Schrmael P, Klose BJ, Hasse S. *Fresenius Z Für Anal Chem* 1982;**310**:209.
224. Schramel P, Lill G, Seif R. *Fresenius Z Für Anal Chem* 1987;**326**:135.
225. Coles BJ, Ramsey MH, Thornton I. *Chem Geol* 1995;**124**:109.
226. Stachel B, Elsholz O, Reincke H. *Fresenius J Anal Chem* 1995;**353**:21.
227. Sturgeon R, Willie SN, Methven BA, Lam JWH, Matusieqicz HI. *Anal At Spectrosc* 1995;**10**:981.
228. Krauss P, Erbsloueh LB, Niedergasaess R, Pepeinik R, Prange A. *Fresenius J Anal Chem* 1995;**353**:3.
229. Li X, Coles BJ, Ramsey MH, Thornton I. *Chem Geol* 1995;**124**:109.
230. Shapiro JB, James WD, Scheiker EA. *J Radioanal Nucl Chem* 1995;**192**:275.
231. Gibson JAE, Willett IR. *Commun Soil Sci Plant Anal* 1991;**22**:1303.
232. Garotti FV, Massaro S, Serrano SHP. *Analysis* 1992;**20**:287.
233. downard AJ, Kipton H, Powell J, Xu S. *Anal Chim Acta* 1992;**256**:117.
234. Schmidt S, Koerdel W, Kloeppel H, Klein W. *J Chromatogr* 1989;**470**:289.
235. Joshi SR. *Appl Radiat Isot* 1989;**40**:691.
236. Livens FR, Singleton DL. *Analyst (London)* 1989;**114**:1097.
237. Chickhaikar S, Sharma K, Patel KS. *Commun Soil Sci Sc Plant Anal* 1995;**26**:625.
238. Asami T, Kubota M, Salto S. *Water Air Soil Pollut* 1992;**62**:349.
239. Lasztity A, Krushevska A, Kotrebai M, Barnes RM, Amarasiriwardena D. *J Anal Spectrosc* 1995;**10**:505.
240. Hwang JD, Huxlry HP, Diomiguardi JP, Vaughn W. *J Appl Spectrosc* 1990;**44**:491.
241. Rurikova D, Beno A. *Chem Pap* 1992;**46**:23.
242. Masscheleyn PH, Delaune RD, Patrick WH. *Environ Sci Technol* 1991;**25**:1414.
243. McGeehan SL, Naylor DV. *J Environ Qual* 1992;**21**:68.
244. Honore Hanse G, Larsen EH, Pritzi G, Cornett C. *J Anal Atom Spectrosc* 1992;**7**:629.
245. Hirsch D, Banin A. *J Anal Qual* 1990;**19**:366.
246. Feng Y, Barratt RS. *Sci Total Environ* 1992;**143**:157.
247. Kim D, Fergusson JE. *Sci Total Environ* 1991;**105**:191.
248. Dubois JP. *J Trace Microprobe Tech* 1991;**9**:149.
249. Mazzucotelli A, Soggia F, Cosma B. *Appl Spectrosc* 1991;**4**:504.
250. Millward CG, Kluckner PD. *J Anal Atom Spectrosc* 1991;**6**:37.
251. Evansm DW, Alberts JJ, Clark RA. *Geochim Cosmochim Acta* 1983;**47**:1041.
252. Kim CS, Smith SJ, Park SW. *J Environ Sci Health* 1996;**A31**:2173.

253. Fawaris BH, Johansen KJ. *Sci Total Environ* 1995;**170**:221.
254. Carbol P, Skarnemark G, Skalberg M. *Sci Total Environ* 1993;**131**:129.
255. Williams TM. *Environ Geol* 1993;**21**:62.
256. Von Gunten HR, Benes P. *Radiochim Acta* 1995;**69**:1.
257. Essington EH, Fowler FB, Polzer WL. *Soil Sci* 1981;**132**:13.
258. Bunzl K, Schimmack W. *Chemosphere* 1989;**18**:2109.
259. Wauters J, Vidal M, Elsen A, Cremers A. *Appl Geochem* 1996;**11**:595.
260. Vidal M, Roig M, Rigol A, Llarado M, Rauret G, Wauters J, et al. *Analyst (London)* 1995;**120**:1785.
261. Thiry Y, Myttenaere C. *J Environ Radioact* 1993;**18**:247.
262. Scheuler B, Sander P, Read DA. *Vacuum* 1990;**41**:1661.
263. Groenewald GS, Ingram JC, McLing T, Giantto AK. *Anal Chem* 1998;**70**:534.
264. Chakraborty R, Das DK, Cervera ML, De la Guardia M. *J Anal Atom Spectrosc* 1995;**10**:353.
265. Milacic R, Stupar J, Kozuh N, Korosin J. *Analyst (London)* 1992;**117**:125.
266. Mesuere K, Martin RE, Fish W. *J Environ Qual* 1991;**20**:114.
267. Gerring LJA, Van der Meer J, Cauet G. *Mar Chem* 1991;**36**:51.
268. Savvin SB, Petrova TV, Ozherayan TG, Reikhshat MM. *Fresenius J Anal Chem* 1991; **340**:217.
269. Hinds MM, Jackson K. *J Anal Atom Spectrosc* 1989;**5**:199.
270. Zhang B, Tao K, Feng J. *J Anal Atom Spectrosc* 1992;**7**:171.
271. Hinds MW, Latimer KE, Jackson KW. *J Anal Atom Spectrosc* 1991;**6**:473.
272. Hinds MW, Jackson KW. *At Spectrosc* 1991;**12**:109.
273. Fernando AR, Plambeck JA. *Analyst (London)* 1992;**117**:39.
274. Wegrzynek D, Holynska B. *Appl Radiat Isot* 1993;**44**:1101.
275. Bermejo-Barrera P, Barciel-Alonso C, Aboal-Somaza M, Bermejo-Barrera A. *J Atom Spectrosc* 1994;**9**:469.
276. Kamburova M. *Talanta* 1993;**40**:719.
277. Sakamoto H, Tommlyasu T, Yonehara N. *Anal Sci* 1992;**8**:35.
278. Azzaria LM, Aftabi A. *Water Air Soil Pollut* 1991;**56**:203.
279. Lee HS, Jung KH, Lee DS. *Talanta* 1989;**36**:999.
280. Bandyopadhyay S, Das AK. *J Indian Chem Soc* 1989;**66**:427.
281. Golimowski J, Orzechowska A, Tykarska A. *Fresenius J Anal Chem* 1995;**351**:656.
282. Robinson L, Dyer FF, Combs DW, Wade W, Teasley NA, Carlton JE. *J Radioanal Nucl Chem* 1994;**179**:305.
283. Welz B, Schlemmeer G, Mudakavi JR. *J Anal Atom Spectrosc* 1992;**7**:499.
284. Baxter DC, Nicol R, Littlejohn D. *Spectrochim Acta Part B* 1992;**47**:1155.
285. Lexa J, Stulik K. *Talanta* 1989;**36**:843.
286. Cherian L, Gupta UK. *Fresenius Z Für Anal Chem* 1990;**336**:400.
287. Saouter E, Campbell PGC, Ribeyre F, Boudon A. *Int J Environ Anal Chem* 1993;**54**:57.
288. Jiao K, Jin W, Metzner H. *Anal Chim Acta* 1992;**260**:35.
289. Rowbottom WH. *J Anal Atom Spectrosc* 1991;**6**:123.
290. Kim CK, Takaku A, Yamammoto M, Kawawamura H, Shiraishi K, Igarashi Y, et al. *J Radioanal Nucl Chem* 1989;**132**:131.
291. Kim CK, Oura Y, Takaku NH, Igarashi V, Ikeda N. *J Radioanal Nucl Chem* 1989;**136**:353.
292. Jia G, Testa C, Desideri D, Mell MA. *Anal Chim Acta* 1989;**220**:103.
293. Green LW, Miller FC, Sparling JA, Joshi SR. *J Am Soc Mass Spectrom* 1991;**2**:240.
294. Holgye Z. *J Radioanal Nucl Chem* 1991;**149**:275.
295. Zhu HM, Tang XZ. *J Radioanal Nucl Chem* 1989;**130**:443.

296. Nakanishi T, Satoh M, Takei M, Ishikawa A, Murata M, Dairyah M, et al. *J Radioanal Nucl Chem* 1990;**138**:321.
297. El-Daoushy F, Garcia-Tenorio R. *J Radioanal Nucl Chem* 1990;**138**:5.
298. Wiliams LR, Leggett RW, Espergren ML, Little CA. *Environ Monit Assess* 1989;**12**:83.
299. Hafez AF, Moharram BM, El-Khatib AM, Adel-Naby A. *Isotopenproxis* 1991;**27**:185.
300. Zhang Y, Moore JN, Frankenberger WT. *Environ Sci Technol* 1999;**33**:1652.
301. McCurdy EJ, Lange JD, Haygarth PM. *Sci Total Environ* 1993;**135**:131.
302. Rojus CJ, de Maroto SB, Valenta P. *Fresenius J Anal Chem* 1994;**348**:775.
303. Cruvinel PE, Flocchini RG. *Nucl Instrum Meth Phys Res Sect B* **B75**: 415.
304. Akcay M, Elik A, Savasci S. *Analyst (London)* 1989;**114**:1079.
305. Stella R, Ganzerli VMT, Maggi L. *J Radioanal Nucl Chem* 1992;**161**:413.
306. Ryabukhin VA, Volynets MP, Myasoedov BF, Rodionova IM, Tuzova AM. *Fresenius J Anal Chem* 1993;**63**:69.
307. Tagami K, Uchida S. *Radiochim Acta* 1993;**63**:69.
308. Morita S, Kim CK, Takaku Y, Seki R, Ikeda N. *Appl Radiat Isot* 1991;**42**:531.
309. Harvey BR, Williams KJ, Lovett MB, Ibbett RD. *J Radioanal Nucl Chem* 1992;**158**:417.
310. De Ruck A, Vandecastelle G, Dams R. *Anal Lett (London)* 1989;**22**:469.
311. Opydo J. *Mikrochim Acta* 1989;**2**:15.
312. Lukaszewski Z, Zembzuski W. *Talanta* 1992;**39**:221.
313. Sagar M. *Mikrochim Acta* 1992;**106**:241.
314. Toole J, McKay K, Baxter M. *Anal Chim Acta* 1990;**245**:83.
315. Shaw TJ, Francois R. *Geochim Cosmochim Acta* 1991;**55**:2075.
316. Parsa B. *J Radioanal Nucl Chem* 1992;**157**:65.
317. Mukhtar OM, Grods A, Khangi FA. *Radiochim Acta* 1991;**54**:201.
318. Mann DK, Oatis T, Wong GTF. *Talanta* 1992;**39**:1199.
319. Lazo EN, Roessier GS, Bervani BA. *J Radioanal Nucl Chem* 1989;**129**:245.
320. Shuktomova II, Kochan IG. *J Radioanal Nucl Chem* 1989;**129**:245.
321. Lazo EN. Report 1988, DOE (Department of Environment)/OR/0033-T 424, Order NO DE89010612 Avail NTIS. 350 pp.
322. Li Z, McIntosh S, Carnride GR, Slavin W. *Spectrochim Acta Part B* 1992;**47B**:701.
323. Boomer DW, Powell MJ. *Anal Chem* 1987;**59**:2810.
324. Noey KC, Liedle SD, Hickey CR, Doane RW. In: *Proceedings symposium on waste management*; 1989. p. 615.
325. D'Silva AP, Bajie SJ, Zamzow D. *Anal Chem* 1993;**65**:3174.
326. Tsukada M, Yammoto D, Endo R, Nakahara H. *J Radioanal Nucl Chem* 1991;**151**:121.
327. Wisburn R, Schechter I, Niessner R, Schroeder H, Kompa KL. *Anal Chem* 1994;**66**:2964.
328. Jerrow M, Marr I, Cresser M. *Anal Proc (London)* 1992;**29**:45.
329. Opydo J. *Water Air Soil Pollut* 1989;**45**:43.
330. Ure AM. *Fresenius Z Für Anal Chem* 1990;**337**:577.
331. Spevackova U, Kucera J. *Int J Environ Anal Chem* 1989;**35**:241.
332. Sagar M. *Tech Instrum Anal Chem* 1992:12 (Hazard Met Environ 133).
333. Knizhnik EI, Prokopenco US, Stolyarov SV, Tokarevskii VV. *At Energy* 1994;**76**:113.
334. Korun M, Martincic R, Pucelj B. *Nucl Instrum Meth Phys Res Sect A* 1991;**A300**:611.
335. Wu D, Landsberger S. *J Radioanal Nucl Chem* 1994;**179**:155.
336. Del Castilho P, Rix I. *Int J Environ Anal Chem* 1993;**51**:59.
337. Haynes RJ, Swift RG. *Geoderma* 1991;**49**:319.
338. Crosland AR, McGarth SP, Lane PW. *Int J Environ Anal Chem* 1993;**51**:153.

339. Ivanova E, Soimenova M, Gentcheva G. *Fresenius J Anal Chem* 1994;**348**:317.
340. Donaldson EM. *Talanta* 1989;**26**:543.
341. Li M, Barban R, Zucchi B, Martinotti W. *Water Air Pollut* 1991;**57–58**:495.
342. Hewitt AD, Reynolds CM. *At Spectrosc* 1990;**11**:187.
343. Paudyn AM, Smith RG. *Can J Appl Spectrosc* 1992;**37**:94.
344. Hewitt AD, Reynolds CM. Report CRR EL—SP—90—19, CETHA—TS—CR—90052. Order No AD—A226 367 (Avail NTIS), 1990.
345. Kratochvil B, Mamba S. *Can J Chem* 1990;**68**:360.
346. Kammin WR, Brandt M. *J Spectrosc* 1989;**4**(49):52.
347. Millward CG, Kluckner PD. *J Anal Atom Spectrosc* 1989;**4**:709.
348. Lo CK, Fung YS. *Int J Environ Anal Chem* 1992;**46**:277.
349. Reynolds AR. In: Isknader IK, Selin HM, editors. *Engineering aspects of metal—waste management*. Boca Raton, Florida: Lewis; 1992. p. 49–61.
350. Kingston HM, Walter PJ. *Spectroscopy* 1992;**7**:20.
351. Krishnamurti GSR, Hunag PM, Van Rees KCJ, Kozak LM, Rostad HPW. *Commun Soil Sci Plant Anal* 1994;**25**:615.
352. Real C, Barreiro R, Carballeira A. *Sci Total Environ* 1994;**152**:135.
353. Van den Akker AH, Van den Heuvel H. *At Spectrosc* 1992;**13**:72.
354. Sanchez J, Garcia R, Millan E. *Analysis* 1994;**22**:222.
355. Pickering WF. *CRC Crit Rev Anal Chem* 1981;**12**:233.
356. Salomons W, Forstner Y. *Environ Technol Lett I* 1980;**1**:506.
357. Welte B, Bles N, Montiel A. *Environ Technol Lett* 1983;**4**:79.
358. Forstner U. *Fresenius Z Für Anal Chem* 1983;**17**:604.
359. Lavkulich LM, Wiens JH. *Soil Sci Am Proc* 1970;**34**:755.
360. Chao TT. *Soil Sci Am Proc* 1972;**36**:764.
361. Tessier A, Campbell PGC, Bisson M. *Anal Chem* 1979;**51**:844.
362. Harrison RM, Laxen DPH, Wilson SJ. *J Environ Sci Technol* 1981;**15**:1378.
363. Shuman LM. *Soil Am J* 1982;**46**:1099.
364. Chao TT, Zhou L. *Soil Sci Am J* 1983;**47**:225.
365. Shuman LM. *Soil Sci Am J* 1983;**47**:656.
366. Maher WA. *Bull Environ Contam Toxicol* 1984;**32**:339.
367. Miller WP, Martens DC, Zelazny LW. *Soil Sci Am J* 1986;**50**:598.
368. Singh JP, Karwasra SPS, Singh M. *Soil Sci* 1988;**146**:359.
369. Chao TT, Sanzolone RF. *Soil Sci Am J* 1989;**53**:385.
370. Sharpley AN. *Soil Sci Am J* 1989;**53**:1023.
371. Griffin TM, Rabenhorst MC, Fannings DS. *Sci Am J* 1989;**53**:1010.
372. Kadama H, Wang C. *Soil Sci Am J* 1989;**53**:526.
373. Guy RD, Chakrabarti CL, Mebain DC. *Water Res* 1978;**12**:21.
374. Tipping E, Hetherington NB, Hilton J, Thompson DW, Bowles E, Hamilton-Taylor J. *Anal Chem* 1985;**57**:1944.
375. Tessier A, Campbell PGC. *Anal Chem* 1988;**60**:1475.
376. Lopez-Sanchez RJF, Rubio R, Rauret G. *Int J Environ Anal Chem* 1993;**51**:113.
377. Shan V, Chen B. *Anal Chem* 1993;**65**:802.
378. Desaules A, Lischer P, Dahinden R, Backmann HJ. *Commun Soil Sci Plant Anal* 1992;**23**:363.
379. Thompson M, Pahlavanpaur B, Walton SJ, Kirkbright GF. *Analyst (London)* 1978;**103**:705.
380. Barei-Funel G, Dalmasso J, Ardisson G. *J Radioanal Nucl Chem* 1992;**156**:83.

381. Burke IT, Mayes WM, Peacock CI, Brown AP, Jarvis AP, Gruiz K. *Environ Sci Technol* 2012;**46**:3085.
382. Sharma P, Ofner JO, Kappler A. *Environ Sci Technol* 2010;**44**:4479.
383. Rauch JN. *Environ Sci Technol* 2010;**44**:5728.
384. Groeneberg JE, Koopmans GF, Comans RNJ. *Environ Sci Technol* 2010;**44**:1340.

Index

Note: Page numbers followed by "f" and "t" indicate figures and tables respectively.

D

E

H

I

L

O

P

R

S

T

U

V

W

X

Y

Z

Printed in the United States
By Bookmasters